2nd International Conference on Electrical Technology and Automatic Control (ICETAC 2021)

Journal of Physics: Conference Series Volume 2121

Chongqing, China
22 – 24 October 2021

ISBN: 978-1-7138-5092-2
ISSN: 1742-6588

Printed from e-media with permission by:

Curran Associates, Inc.
57 Morehouse Lane
Red Hook, NY 12571

proceedings.com

Some format issues inherent in the e-media version may also appear in this print version.

Institute of Physics
Dirac House, Temple Back
Bristol BS1 6BE UK

Phone: 44 1 17 929 7481
Fax: 44 1 17 920 0979

techtracking@iop.org

Additional copies of this publication are available from:

Curran Associates, Inc.
57 Morehouse Lane
Red Hook, NY 12571 USA
Phone: 845-758-0400
Fax: 845-758-2633
Email: curran@proceedings.com
Web: www.proceedings.com

TABLE OF CONTENTS

Preface

The 2021 2nd International Conference on Electrical Technology and Automatic Control (ICETAC 2021) was held on October 23, 2021 as a virtual conference due to the growing concerns over the coronavirus outbreak (COVID-19), and in order to protect the well-being of our attendees, partners, and staff as our number one priority.

ICETAC 2021 brings together more than 70 national and international researchers from industry, government, and academia. We invited submissions of papers and abstracts on all topics related to Electrical Technology and Automatic Control. The conference provides networking opportunities for participants to share ideas, designs, and experiences on the state-of-the-art and future direction of Electrical Technology and Automatic Control. ICETAC 2021 will feature a high-quality technical & experiential program dealing with a mix of traditional and contemporary hot topics in paper presentations and high-profile keynotes.

During the conference, the conference model was divided into three sessions, including oral presentations, keynote speeches, and online Q&A discussion. In the first part, some scholars, whose submissions were selected as the excellent papers, were given about 5-10 minutes to perform their oral presentations one by one. Then in the second part, keynote speakers were each allocated 30-45 minutes to hold their speeches.

In the second part, we invited four professors as our keynote speakers. Prof. Yuanchang Zhong, our first keynote speaker, from Chongqing University, China. His mainly research focus on Electronic Technology, Electronic Measurement Technology, System Engineering etc. And then we had Prof. Qinmin Yang, from Zhejiang University, China. He presented an insightful speech: *Theoretical research and practice in intelligent control design for wind energy*. In this talk, challenges faced by control community for high reliable and efficient exploitation of wind energy are discussed. Prof. Badrul Hisham Ahmad, Universiti Teknikal Malaysia Melaka, performed a speech: *Switchable Absorptive Bandstop to Bandpass Filter using Parallel-coupled Resonator*. Our finale keynote speaker, Prof. He Yan, from Chongqing University of Technology, China. He presented an insightful speech: *Research on Key Technologies of Semantic SLAM for Mobile Robot in Dynamic Scene*. Their insightful speeches had triggered heated discussion in the third session of the conference. Every participant praised this conference for disseminating useful and insightful knowledge.

The proceedings are a compilation of the accepted papers and represent an interesting outcome of the conference. Topics include but are not limited to the following areas: Electric Vehicle Technology, Intelligent Control System, Electrical Materials, Sensor Network System, Acoustics and Noise Control and more related topics. All the papers have been through rigorous review and process to meet the requirements of International publication standard.

We would like to acknowledge all of those who supported ICETAC 2021. The help and contribution of each individual and institution was instrumental in the success of the conference. In particular, we would like to thank the organizing committee for its valuable inputs in shaping the conference program and reviewing the submitted papers.

The Committee of ICETAC 2021

Journal of Physics: Conference Series **2121** (2021) 011001 doi:10.1088/1742-6596/2121/1/011001

Committee member

General Conference Chairs
Prof. Yuanchang Zhong, Chongqing University, China.
Prof. Mohamed Benbouzid, Université de Bretagne Occidentale, France

Technical Program Committee Chair
Prof. Yonghui Li, the University of Sydney, Australia.

Local Committee Chairs
Prof. YinQuan Yu, East China Jiaotong University, China.
Prof. Goh Hui Hwang, Institution of Engineering and Technology (IET), Malaysia

Publication Chair
Assoc. Prof. GHAMGEEN IZAT RASHED, Wuhan University-School of Electrical Engineering and Automation, China

Technical Program Committees
A. Prof. Hong Ki An, Huaiyin Institute of Technology, China
A. Prof. Da-Jie Lin, Feng Chia University, Taiwan, China.
A. Prof. Bing Wu, The Intelligent Transportation System Center, WUT, China.
Dr. MOHAMED EL MALKI, Executive Ofecier of Eudoxia Research Centre, India
Dr. Shah Fahad, Electrical Engineering Department of Zhejiang University, Pakistan
Dr. G. Sunitha, Sree Vidyanikethan Engineering College, India
Dr. F. M. Chen, The Fourth Military Medical University, China

Peer review declaration

All conference organisers/editors are required to declare details about their peer review. Therefore, please provide the following information:

- Type of peer review: Double-blind
- Conference submission management system: AI Scholar System
- Number of submissions received: 91
- Number of submissions sent for review: 86
- Number of submissions accepted: 45
- Acceptance Rate (Number of Submissions Accepted / Number of Submissions Received X 100): 49.5%
- Average number of reviews per paper: 2
- Total number of reviewers involved: 30
- Any additional info on review process:
- Step 1. Each of selected paper will be reviewed by two/three professional experts in the related subject area.
- Step 2. Review Reports received from the experts will be judged by one of the editors either Review Reports are logical or not?
- Step 3. If not logical, then editor can assign new reviewer or can also judge at his/her own.
- Step 4. If logical, then Review Reports will be sent to authors to modify the manuscript accordingly.
- Step 5. Authors will be required to revise their papers according to the points raised.
- Step 6. Revised version will then be evaluated by the editor for the incorporation of the points raised by the reviewers.
- Step 7. Then the editor will send the revised manuscript to the reviewers again for re-evaluation.
- Step 8. If the reviewers approve the revise version of the manuscript, then will be accepted for publication.

- Contact person for queries:
  Xuexia Ye
  publication@keoaeic.org
  AEIC Academic Exchange Information Centre

Reliability Design and Simulation of High Density Packaging T/R Module

Hailong Zhao*, **Yu Zhang, Honglei Ran, Hao Peng, Kui Zhang and Lijing Yin**

The 13th Research Institute, CETC, Shijiazhuang, Hebei, 050051, China

*Corresponding author's e-mail: zhaohl@cetc13.cn

Abstract. The requirements, development, packaging characteristics and existing reliability problems of high density packaging T/R modules are introduced, and the importance of reliability design is recognized. Aiming at the reliability problems of Electro Magnetic Compatibility (EMC), heat dissipation and electrostatic discharge, the reliability design and simulation scheme are given, and some cases are provided. Through design analysis and improvement, the reliability of 3D high density packaging T/R modules can be effectively improved.

1. Introduction

In recent years, phased array has been widely used in radar communication and electronic countermeasure, and T/R module is the key module of active phased array system. Limited by specific use environment and conditions, T/R modules have strict requirements on volume, weight, vibration resistance and impact resistance [1]. At present, the widely used microwave Multi-Chip Modules (MCM), whose microwave module assembled by its 2D plane assembly process technology cannot meet the characteristics of miniaturization, lightweight, conformal and stealth, are facing another breakthrough in process integration technology [2]. In recent years, the technology hotspot of 3D integrated packaging process is an effective way to achieve the integration, reduction, lightweight, high density and excellent electrical performance of microwave modules.

3D-MCM refers to a 3D structure module obtained by aligning and stacking a plurality of packaged 2D-MCM through a partition or other physical structure in the vertical direction in addition to expanding the electronic components and layout on the horizontal plane. Compared with conventional MCM, 3D-MCM can make the chip interconnection shorter, the transmission delay less, the working speed of modules faster, greatly reduce the physical size of the system and improve the assembly efficiency, so as to achieve higher chip integration [3].

With the development trend of miniaturization, high integration, high performance and multi-channel of T/R modules, tile T/R modules developed on the basis of 3D high density integration technology [1,4] have incomparable advantages in reducing module cost, volume and weight compared with current brick T/R modules, and are easier to array integration. At the same time, in the 3D tile multi-channel T/R module with higher integration, due to the denser arrangement of various devices, it is easier to produce reliability problems than brick T/R module. Reliability design and simulation technology is an important mean to solve the reliability problems of T/R module.

2. High density packaging features

Generally, brick structure and tile structure are the most common T/R modules. In the brick module, each transceiver channel is integrated into a unit module, and finally spliced into a complete subarray. The integration mode is relatively simple and the process flow is relatively mature. Because its longitudinal size is too large, it is usually suitable for the research and design of single channel transceiver modules. This form of structure has low integration and occupies a large transverse area, which cannot meet the application requirements such as array conformal. In addition, because the receiving branch and transmitting branch are located in the same plane, it is easy to form a closed-loop structure, resulting in mutual interference between the transmitting and receiving branches, so there is a very serious problem of Electro Magnetic Compatibility (EMC).

The tile T/R module is to layer a complete subarray according to functions. Each layer has some specific function of all channels of the subarray. Finally, all functional layers are connected together through 3D interconnection to form a fully functional phased array front-end. Compared with traditional brick modules, tile T/R modules have the advantages of high integration, small volume, flexible use and strong reconfigurability. In the development tide of miniaturization, reconfiguration, integration of electronic systems and the extension of the use frequency band of phased array systems to W-band or even terahertz band, tile T/R modules will replace brick modules and become the mainstream design direction [5]. Figure 1 shows the structural diagram of a typical 3D integrated tile multi-channel T/R module [6].

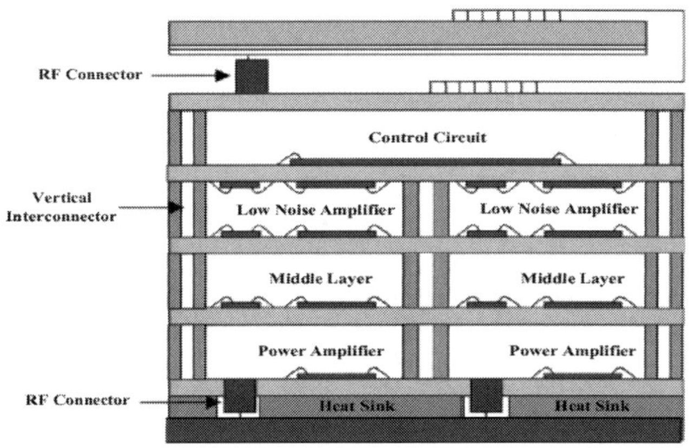

Figure 1. Structural diagram of 3D integrated tile T/R module

As can be seen from figure 1, the chips with various functions are assembled in layers, and the signal transmission between different layers is realized through vertical interconnection holes, which greatly reduces the longitudinal size and improves the integration. The tile T/R module is 20% ~ 80% smaller than the existing brick T/R module, the thermal path is shortened, the heat dissipation efficiency is improved accordingly, and has higher reliability. Although it is more difficult to realize miniaturized tile T/R modules, it has great advantages in reducing production cost, reducing size and weight, and is easy to realize large-scale multi-channel module array [7]. Based on the above advantages and the practical application requirements of modern radar array, T/R module is gradually developing from brick structure to tile structure [8].

3. Reliability design criteria

The number of T/R modules used in phased array radar is generally very large. In order to improve the reliability of the whole system, the reliability of core modules must be improved. Each function of the T/R module depends on a variety of different electronic circuits, so the cognition of circuit design reliability must be improved. According to experience, reliable circuit design technology mainly includes the following aspects: simple scheme, mature technology, inheritance of experience, correct selection and use of components, over stress design, tolerance design, overvoltage, overcurrent and overheat protection, EMC design, thermal design, etc.

This paper does not discuss the widely used circuit design requirements such as simplified design, redundant design and derating design, but focuses on reliability design and simulation technologies such as EMC design, thermal design and antistatic design. It is designed from the aspects of EMC, thermal stress and antistatic, and combined with simulation, test verification and other methods to improve the reliability of tile T/R module.

4. Reliability design scheme

4.1. EMC design

The early design of microwave modules mainly focused on logic and function. Even if there were EMC problems, they could be simply solved by isolation, filtering and grounding. However, there are EMC problems in the layout, wiring, through-hole distribution and packaging mode of high density packaging T/R modules. The EMC of products must be fully considered in the design work. Due to the complexity of EMC problems, the traditional EMC problems are mostly found by the test after product design and production, and then corrected afterwards, resulting in the improvement of design cost and the growth of research and development cycle. Serious EMC problems will even lead to irreparable product design failure.

With the development of T/R module technology, the EMC problem of modules is becoming more and more complex. The structure of 3D high density integrated modules is more complex, the circuit layout is more compact, and the problem of EMC is more prominent, which needs further analysis and design. Especially in the 3D integrated tile T/R module, the signal transmission between module substrates adopts interlayer vertical interconnection, the high-frequency transmission performance is not easy to guarantee, and the connection reliability is poor. The vertical interconnection between microwave signal layers not only needs to achieve the integrity of microwave signal transmission, but also needs to ensure the structural integrity, do not produce high-order modes and radiation effects, or do not inhibit its ability to spread to the outside world, so as to protect the normal operation of other devices in the module. Therefore, vertical interconnection is one of the key technologies to realize tile multi-channel T/R modules. Table 1 shows several common millimeter wave vertical interconnection methods and performance comparison [9].

Table 1. Comparison of various microwave vertical interconnection technologies

Transmission line type	Irregular waveguide	Quasi coaxial line	Plane transmission line	Planar line coupler
High frequency transmission performance	Excellent	Good	Normal	Poor
EMC	Excellent	Good	Poor	Bad
Contact mode	Mechanical contact	Mechanical contact, elastic crimping or bonding of outer conductor; Elastic crimping or bonding of inner conductor	Elastic crimping or bonding	Mechanical contact of ground plane, non-contact of conduction band
Connection reliability	Excellent	Poor	Poor	Good
Miniaturization	Bad	Excellent	Good	Normal
Mass production	Excellent	Poor	Normal	Excellent

It can be seen from table 1 that various vertical interconnection modes cannot fully meet the integration requirements of tile modules. Due to the influence of spatial coupling and surface wave of dielectric substrate, the EMC of open planar transmission line structure is poor. The closed structure of coaxial type is conducive to improve the EMC of signal transmission. Therefore, in the design of vertical interconnection between tile T/R module layers, in addition to good high-frequency performance, simple structure and process and easy realization, metal shell should be selected for shielding as much as possible to maintain good EMC performance, so as to ensure that the signal is not subject to external interference and leakage. In T/R module, the discontinuity of transmission line will be amplified and propagated through cavity resonance effect, which is easy to cause electromagnetic interference to various devices in the module. Therefore, improving the transmission of microwave signal and restraining the propagation of high-order modes have very positive significance for the design of module EMC.

With the development of computer technology and the improvement of electromagnetic simulation analysis software, the data obtained by electromagnetic simulation have been very consistent with the actual experimental data. The electromagnetic simulation results have high guiding value for the design. Using electromagnetic simulation software, the EMC problem of the product can be roughly estimated in the product design stage, which is very helpful to solve the EMC problem, so as to improve the product performance, save the cost, shorten the research and development time and improve the success rate of product development. Using electromagnetic simulation software to simulate and analyse the EMC of newly designed products has become a necessary step in the process of product design. Since the modelling and simulation of T/R module electromagnetic microwave related disciplines involve different professional technologies and methods, it is necessary to pay attention to the characteristics of different technologies and the coordination of processes from the basic workflow. The general electromagnetic simulation workflow is shown in Figure 2, and the main links are as follows:

(1) According to the simulation input data (such as PCB layout file, cavity view, etc.), build the basic 3D physical model of T/R module. On this basis, make professional processing of the model based on the professional characteristics of electromagnetic microwave simulation, and build the 3D electromagnetic model;

(2) For the 3D electromagnetic model, set the necessary excitation, boundary conditions and simulation frequency, and perform iterative optimization of mesh dissection;

(3) Perform 3D electromagnetic simulation of T/R module, and obtain the electromagnetic field distribution and other data of different frequency points in the module cavity under signal excitation;

(4) According to the simulation input data (such as circuit system principle, device index, S parameters, etc.), build the basic T/R module circuit system model, and combined with the passive 3D structure S parameter model obtained by the 3D electromagnetic simulation of T/R module, build the field-circuit combined T/R module circuit system model;

(5) Perform field-circuit combination collaborative simulation of active-passive/3D structure, and obtain the index data of T/R module microwave circuit system considering the actual 3D electromagnetic characteristics.

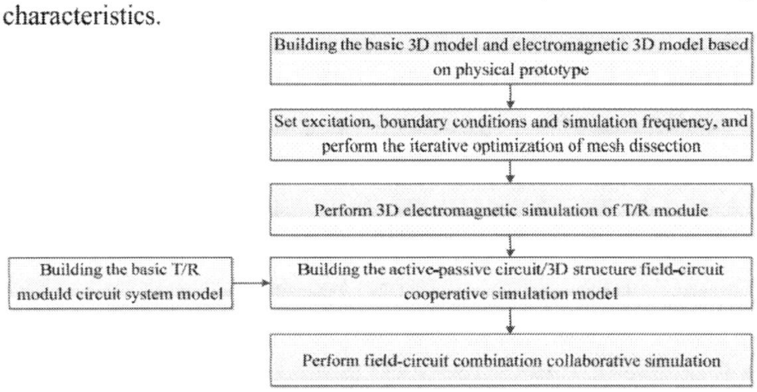

Figure 2. Electromagnetic simulation workflow

4.2. Thermal design

The thermal design of micro tile T/R module is an important link throughout the whole process of reliability design. The purpose of thermal design research is to control the product temperature within the specified range, minimize the temperature difference inside the product and ensure the normal operation of the product. By selecting materials and devices and taking reasonable process measures, the heat generated by components is effectively transmitted to the shell to ensure that the junction temperature or maximum temperature of power devices with large calorific value is controlled within the allowable temperature range, so as to make the heat source layout uniform and avoid local heat source concentration. At the same time, minimize the impact of environmental temperature change on product performance, improve product life, eliminate quality hidden dangers and improve reliability.

After the thermal design is completed, the analysis model is built by using the finite element analysis software for thermal simulation. The junction temperature is one of the important indicators to judge the thermal performance of the chip. Untimely chip heat dissipation will lead to excessive junction temperature [10]. Through the thermal design of the circuit, reducing the junction temperature of the semiconductor chip can better improve the reliability of the circuit.

For example, the GaN chip parameters of a T/R module are as follows: the thermal resistance is about 4℃/W (as shown in table 2), the working voltage is +28V, the pulse width is 200μs, the duty cycle is 30%, and the test platform temperature is +70℃. The process ensures that the lifetime of GaN chip is 1 million hours at 225℃ junction temperature; the junction temperature below 150 ℃ is recommended. When the module works at +28V, the output power is about 3.5W and the dynamic working current is 125mA (30% duty cycle). Therefore, the thermal power consumption of GaN chip is calculated as: 28V*0.125A-3.5W*30%=2.45W; GaN chip temperature rise: 4℃/W*2.45W=9.8℃.

Table 2. Material properties of T/R modules

Name	Material Science	Thermal Conductivity (W/m·K)
Chip	GaN	Thermal resistance 4℃/W
Shim	MoCu	170
Box body	6061Al	154
AuSn Solder	$Au_{80}Sn_{20}$	4 (worst condition)
PbSn Solder	$Pb_{63}Sn_{37}$	50

For the above thermal design, the thermal design method is verified by the thermal simulation software to make the thermal design of the product meet the product requirements. The ambient temperature is 70℃, and the thermal simulation is carried out according to the material properties in table 2, the simulation results are shown in Figure 3. The maximum temperature of the chip bottom is 78℃, and the maximum temperature rise of the chip is 8℃, which is lower than the design temperature rise, indicating that the structure of the T/R module has good heat dissipation capability.

Figure 3. Thermal simulation diagram of T/R module

4.3. Antistatic design

The damage of electrostatic discharge to T/R modules is often hidden and developmental, that is, electrostatic damage is sometimes difficult to find, and will be exposed after a certain time of accumulation, resulting in the complete failure of components on the module circuit board. Therefore, it is particularly necessary to take antistatic measures in advance for antistatic design. The antistatic design consists of a series of links, including components design, circuit design, components selection, procurement, packaging, transmission, storage, transportation, testing, assembly, operation and use, and each link should be considered.

For example, in addition to meeting the conventional anti-static requirements for a certain type of T/R module, the following measures are taken in circuit design and structural design:

(1) Devices with high electrostatic sensitivity shall be selected as far as possible during device selection, and the protection grade shall be at least 1A;

(2) Maintain well connection between the module power ground and the box body to provide electrostatic discharge circuit;

(3) Protect internal devices by controlling input port connected in series with current limiting resistor, use decoupling capacitor to fully filter in the power input terminal, and add diodes to improve the antistatic ability;

(4) The bottom layer of the circuit board is a large area metal layer, which is connected with the box body to form an effective electrostatic discharge path;

(5) Place the electrostatic sensitive devices between interstage circuits in the circuit layout, and avoid placing the electrostatic sensitive devices at the input, output and power terminals;

(6) Rationalize the structural design and layout, and avoid the design of sharp corners, narrow seams, shell openings, etc.

5. Conclusion

Active phased array radar and communication system are developing towards miniaturization, high working frequency, high integration and low cost, which puts forward higher and higher requirements for the development of T/R module. However, the miniaturization and high integration of T/R modules have brought more serious and complex reliability problems such as EMC and heat dissipation, which seriously restricts the development of T/R modules. Therefore, it is of great research significance to analyse and improve the reliability of 3D high density packaging T/R modules.

References

[1] Hauhe M S, Wooldrige J J. (1997) High Density Packaging of X-Band Active Array Modules. IEEE Trans on Modules, Packaging, and Manufacturing Technology, 20(3): 279-291.

[2] Mao X H. (2009) High Density Integrated Technologies and Miniature in Electronic Equipment. Electronic Information Warfare Technology, (4): 68-71.

[3] Du S, Zhou D L, (2010) Chen F. Structure and fabrication of 3D MCM for high density multilayer substrate. Journal of Integrated Circuit, 22(3): 28-32.

[4] Schreiner M, Leier H, Menzel W, et al. (2003) Architecture and Interconnect Technologies for a Novel Conformal Active Phased Array Radar Module. IEEE MTT-S Microwave Symposium Digest, pp: 567-570.

[5] Tong H M, Lai Y S, Wong C P. (2013) Advanced Flip Chip Packaging. Springer, Heidelberg.

[6] Barbier T, Mazel F, Reig B, et al. (2005) A 3D wideband package solution using MCM-D BCB technology for tile TR module. In: Gallium Arsenide and Other Semiconductor Application Symposium, 2005. European. pp: 549-552.

[7] Yan W, Jiang W Z, Yu S L. (2009) The Microassembly Technology for Miniature and High Density Microwave Modules and Its Applications. Defense Manufacturing Technology. 5(10): 43-47.

[8] Yang Y L. (2018) Research on X-Band Planar Phased Array T/R Module Microsystem. School of Electronic Science and Engineering, Chengdu.

[9] Huang J, (2011) Crucial Integration Technology of T/R Module for Millimeter-wave Active Phased Array. Telecommunication Engineering, 51(2): 1-6.

[10] Lai Y, Cordero N, Barthel F, et al. (2009) Liquid cooling of bright LEDs for automotive applications. Applied Thermal Engineering, 29(5-6): 1239-1244.

Design of a Ku-band Tile T/R Module

Yu Zhang, Hailong Zhao*, Xuefeng Song and Zenglu Li

The 13th Research Institute, CETC, Shijiazhuang, Hebei, 050051, China

*Corresponding author's e-mail: zhaohl@cetc13.cn

Abstract. This paper introduces a 16 channel Ku-band tile T/R module design in 3D highly integrated packaging. The technical indicator design of T/R module is completed by using four channel amplitude phase multifunctional chip and various design methods. Through reasonable indicator distribution and structure layout, the miniaturization design of T/R module is realized by using high-density substrate and 3D interconnection structure. Through process optimization, structural design and simulation verification, the reliability and producibility of T/R module is studied, and the consistency of products is improved. The test results show that the design has good microwave performance and can meet the requirements of high performance, low cost and batch production of Ku-band T/R module.

1. Introduction

T/R module is an important part of phased array radar, and its performance indicator directly affects the performance of the whole phased array radar [1]. Modern military and civil electronic equipment, especially active phased array radar and communication system, is developing towards miniaturization, high working frequency, high integration and low cost, which puts forward higher and higher requirements for T/R modules.

With the rapid development of integrated circuit, the integration of T/R module used in active phased array radar becomes higher and higher [2]. Conformal antenna requires T/R module to have small size and ultra-thin structure. At present, there have been tile T/R modules integrating antennas on T/R modules [3]. The lightweight of tile T/R modules can better meet the needs of radar and is the inevitable result of the development of radar system. This T/R module adopts 3D assembly technology to integrate the antenna unit and module circuit into one circuit module, which avoids the connector interconnection between the T/R module circuit and antenna unit, greatly reduces the cost of T/R module, improves the integration, reduces the volume and weight, improves the stability of radar in complex environment [4], and meets the development requirements of phased array radar miniaturization and lightweight [5]. It is an important direction of T/R module development at present.

According to the requirements of phased array radar for micro tile T/R module, a new highly reliable Ku-band tile T/R module with 16 channels and 4*4 structure is designed in this paper.

2. Overall indicator design

Due to the complexity of T/R module system, many problems are involved, including the internal details of the module, the relationship between the module and the outside, signal quality, delay, distribution, noise, etc. The overall indicator design should be considered in many ways. One of the key problems is the layout and wiring of components and the signal interconnection between layers. It is necessary to comprehensively consider the crosstalk, noise, electrical path radiation and other problems between the signals of each chip and component on the substrate under the small cavity body.

The second key problem is the design of passive device in the carrier, which needs to comprehensively consider the limitations (accuracy) of the passive device process, quality factor (Q), resonance frequency, etc. With the increase of module complexity and the decrease of module size, the difficulty of system design will increase. In addition to using the design software, the numerical simulation of system performance should also participate in the design process, such as high frequency electromagnetic field simulation, heat transfer simulation, and reliability.

The design ideas of T/R module in this paper are as follows: complete the technical indicator design of T/R module through various design means, solve the problem of rationality of indicator distribution, and solve the problem of realization of technical indicator and optimum process route; through the redundancy analysis of process error, solved the influence of process deviation on technical indicators and device compatibility; through the time domain analysis of power supply and wave control devices, realize the accurate control of power management and wave control timing; through the performance analysis of the whole module, solve the problem of matching with the technical indicators of the whole machine; through testing and debugging, solve the problem of technical indicator compliance, and solve the problems of signal crosstalk, noise, circuit radiation, etc. The microwave link block diagram of T/R module is shown in Figure 1.

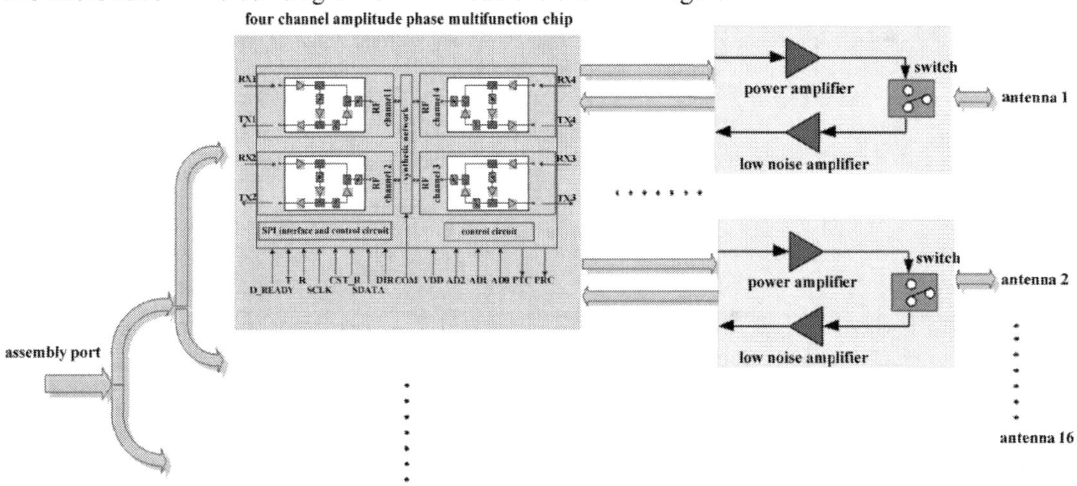

Figure 1. Microwave link block diagram of T/R module

After receiving from the antenna, the receiving branch signal is amplified by the switch and low noise amplifier, then phase shifted and attenuated by four channel amplitude phase multifunctional chip, and output through the quasi coaxial transmission via or connector. NC (Numerical Control) attenuator digits, NC attenuation amount, NC attenuation step, NC attenuation accuracy, NC phase shifter digits, NC phase shift range, NC phase shift step, phase shift accuracy RMS (Root Mean Square) error and other indicators can be guaranteed by the device. The main indicators are designed as follows:

(1) Noise figure, R-branch gain

The theoretical calculated value of noise coefficient of the whole R-branch is about 4dB, and the theoretical value of module receiving gain is about 12dB.

(2) Gain flatness design of R-branch

In order to improve the gain flatness of R-branch, two measures are taken in the design. First, the flatness of the devices used is optimized. Secondly, on the RF (Radio Frequency) transmission path, especially in 3D interconnection, ensure that the in-band insertion loss curve is relatively smooth. Through the above measures, the gain flatness of R-branch can be controlled within ±1dB at room temperature.

(3) Design of phase shifting accuracy and attenuation accuracy of NC phase shifter

The NC phase shifter chip in the amplitude phase multifunction chip is 6 bits, and the minimum step is 5.625°. Within the module working frequency band, the 64 state RMS phase shift error of the

NC phase shifter is ≤2°. The phase shifter is installed inside the module. Due to the influence of the cavity and the port stationary wave, the indicator will deteriorate by about 1°, and the RMS phase shift error can be controlled at about 3°.

The digits of NC attenuator are 6, the step is 0.5dB, and the total attenuation is 31.5dB. The attenuation accuracy of the NC attenuator is determined by the attenuation accuracy of the selected NC attenuator itself. The attenuation accuracy of 64 attenuation states of the NC attenuator selected in this paper have is ≤0.5dB.

(4) P_{-1} design of R-branch input

The input P_{-1} at the receiving end of the four channel amplitude phase multifunction chip is -5dB, the gain of the low noise amplifier is 26dB, and the loss of connector and switching is about 1dB, then the module receiving input P_{-1} is: 1-26-5=-30dBm.

(5) Design of port stationary wave

The receiving input port is the antenna port, the microwave connector is the standard SSMP connector, and the stationary wave coefficient is 1.25. Cascaded with the connector is a GaN multifunctional device with a port stationary wave of 1.6. After reasonable matching design, it can ensure that the stationary wave of the receiving input port is within 1.8. The output port of R-branch is SSMP connector and power divider chip, and the chip stationary wave is 1.3, which meets the requirements of receiving and output stationary wave ≤1.8.

(6) Design of saturated output power and output power flatness of T-branch

The transmitted signal enters from the assembly port of the module, passes through the one point four power divider, and then reaches the port of the four channel amplitude phase multifunction chip. The signal is phase shifted by the amplitude phase multifunction chip, and then saturated outputs by power amplify. After the signal passes through the port insulator or connector, one point four power divider and four channel amplitude phase multifunction chip, the loss is about 0.5dB, and the power gain of the power amplifier is ≥20dB. Therefore, when the input power is greater than 14dBm, the power amplifier can approach the saturated output, and the saturated output power is about 34dBm. Minus the loss of the output port connector, the output power of the transmission channel is about 33dBm.

The saturated output power of GaN power amplifier chip is within the working bandwidth and the flatness is within ±0.5dB.

(7) Top-down design of launching pulse

The indicator requires that the top-down of transmitted pulse is ≤0.5dB (pulse width 200us, 30% duty cycle). If the power modulator is close to the amplifier, and the transmission length of the circuit board and wire is reduced, the top-down of the transmitted pulse can be controlled at about 0.4dB.

3. Miniaturization design

The function of T/R module is complex, the number of channels is large, and there are many microwave chips and control chips. It is very difficult to develop it to achieve certain power and sensitivity in order to realize the transmission and reception of RF signals in a very small volume. This design reduces the chip space through reasonable indicator allocation and improving the multi-function chip; through reasonable structural layout, a 3D interconnection structure of RF and control lines is formed to reduce the size and weight of modules; the miniaturization and high performance of T/R module are realized by using high-density substrate. The mixed signal design of microwave transmission line, logic control line and power line is integrated into the same 3D microwave transmission structure, which minimizes the volume and weight of the T/R module.

PCB substrate is used in this design. PCB substrate has low dielectric constant, low RF loss, low coefficient of thermal expansion and high glass transition temperature (T_g). It can realize the interconnection of any layer, and has extremely dense wiring and low processing cost, which is the preferred plate to realize high-density assembly products. PCB substrate can achieve thinner thickness, which is more suitable for tile modules with strict size requirements.

There are 4 layers of copper foil on the substrate, and the overall plate thickness is designed to be 0.3mm. In the design, the substrate stack is designed as follows: the top layer is RF microwave layer, mid1 layer is microwave ground layer, mid2 layer is power supply and controlling wire and the bottom layer is GND layer. Most of the lines that can be fully utilized are in mid2 layer, and mid1 layer is only used in areas other than microwave flooring. The wiring density of circuit board is very high and the requirements are higher. In the design, the continuity of microwave ground shall be fully considered to ensure the RF performance of the module; isolation between low frequency and RF signals shall also be carried out to avoid crosstalk.

A variety of vertical interconnection modes such as substrate vertical interconnection, insulator or transition block interconnection, BGA ball planting, fuzz button are compared. In this design, microwave multilayer substrate is used to solve microwave interconnection. Microwave vertical interconnection transmission between substrates is the simplest and easy way of 3D transmission of microwave signals. It is mostly realized by multi-layer microwave boards, which takes up a small space. Mature process assembly methods can be used for module assembly, which will not increase the difficulty of assembly. After the signal is vertically transmitted, the connection part of the microstrip or stripline should be well matched and designed, which can be applied to higher frequencies. Because there are microwave circuits, power supply and control circuits on the whole circuit board, attention should be paid to avoid crosstalk between signals during design.

4. Reliability design

Due to the 3D integration mode, tile T/R modules face greater challenges than "brick" in the design of signal transmission, module coupling, heat dissipation performance and power integrity. This paper mainly optimizes the reliability from the aspects of electromagnetic compatibility design, heat dissipation design and sealing design.

Because T/R module encapsulates all RF chips and components in one cavity, it is easy to produce coupling effect, resulting in mutual interference between different modules [6]. This design mainly analyses and solves this problem from the following aspects: ① In the design, appropriate filtering and shielding are added between the connection of microwave circuit, wave control circuit and power supply, and reasonable layout is adopted, avoiding long distance parallel lines of signals; ② The multi-channel local direct cavity division design is adopted in the structural design, which has good channel isolation, small crosstalk and high circuit stability; ③ A whole sheet of absorbing material is pasted on the box cover to absorb the space radiation of the transmission line; ④ In the 3D microwave transmission structure, the quasi coaxial and coplanar waveguide mode is adopted, which has good stability, and realizes the maximum possible spatial isolation and shielding measures with the wave control and power supply circuit to avoid crosstalk.

When the T/R module works, the shell structure will affect the heat dissipation performance of the module. When the output power of the module is high, it will obviously heat up. When the module temperature exceeds the normal working temperature of the chip, the chip and device may fail. This paper considers heat dissipation from two aspects: ① For internal device heat dissipation, power amplifier devices (including GaN power amplifier and circulator, et al.) are sintered on molybdenum copper carrier with gold-stannum solder, and molybdenum copper carrier is sintered on the box body to ensure sufficient heat dissipation; ② For external heat dissipation of module, when the system is integrated, the modules are first fixed to the cold plate, and the heat is quickly exported through the cold plate. The contact surface between the modules and the cold plate is fully considered in the structural design.

The function of T/R module is complex, there are many chips, and there are aluminium pressure point chips that need aluminium wire bonding, which has strict requirements on the sealing of the environment. This paper solves the sealing problem of T/R module from the aspects of packaging material, packaging structure and packaging technology: ① SSMP connector and micro torque low frequency connector are gas sealing structure; ② Microwave connector and low frequency connector are welded on the box body at one time by welding process; ③ The box body is sealed by laser sealing welding, so as to meet the sealing requirements of modules.

5. Producibility design

In order to meet the requirements of batch production of tile T/R modules, it is necessary to design the modules with low cost and producibility: through reasonable process design, select mature process routes as far as possible to solve the problem of high density assembly of tile modules; solve the problem of testing and installation of modules by using the opposite plug connector and setting the installation hole at the appropriate position; control the cost of equipment and outsourcing, and optimize the design to make the modules free of debugging or micro debugging, so as to reduce the processing cost of modules. Producibility design mainly includes: process optimization design, free (micro) debugging design and automatic production.

5.1. Process optimization design

The traditional T/R module all adopts serial operation, which restricting the production cycle of T/R module. By optimizing the structure layout and adopting parallel modular design, the production cycle can be reduced and the production efficiency can be improved. At the same time, due to the modular design of the cavity, the performance of the modules has been improved to a certain extent, and it is convenient for module maintenance to ensure that T/R modules can realize mass and efficient production.

5.2. Free (micro) debugging design

There are many test indicators of T/R modules, which bring huge workload to debugging and production. Therefore, in order to implement producibility, it is necessary to be free (micro) debugging. 3D electromagnetic field software simulation can effectively simulate the transmission characteristics of microwave signals. By using 3D electromagnetic field software to simulate interconnection matching, cavity effect and system cascade, it can effectively guide the design layout and reduce the debugging workload in the early stage of design. At the same time, the first piece design is carried out according to the simulation result, and then the software simulation is fed back according to the first piece debugging result, and the joint design is carried out again, so that the batch product is free of (micro) debugging as much as possible.

5.3. Automatic production

Due to the complex structure of T/R components, the variety and quantity of chips, and the limited space and volume, to a certain extent, the utilization rate of automatic assembly and automatic bonding of T/R modules is reduced to a certain extent. Therefore, it is hoped to formulate design specifications for T/R automatic assembly, optimize cavity structure design process, and design corresponding tooling for T/R automatic assembly through this project, so as to improve the automation rate of assembly bonding and improve product consistency.

6. Conclusion

Process implementation was carried out after the design was completed. The test results show that the technical indicators of this design meet the requirements, and the microwave performance is good, which simplifies the production process, reduces the production cost, and can meet the requirements of high performance, low cost and batch production of T/R modules. After the implementation of the project, it can realize technological leapfrog development, promote the engineering application of a new generation of miniaturized T/R modules, provide conditions for the development of a new generation of phased array radar, promote a significant leap in the performance of T/R modules for Ku-band radar, and meet the requirements of conformal phased array radar for Ku-band high power and small size.

References

[1] Chappell W, Fulton C. (2010) Digital array radar panel development. In: 2010 IEEE International Symposium on Phased Array Systems and Technology. Waltham. pp: 50-60.
[2] Schuh P, Rieger R, (2009) Fleckenstein A, et al. T/R-module technologies today and possible evolutions. In: 2009 International Radar Conference "Surveillance for a Safer World" (RADAR 2009). Bordeaux. pp: 1-5.
[3] Yang Y L. (2018) Research on X-Band Planar Phased Array T/R Module Microsystem. School of Electronic Science and Engineering, Chengdu.
[4] Fu S Q, Peng Y, Xue Y. (2018) Analysis of the adaptability test requirements of radar equipment complex electromagnetic environment. Journal of Terahertz Science and Electronic Information Technology, 16(5): 868-870.
[5] Peng X L. (2010) Integration technology of phased array antenna. Telecommunication Engineering, 50(10): 112-117.
[6] Wang Y Q. (2019) Research on Key Technology of X-band High Power T/R Module Based on LTCC Technology. Xidian University, Xian.

A Day-Ahead Scheduling Model of Electricity-Gas Integrated System with Power-to-Gas

Jiacheng Ruan[a], Yongji Cao[b]

school of electrical engineering, Shandong University, Jinan, China

[a]sdurjc@163.com, [b]CaoYongji@sdu.edu.cn

Abstract—In recent years, with the increasing of gas-fired power plants and the development of power-to-gas (P2G) technology, and the interdependence between the power system and the natural gas system has gradually deepened. This paper proposes a day-ahead optimal scheduling model for an electricity-gas integrated system. Based on P2G, it can realize the two-way movement of energy between the power system and the natural gas system, promote the coordinated and optimized operation of the two energy systems, and improve energy utilization. The model proposed in this paper minimizes the total cost of both systems. Firstly, the power system sub-model and the natural gas system sub-model in the collaborative optimization model are studied separately, and the constraints between the two systems are considered. Then, the piecewise linearization method and DC power flow simplification method are adopted, the nonlinear problem is transformed into a mixed integer linear programming problem. Finally, the load forecast value of the day-ahead dispatch is loaded, and the 24-hour dispatch result is obtained through the simulation platform, and the P2G is used for joint dispatch.

1. Introduction

With the increasing in the scale of new energy, the randomness, volatility of wind power have caused a lot of energy waste. The gradual development of power-to-gas (P2G) technology provides a new solution, making it possible for the two-way flow of energy between the power system and the natural gas system, becoming one of the options that can solve the problem of long-term power storage. And in theory, it can alleviate grid congestion [1]. In recent years, the proportion of natural gas power generation capacity has been increasing. Compared with traditional thermal power plants, gas power generation is in line with the general trend of low-carbonization. It is estimated that by 2035, Chinese total natural gas power generation is expected to increase to 760 billion kWh. With the large-scale coverage of gas networks, natural gas power generation will become an important part of the future energy system[2][3]. The combination of power system and natural gas system through P2G is an optimized configuration of energy. Compared with traditional energy storage equipment, the electricity-to-gas technology has large storage capacity and long discharge time [4][5], which can effectively absorb large-scale wind power and realize long-term, large-scale space-time translation of energy [6].

In recent years, research has mainly focused on the impact of the joint operation of power and natural gas networks on the safety and steady state of the power system. In [21][22] , a steady-state flow model of natural gas networks is proposed, through the natural gas system and gas turbine generator sets. Perform short-term combination to analyze the impact of natural gas system on the short-term dispatch of power system unit combination. In [23] the increasing degree of coupling between the power system and the natural gas system is analyzed, and corresponding countermeasures to the potential security threats caused by the production scheduling and safe operation of the power system is proposed. In addition, most of the existing papers are limited to the analysis of the impact of electricity-to-gas devices on the power system and natural gas system[25], the economic analysis of electricity-to-gas devices[26], the capacity setting and selection of electricity-to-gas facilities site[26][27], and comparison of electricity-to-gas devices and other energy storage methods [28].

In summary, it is possible to consider the integration of electrical systems and natural gas systems, and the introduction of wind power and electricity-to-gas facilities has less research on optimization scheduling. Therefore, how to maximize the use of new energy, improve energy utilization, and achieve economic optimization under the constraints of the power grid and gas grid, is an urgent problem to be solved.

2. Principles and Advantages of P2G Technology

P2G technology includes two types of electricity to hydrogen and electricity to methane. The chemical reaction principle of electro-hydrogen conversion is electrolysis of water; the chemical reaction principle of electro-conversion of methane is electrolysis of water and methanation.

The water electrolysis reaction is to electrolyze water to generate H_2 and O_2. There are three main methods for producing hydrogen from electrolyzed water, which are alkaline electrolysis of water to produce hydrogen, solid polymer electrolysis of water to produce hydrogen, and high temperature solid oxide electrolysis of water to produce hydrogen. At present, the efficiency of the electrolysis water reaction is about 56% to 73%[7][8].

The methanation reaction in which CO_2 and hydrogen are added to the catalyst is one of the effective ways to utilize CO_2. The main reaction products are CH_4 and H_2O. At present, the efficiency of the methanation reaction is about 75% to 80% [9].

The electro-hydrogen conversion only requires the electrolysis of water, which avoids the energy loss caused by the subsequent methanation reaction and reduces the equipment construction costs related to the methanation reaction. However, the injection of hydrogen into the natural gas pipeline will cause risks in the transportation pipeline. At present, the maximum allowable volume fraction of hydrogen in natural gas pipelines is about 10% to 15%, and the maximum allowable volume fraction of hydrogen in gas turbine fuel is about 5% [12].The whole process of electric conversion to methane includes two steps of electrolysis of water reaction and methanation reaction, the efficiency is about 42% to 58% [13], and the efficiency is lower than that of electric conversion to hydrogen. But the most important advantage of methane is that it can be directly injected into existing natural gas pipelines and storage devices for direct production and consumption.

The current power system development has many systems for storing electrical energy, all of which have different storage principles, and there are big differences in capacity, energy storage efficiency, construction and operation costs, etc. [14]. Compared with other energy storage methods, we can see from Fig. 1 that P2G technology has obvious advantages in terms of capacity. The rated power can reach the gigawatt level, and the discharge time can last for several days or even months.

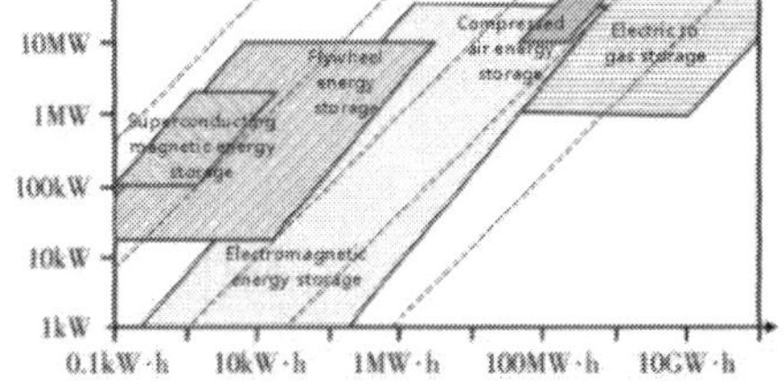

Fig.1. Comparison of energy storage capacity.

 2121 (2021) 012003 doi:10.1088/1742-6596/2121/1/012003

3. Optimization Model

Electricity-gas integrated system proposed in this chapter not only considers the economic and technical constraints of power system dispatch, but also considers the economic and technical constraints of natural gas system dispatch, so as to realize the optimal day-ahead dispatch of the joint system.

3.1 Nomenclature

a, g, j, s stand for P2G facility, natural gas load, gas supply point, and gas storage facility, respectively.
d, i, t, w represent power load, power generation unit, time and wind power plant, respectively.
m, n represents the index of the natural gas network node.
F_i^c is cost curve of power generation unit.
G_{jt}^g is the gas production at gas supply point.
I_{it}^b, $I_{at}^{b,ptg}$ are the status of the facility, 0 means closed, 1 means open, respectively.
P_{it}^b, P_{wt}^b, $P_{at}^{b,ptg}$ are the output active power of each units, respectively.
SU_{it}^b, SD_{it}^b are start and stop costs of unit, respectively.
X_{it}^{on}, X_{it}^{off} are on/off time counter with unit *i* at time *t, respectively.*
E_{st} is the gas storage capacity of *s* at time *t*.
Pr_{mt} is the pressure of the natural gas system.
$\widetilde{Pr}_{mn,t}$ is the average pressure of *m* and *n* at time *t*.
$Q_{mn,t}^{in}$, $Q_{mn,t}^{out}$ are the amount of gas in and out of pipe *mn, respectively.*
Q_{st}^{in}, Q_{st}^{out} are the input and output reactive power of gas, respectively.
$\tilde{Q}_{mn,t}$ is the average gas flow rate of pipeline *mn*.
G_{it}^{gu} is the gas consumption of gas-fired units.
$G_{at}^{g,ptg}$ is the gas production of P2G facility a at time *t*.
C_i^{fuel}, C_w^b, C_j^G, C_s^G are each corresponds to the fuel price of power generation unit *i*, the cost of wind power overflow, the price of natural gas produced at gas supply points, and the price of natural gas stored at gas storage point *s*.
$P_{f,wt}^b$, P_{dt}^b, G_{gt} correspond to the predicted output, power load and gas load of the wind power plant at *t*, respectively.
SR_t is the reserve at time *t*.
T_{it}^{on}, T_{it}^{off} is the minimum on/off time of unit *i*.
UR_i, DR_i are ramp up/down rate of power generation unit *I, respectively.*
K_{mn}^{gf}, K_{mn}^{lp} are constant of pipeline flow characteristic and pipeline characteristic.
η_a^{ptg} is the efficiency of P2G equipment *a*.
GU is the gas unit equipment.
K_a, K_d are relation matrix of P2G equipment and load, respectively.
K_P, K_w are the relation matrix of power generation unit and wind power, respectively.
$\mathrm{PL}^{\max}$ is the maximum power flow (vector).
SF is the transfer factor matrix.

3.2 Objective Function

The model aims to minimize the overall energy cost, which includes non-gas turbine operating costs, wind spillover costs, natural gas production costs, and natural gas storage costs. The objective function is shown as follows.

$$\min = \sum_{i\in GU} C_i^{fuel}[F_i^c\left(P_{it}^b\right) + SU_{it}^b + SD_{it}^b]$$
$$+\sum C_w^b\left(P_{f,wt}^b - P_{wt}^b\right) + \sum_j^w C_j^G \cdot G_{jt}^g + \sum_s C_s^G \cdot Q_s^{out} \quad (1)$$

3.3Model of Power System

$$\sum_i P_{it}^b + \sum_w P_{wt}^b - \sum_a P_{at}^{b,ptg} = \sum_d p_{dt}^b \tag{2}$$

$$P_i^{min} \cdot I_{it}^b \le P_{it}^b \le P_i^{max} \cdot I_{it}^b \tag{3}$$

$$0 \le P_{at}^{b,ptg} \le P_a^{max} \cdot I_{at}^{b,ptg} \tag{4}$$

$$0 \le P_{wt}^b \le P_{f,wt}^b \tag{5}$$

$$\left(X_{i(t-1)}^{on} - T_{it}^{on}\right) \cdot \left(I_{i(t-1)}^b - I_{it}^b\right) \ge 0 \tag{6}$$

$$\left(X_{i(t-1)}^{off} - T_{it}^{off}\right) \cdot \left(I_{it}^b - I_{i(t-1)}^b\right) \ge 0 \tag{7}$$

$$SU_{it}^b \ge su_i \cdot \left(I_{it}^b - I_{i(t-1)}^b\right), SU_{it}^b \ge 0 \tag{8}$$

$$SD_{it}^b \ge sd_i \cdot \left(I_{i(t-1)}^b - I_{it}^b\right), SD_{it}^b \ge 0 \tag{9}$$

$$P_{it}^b - P_{i(t-1)}^b \le UR_i \cdot I_{i(t-1)}^b + P_i^{min} \cdot \left(I_{it}^b - I_{i(t-1)}^b\right) + P_i^{max}\left(1 - I_{it}^b\right) \tag{10}$$

$$P_{i(t-1)}^b - P_{it}^b \le DR_i \cdot I_{it}^b + P_i^{min} \cdot \left(I_{i(t-1)}^b - I_{it}^b\right) + P_i^{max}\left(1 - I_{i(t-1)}^b\right) \tag{11}$$

$$\sum_i P_i^{max} \cdot I_{it}^b \ge \sum_d P_{dt}^b + SR_t \tag{12}$$

$$-\boldsymbol{PL^{max}} \le \boldsymbol{SF} \cdot \begin{pmatrix} \boldsymbol{K_P} \cdot \boldsymbol{P_t^b} + \boldsymbol{K_w} \cdot \boldsymbol{P_{wt}^b} \\ -\boldsymbol{K_a} \cdot \boldsymbol{P_{at}^{b,ptg}} - \boldsymbol{K_d} \cdot \boldsymbol{P_{dt}^b} \end{pmatrix} \le \boldsymbol{PL^{max}} \tag{13}$$

$$I_{it}^b - I_{at}^{b,ptg} \le 1, \ \ if \ K_p(e,i) = K_a \ \ (e,a) = 1 \tag{14}$$

The above are the basic constraints of the power system, including the system load balance generator set, the power limit of the electric-to-gas device and the wind farm, the minimum on/off time limit, the startup and shutdown cost, the uphill and downhill limit, and the system rotation reserve requirements , DC power flow constraint.

Constraint (14) ensure that the gas-fired unit and the P2G mechanism will not operate at the same time, ensuring operational safety.

3.4Model of Natural Gas System

In this section, the gas network model is used to describe the dynamic characteristics of the natural gas system, and the gas flow rate and line package speed are considered by changing the input and output gas flow. Constraint (15) represents the gas network node balance, which means that the total gas flow injection is equal to the total extraction amount at each node.

$$\sum_{n \in G(m)} \left(Q_{mn,t}^{out} - Q_{mn,t}^{in}\right) + \sum_{s \in G(m)} \left(Q_{st}^{out} - Q_{st}^{in}\right) + \sum_{j \in G(m)} G_{jt}^g$$
$$+ \sum_{a \in G(m)} G_{at}^{g,ptg} = \sum_{i \in G(m)} G_{it}^{gu} + \sum_{g \in G(m)} G_{gt} \tag{15}$$

$$G_j^{min} \le G_{jt}^g \le G_j^{max} \tag{16}$$

$$Pr_m^{min} \le Pr_{mt} \le Pr_m^{max} \tag{17}$$

$$E_{st} = E_{s(t-1)} + Q_{st}^{in} - Q_{st}^{out} \tag{18}$$

$$E_{st}^{min} \le E_{st} \le E_s^{max} \tag{19}$$

$$Q_s^{in,min} \le Q_{st}^{in} \le Q_s^{in,max} \tag{20}$$

$$Q_s^{out,min} \le Q_{st}^{out} \le Q_s^{out,max} \tag{21}$$

Constraint (16) shows the maximum and minimum amount of gas produced by the gas well, and constraint (17) represents the pressure limit of each gas node, constraint (18) represents the balance of the storage capacity of the gas tank, and constraint (19) represents the maximum and minimum storage capacity.

$$\tilde{Q}_{mn,t} = sgn(Pr_{mt}, Pr_{nt}) \cdot K_{mn}^{gf} \cdot \sqrt{|Pr_{mt}^2 - Pr_{nt}^2|} \quad (22)$$

$$\tilde{Q}_{mn,t} = \frac{Q_{mn,t}^{out} + Q_{mn,t}^{in}}{2} \quad (23)$$

$$sgn(Pr_{mt}, Pr_{nt}) = \begin{cases} 1, Pr_{mt} \geq Pr_{nt} \\ -1, Pr_{mt} \leq Pr_{nt} \end{cases} \quad (24)$$

(22)-(24) are low simulation in natural gas pipeline.

$$LP_{mn,t} = K_{mn}^{lp} \cdot \widetilde{Pr}_{mn,t} \quad (25)$$

$$LP_{mn,t} = LP_{mn,t-1} + Q_{mn,t}^{in} - Q_{mn,t}^{out} \quad (26)$$

$$\widetilde{Pr}_{mn,t} = \frac{Pr_{mt} + Pr_{nt}}{2} \quad (27)$$

LP represents the amount of natural gas contained in the pipeline, and LP can be regarded as proportional to the average pressure of the gas pipeline and pipeline characteristics. Since the uncertainty of real-time gas load can actually be balanced by LP, the uncertainty of natural gas load is not considered in the proposed joint day-ahead optimal scheduling model.

3.5 Joint Constraint

$$G_{it}^{gu} = [F_i^c(P_{it}^b) + SU_{it}^b + SD_{it}^b] / HHV, \ \forall i \in GU. \quad (28)$$

$$G_{it}^{g,ptg} = \phi \cdot P_{at}^{b,ptg} \cdot \eta_a^{ptg} / HHV, \forall a \in PtG \quad (29)$$

Gas generators and P2G facilities establish a connection between the power system and the natural gas system. Constraint (28) is the relationship between gas power generation dispatch and gas consumption unit. Constraint (29) represents a P2G facility, where the higher heating value (HHV) is 1.026 MBtu/kcf and the energy conversion factor ϕ is 3.4MBtu/MWh.

Finally, the optimal model of electricity-gas integrated system is established.

4 Strategy of Solution

4.1 Linear Method

The piecewise linearization method is utilized to simplified the objective function (1), whose principle is shown in Fig. 2.

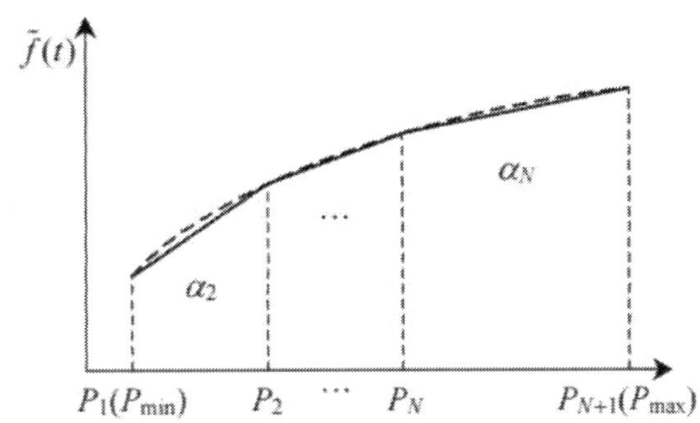

Fig.2. Piecewise linearization of fuel cost

Linearized model of fuel costs is shown as follows:

$$\bar{f}(t) = \sum_{i=1}^{N} (\alpha_i P_i(t) + \beta_i B_i(t)) \quad (30)$$

$$\alpha_i = \frac{f(P_{i+1}) - f(P_i)}{P_{i+1} - P_i} \quad (31)$$

$$\beta_i = f(P_i) - P_i \alpha_i \quad (32)$$

$$P_i B_i(t) \leq P_i(t) \leq P_{i+1} B_i(t) \tag{33}$$

$$U(t) = \sum_{i=1}^{N} B_i(t) \leq 1 \tag{34}$$

$$-M(1-\delta_{mn,t}) \leq \tilde{Q}_{mn,t}^{\ 2} \text{-} K_{mn}^{gf^2} \ (Pr_{mt}^2 - Pr_{nt}^2) \leq M(1-\delta_{mn,t}) \tag{35}$$

$$-M\delta_{mn,t} \leq -\tilde{Q}_{mn,t}^{\ 2} \text{-} K_{mn}^{gf^2} \ (Pr_{mt}^2 - Pr_{nt}^2) \leq M\delta_{mn,t} \tag{36}$$

$$-M(1-\delta_{mn,t}) \leq Pr_{mt} + Pr_{nt} \leq M(1-\delta_{mn,t}) \tag{37}$$

The linearization of A is completed through (35)-(37), and the linearization of Pr_{mt}^2、Pr_{nt}^2 are completed in the same way.

4.2Power Transfer Distribution Factor

$\mathbf{B_{bus}}$ represents the non-full rank node admittance matrix, $\boldsymbol{\theta}$ represents the voltage phase angle matrix, $\mathbf{P_{flow}}$ represents the node injection active column vector, $\mathbf{C}$ represents the branch-node correlation matrix, and $\mathbf{B_{branch}}$ represents the non-full rank branch susceptance matrix.

$$\mathbf{P_{inj}} = \mathbf{B_{bus}}\,\boldsymbol{\theta} \tag{38}$$

$$\mathbf{P_{flow}} = \mathbf{B_{branch}}\mathbf{C}\boldsymbol{\theta} \tag{39}$$

$$\mathbf{P_{flow}} = \mathbf{P_f^T}\mathbf{P_f}\mathbf{P_{flow}} = \mathbf{P_f^T}\begin{bmatrix}\mathbf{P_{flow}^{int}}\\ \mathbf{P_{flow}^{ext}}\end{bmatrix} \tag{40}$$

$$\mathbf{P_{inj}} = \mathbf{P_g^T}\mathbf{P_g}\mathbf{P_{inj}} = \mathbf{P_g^T}\begin{bmatrix}\mathbf{P_{G1}}\\ \cdot\\ \cdot\\ \cdot\\ \mathbf{P_{Gn}}\end{bmatrix} = \mathbf{P_g^T}\mathbf{P_{injection}} \tag{41}$$

Where, $\mathbf{P_f}$ represents the permutation matrix sorted by intra-class road flow, $\mathbf{P_g}$ represents the permutation matrix sorted by injection classification, and $\mathbf{P_{injection}}$ represents the active injection vector arranged according to the classification order [21].

5 Case Analysis

The six-bus power system on the left includes three generators G2-G4, two gas turbines G1, G5, a P2G device, a wind farm, seven transmission lines and three loads, which is shown in Fig. 3. The P2G equipment has a capacity of 37.5 MW and an efficiency of 0.64 [22]. The fuel price of G2-G4 is \$2.5/MBtu. The seven-node gas system includes two gas wells W1 and W2, a gas storage S1, six pipelines PL1-PL6, and five gas loads. Both the initial and final bus packages are set to 106800 kcf. And the forecast values of wind power an loads in next 24 hours are shown in Fig. 4.

The optimal result of output active power of P2G devices is shown Table I according to the day-ahead optimal dispatch model proposed in this paper.

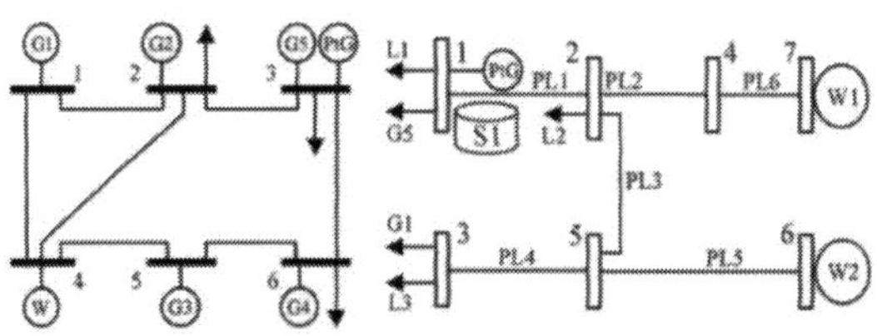

Fig.3. 6-node power system/7-node natural gas system.

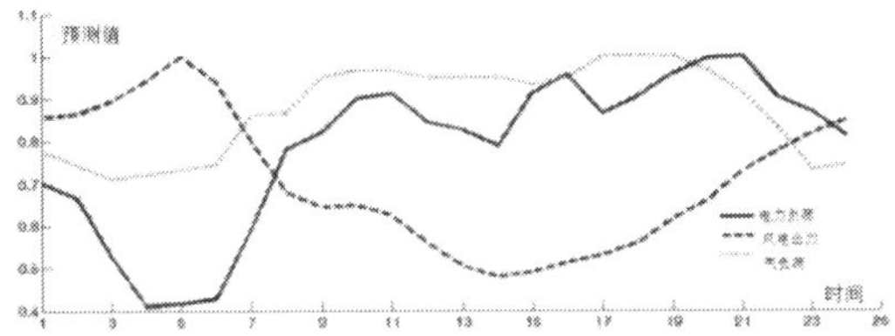

Fig.4. Forecast values of wind power and loads in the next 24 hours.

TABLE I. OUTPUT ACTIVE POWER OF P2G DEVICES (MW)

	t4	t5	t6
m1	77.481	79.532	6.676
i3	36.533	37.500	4.148

Table I shows that the P2G equipment at node 1 of the natural gas system has consumed 77.471 MW, 79.532 MW and 6.676 MW of wind power during t4, t5, t6, respectively. The P2G equipment at node 3 of the left power system also consumed 36.533 MW, 37.500 MW, 3.148 MW of wind power during t4, t5, t6. The P2G equipment converts this wind energy into natural gas and stores it in the natural gas system when the power load is low, as shown in Fig 4 It can also be seen that the power load from t3 to t6 is at a trough, the output of wind farm is at its peak at this time, and the generated electric energy cannot be consumed by the electric load, so it is converted into natural gas through the P2G equipment and enters the natural gas system.

In the time period from t3 to t6, the gas load increased by about 250kcf. Firstly, if the natural gas system is taken out independently, the increase in the system gas load will inevitably increase the amount of natural gas entering the system from the gas wells W1 and W2, or the gas storage tank Gas enters the system, and in this model, the gas output of the gas storage tank is 0 during the period from t3 to t6, and the intake of the gas wells W1 and W2 into the only channels PL5 and PL6 of the natural gas system does not show much fluctuation , The natural gas volume of the PL5 pipeline was maintained at 2000kcf, and the intake volume of the PL6 pipeline did not increase but decreased, from 4148.443kcf to 3939.483kcf. Combined with the previous analysis, it can be seen that during this period of time, the natural gas produced by the P2G facility enters the natural gas system, which offsets the increase in gas load and reduces the speed requirements of the gas well in response to load fluctuations.

Combining Fig.4 and Simulation results from GAMS, we find that when natural gas load is low and the electric load rises(t14-t16), the electric load increases by about 90 MW. At same time, the operation result of the optimal dispatch model shows that the gas turbine G5 at i3 increased its output from 10 MWh to 50 MWh. The output of the gas turbine at i1 increased by 12 MWh. At this time, the output of non-gas turbine G2 at i2 and non-gas turbine G4 at i6 increased by 35 MWh.

If the power system and the natural gas system are completely independent at this time, the additional load about 90 MW needs to be fully borne by the non-gas turbine unit, which will greatly increase the power generation cost of the non-gas generator unit, leading to the activation of the backup generator unit , Resulting in a further increase in power generation costs.

The operation results show that the minimum value of the objective function, that is, the lowest comprehensive energy cost is 537,995.392 USD. If the P2G equipment are removed from the model, the energy cost will be 597,756.862 USD, By contrast, P2G saves about 10% of economic costs and creates a good Economic benefits.

P2G technology plays a significant role in eliminating wind power, it can reduce the impact of load fluctuations on both power system and natural gas system, can save power generation costs.

6 Conclusion

The electricity-gas integrated system day ahead scheduling model makes the time characteristic curve of power load and gas load more stable, effectively reducing the difficulty of responding to load fluctuations in the power system, alleviating the pressure of power dispatch, reducing the cost of power generation. at the same time, it also effectively relieves the congestion of the natural gas network, and has a very positive impact on the optimal dispatch of the natural gas system. The model in this paper realized optimal economy and efficient use of energy. The subject has great research value in the future.

References

[1] Y. Li *et al.*, "Day-ahead schedule of a multi-energy system with power-to-gas technology," *2017 IEEE Power & Energy Society General Meeting*, Chicago, IL, 2017, pp. 1-5,
[2] DONG Zhaoyang, ZHAO Junhua, WEN Fushuan, et al. From Smart Grid to Energy Internet:Basic Concept and Research Framework[J]. Automation of Electric Power Systems, 2014, 38(15):1-11.
[3] PPG highlights solutions for wind power market at China Wind Power 2014[J]. Focus on Powder Coatings,2015,2015(1).
[4] An Analysis of Prospects for Application of Large-scale Energy Storage Technology in Power Systems[J]. Automation of Electric Power Systems, 2013, 37(1):3-8..
[5] WANG Chengmin, SUN Werqing, YI Tao, et al.Review on energy storage application planning and benefit evaluation methods in smart grid [J]. Proceedings of the CSEE, 2013, 33(7): 33-41.
[6] Wei Zhinong, Zhang Side, Sun Guoqiang, et al. Power-to-gas considered peak load shifting research for integrated electricity and natural-gas energy systems[J]. Proceedings of the CSEE, 2017, 37(16) : 4601-4609(in Chinese).
[7] KUDO A , MISEKI Y. Heterogeneous photocatalyst materials for water splitting [J] . Chemical Society Reviews , 2009 ,38 (1): 253-278.
[8] LIU Ming-yi, YU Bo, XU Jing-ming. Efficiency of solid oxide water electrolysis system for hydrogen production [J] . Journal of Tsinghua University:Science and Technology, 2009, 49(6):868 —871. (in Chinese).
[9] Lombardi P, Sokolnikova T, Suslov K, et al. Power to Gas as an alternative energy storage solution to integrate a large amount of renewable energy: economic and technical analysis[J]. system, 2011, 11: 12.
[10] Schaaf T, Grünig J, Schuster M R, et al. Methanation of CO 2-storage of renewable energy in a gas distribution system[J]. Energy, Sustainability and Society, 2014, 4(1): 2.
[11] LIU C J, GUO Q T, YE J Y, et al. Perspective on catalyst investigation for CO 2 conversion and related issues[J]. CIESC Journal, 2016, 67(1): 6-13..
[12] HUANG Ming, WU Yong. , WEN Xizhi, et al. Feasibility a- nalysis of hydrogen transport in natural gas pipeline [J] . Gas & Heat, 2013, 4(33): 39 —42. .
[13] Judd R, Pinchbeck D. Hydrogen admixture to the natural gas grid[M]//Compendium of Hydrogen Energy. 2016: 165-192.
[14] International Electrotechnical Commission. Electrical energy storage white paper[J]. Geneva, Switzerland: International Electrotechnical Commission, 2011: 1-78.
[15] YAN Xiao-hui, XU Yu-jie, JI Lv, et al. Forcasting and analysis on large-scale energy storage technologies in China [J]. Electric Power, 2013, 46(8): 22-29.
[16] YUAN Xiaoming, CHENG Shijie, WEN Jinyu. Prospects Analysis of Energy Storage Application in Grid Integration of Large-scale Wind Power[J]. Automation of Electric Power Systems, 2013, 37(1):14-18.
[17] LUO C X. Application of power-to-gas technology for low carbonation [J]. Sino-Global Energy, 2017, 22(4): 20-26.

[18] Tomasgard A, Rømo F, Fodstad M, et al. Optimization models for the natural gas value chain[M]//Geometric modelling, numerical simulation, and optimization. Springer, Berlin, Heidelberg, 2007: 521-558.
[19] Wu Xiong, Wang Xiuli, Wang Jianxue, et al. Economic generation scheduling of a microgrid using mixed integer programming [J]. Proceedings of the CSEE, 2013, 33(28): 1-9
[20] Yin Zili, Chen Yuxing. A power flow computation method for reduction grid based on power transfer distribution factor [J]. Power System Protection and Control, 2016, 44(17) : 25-30.
[21] Li T, Eremia M, Shahidehpour M. Interdependency of natural gas network and power system security[J]. IEEE Transactions on Power Systems, 2008, 23(4): 1817-1824.
[22] Liu C, Shahidehpour M, Fu Y, et al. Security-constrained unit commitment with natural gas transmission constraints[J]. IEEE Transactions on Power Systems, 2009, 24(3): 1523-1536.
[23] Saldarriaga C A, Hincapié R A, Salazar H. A holistic approach for planning natural gas and electricity distribution networks[J]. IEEE transactions on power systems, 2013, 28(4): 4052-4063.
[24] Jentsch M, Trost T, Sterner M. Optimal use of power-to-gas energy storage systems in an 85% renewable energy scenario[J]. Energy Procedia, 2014, 46: 254-261.
[25] WANG Yelei, ZHAO Junhua, WEN Fushuan, et al. Market Equilibrium of Multi-energy System with Power-to-gas Functions[J]. Automation of Electric Power Systems, 2015, 39(21):1-10..
[26] Tian Yuanyuan, Liao Qingfen, Liu Dichen, et al. Planning of urban distribution network considering the integrated energy supply-side reform[J].Power System Technology, 201, 40(10): 2924-2933(in Chinese).
[27] HUANG Guori, LIUWeijia, Wen Fushuam, et al Collaborative planning of integrated electricity and natural gas energy systems with power -to-gas stations[J].Electric Power Construction, 2016, 37(9):1-13..
[28] LIU Kaijun. Development path exploration of energy internet [J] Electric Power Construction, 2015, 36(10) : 5-10.

An iterative identification algorithm and its convergence analysis of closed-loop power system based on ambient signals

Weijie Du [1], Miao Yu [1,2,*], Jinglin Li [1], Shouzhi Zhang [1], and Jingxuan Hu [1]

[1] School of Mechanical and Vehicle Engineering, Beijing University of Civil Engineering and Architecture, Beijing, 100044, China

[2] Beijing Key Laboratory of Service Performance of Urban rail Transit Vehicles, Beijing University of Civil Engineering and Architecture, Beijing, 100044, China

*Corresponding author's e-mail: yumiao@bucea.edu.cn

Abstract. Identification and control are important problems of closed-loop power system.At present, most studies are separate identification methods. This paper studies an online and real-time integrated identification method, which can solve the problems of model set and controller design of closed-loop power system.This paper investigates a new iterative identification algorithm and its convergence problem of closed-loop power system based on ambient signals. Firstly, the whole algorithm procedure is given. This algorithm uses the iterative process under the closed-loop condition, which combines system model identification with controller design. Then the complementary of model identification and control design has been realized. Secondly, because of the dynamic performance of the iterative identification algorithm, it has characteristics described from the perspective of a partitioned dynamic system. Regard each iterative identification step as a state node. In this situation, the algorithm guarantees all the state nodes converge to the Lyapunov stable equilibrium. Finally, the simulation results show the correctness and effectiveness of the proposed method through the simulation of a power system with four-machine-two-region.

1. Introduction

The power system model is important to help design a correct and effective controller and also to analyze the dynamic characteristics of the power system, the controller can help the power system to maintain stability[1]. Recently, more and more scholars have started to study the problem of identification of control[2]. Nie Yonghui et. al. proposed a new method to design wide area damping controller based on artificial immune system theory. The improved fireworks algorithm was used to optimize the parameters of immune controller, which could improve the stability of power system[3]. Li Peiping et al. proposed an adaptive wide-area coordinated damping controller based on the objective representation heuristic dynamic programming algorithm, which could suppress the low frequency oscillation in the weak damping interval of the system and improved the transient stability of the system[4]. Zhang Yan et. al. analyzed the damping characteristics of doubly fed induction generator (DFIG) grid-connected system based on energy method, and discussed the relationship between damping coefficient and system damping, which characterized oscillation energy dissipation, from the physical level[5].Yang Jing et al. designed a subsynchronous oscillation robust damping controller based on the additional control of direct-drive wind turbine, which could effectively improve the adaptability of subsynchronous oscillation damping control[6]. In this paper, a new control method is designed based on iterative identification, which can solve the problems of model design and controller design that are difficult to solve in current power systems by analyzing the convergence of algorithms for closed-loop power systems. The first chapter of this paper gives the introduction, and the second chapter gives the model to identify the environmental signals and the mathematical model of the closed-loop power system. A new control and iterative identification method is proposed in the third chapter. The simulation results and effect analysis of the identification are given in Section IV. Chapter 5 is the conclusion and outlook.

2. Problem formulation and preliminaries

2.1. Problem formulation

The closed-loop system environment signal model is shown in Figure 1, where we can add disturbance signals to the power system to better simulate the disturbance signals of random nature. The input signal in Figure 1 is u and i is the output signal. The random interference signals are e_{10} and e_{20}, and their variances can be represented by λ_{10} and λ_{20} with a mean value of zero. Subject to an excitation signal with an external r. The filters are H_{10} and H_{20}, which are stable and reversible. W_0 is the feedback controller. In addition, Fig.1(b) represents the closed-loop identification model, y represents the system input signal, and u represents the output signal of the system. $G(q,\theta)$ is the control system parameter of positive feedback, and $H(q,\theta)$ represents the disturbance filtering model above the positive direction. From Fig.1(b) can Identify the mathematical equation as shown in 7[7]:

$$y(t) = G(q,\theta)u(t) + H(q,\theta)e(t)$$
$$u(t) = -W(q,\theta)y(t) + r(t) \tag{1}$$

where $H(q,\theta) = \dfrac{F(q,\theta)}{E(q,\theta)}, G(q,\theta) = \dfrac{B(q,\theta)}{A(q,\theta)}, W(q,\theta) = \dfrac{R(q,\theta)}{P(q,\theta)}$

the model used for recognition is represented:

$$\begin{bmatrix} y \\ u \end{bmatrix} = \begin{bmatrix} \dfrac{G}{1+GW} & \dfrac{H}{1+GW} \\ \dfrac{1}{GW+1} & -\dfrac{WH}{GW+1} \end{bmatrix} \begin{bmatrix} r \\ e \end{bmatrix} = \Xi(G,H,W)\begin{bmatrix} r \\ e \end{bmatrix} \tag{2}$$

$$\begin{bmatrix} \hat{y} \\ \hat{u} \end{bmatrix} = \Xi(\hat{G},\hat{H},W)\begin{bmatrix} r \\ e \end{bmatrix}$$

The uncertainty set for error prediction for each iteration of the identification set from the identification model in the power system represented in Figure 1 is given by the following equation:

$$\zeta_{PE}(G_k, \Upsilon_N, \Upsilon_D, R) \triangleq \left\{ G(q,\theta) \middle| \begin{array}{l} G(q,\theta) = \dfrac{G_k + \Upsilon_N\theta}{1+\Upsilon_D\theta}, \theta \in U_\theta \\ U_\theta = \{\theta | (\theta-\hat{\theta})^T R(\theta-\hat{\theta}) \le \chi_a \end{array} \right\} \tag{3}$$

where $\theta \in R^{k\times 1}$ is a power system model parametrization, G_k is a known transfer function, $R \in R^{k\times k}$ is the definite matrix of symmetric positive, the transfer function vectors are Υ_N, Υ_D,and the estimated value is $\hat{\theta}$. Uncertainty sets are easily obtained in closed-loop power systems using prediction error identification methods.

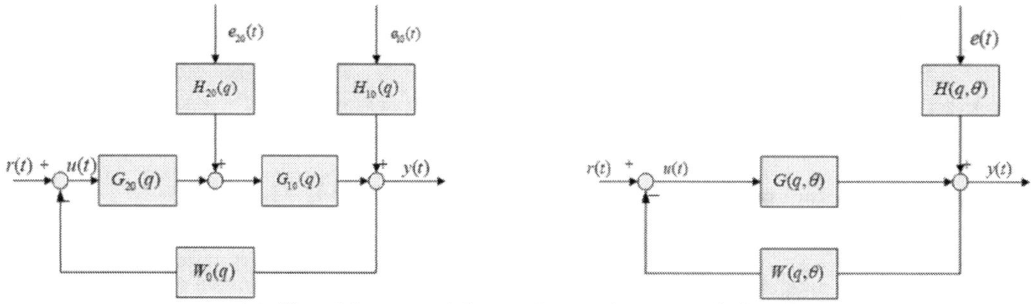

(a) Closed-loop model contains environmental signals
(b) The identification model corresponding to model (a)
Fig.1 Power system closed-loop model includes environmental signals

2.2. Preliminaries

Table 1 indicates a summary of some notations, and directed graphs are used to indicate dynamic networks with a graphical topology and dynamic networks with unknown random time delays.

Table 1 Explanation of symbols

Symbols	Meaning
$\mathbb{R}^n$	n-dimensional Euclidean space
$\overline{\mathbb{R}}^n_+$	nonnegative orthants of $\mathbb{R}^n$
$\mathbb{R}^n_+$	positive orthants of $\mathbb{R}^n$
$x \geq\geq 0, x \in \mathbb{R}^n$	every component of x is nonnegative
$x \gg 0, x \in \mathbb{R}^n$	every component of x is positive
$A \geq\geq 0, A \in \mathbb{R}^{n\times m}$	nonnegative matrix
$A \gg 0, A \in \mathbb{R}^{n\times m}$	positive matrix
$A \geq 0, A \in \mathbb{R}^{n\times n}$	nonnegative definite matrix
$(\bullet)^{\mathrm{T}}$	matrix transpose
$(\bullet)^{\mathrm{D}}$	Drazin generalized inverse
$\lVert\bullet\rVert$	Euclidean vector norm
$\mathcal{G} = (\mathcal{V}, \mathcal{E}, \mathcal{A})$	weighted directed graph
$\mathcal{V} = \{1,\ldots,n\}$	set of vertices
$\mathcal{E} \subseteq \mathcal{V} \times \mathcal{V}$	direction of information flow
$\mathcal{A}_{(i,j)} = a_{ij} > 0$	weighted adjacency matrix
a_{ij}	weight of each edge

The basic definition of convergence analysis of iterative recognition algorithms is given below.

Definition 1. The dynamic network $\mathcal{G}$ with timedelay an be expressed in the following form:

$$\dot{\xi}(t) = f(\xi(t)) + f_d(\xi(t-\tau_1),\ldots,\xi(t-\tau_{nd}))$$
$$\xi(\theta) = \phi(\theta), -\tau^* \leq \theta \leq 0, t \geq 0 \tag{4}$$

where $f_d : \mathbb{R}^n \times \cdots \times \mathbb{R}^n \to \mathbb{R}^n$ is locally Lipschitz continuous, $f_d(0,\ldots,0) = 0, \tau^* = max_{i\in\{1,\ldots,n_d\}}\tau_i, i = 1,\ldots,n_d$ and the vector $\xi(t) \triangleq [\xi_1(t),\ldots,\xi_n(t)]^{\mathrm{T}}$ is the state of algorithm nodes; $\phi(\bullet) \in C = C([-\tau^*,0],\mathbb{R}^n)$ is a continuous vector-valued function and satisfies the initial state of (4).

Definition 2. The (4) in a nonlinear time-delay dynamic system is a compartmentalized dynamic system if $F(\xi) \triangleq f(\xi) + f_d(\xi,\xi,\ldots,\xi)$ and $F(\bullet)$ is compartmental, where $f(\bullet)$ and $f_d = [f_{d1},\ldots,f_{dn}]^{\mathrm{T}}$ are given by

$$f_i(\xi(t)) = -\sum_{j=1, j\neq i}^{n} a_{ji}(\xi(t))$$
$$f_{di}(\xi(t-\tau_1),\ldots,\xi(t-\tau_{n_d})) = \sum_{j=1, j\neq i}^{n} a_{ij}(\xi(t-\tau_{ij})) \tag{5}$$

$\tau_{ij}, i \neq j, i, j \in \{1,\ldots,n\}$ means the logistics transfer time from compartment j to i, and $a_{ij}(\xi(\bullet)) \geq 0, \xi(\bullet) \in C_+, i \neq j, i, j \in \{1,\ldots,n\}$ means the instantaneous logistics rate from compartment j to i. From the above, the form of nonlinear dynamic network with G time lag can be written as

$$u(t) = \dot{\xi}(t) = f(\xi(t)) + \sum_{i=1}^{n_d} f_{d_i}(\xi(t-\tau_i))$$
$$\xi(\theta) = \phi(\theta), -\tau^* \leq \theta \leq 0, t \geq 0 \tag{6}$$

where $f: \bar{\mathbb{R}}_+^n \to \bar{\mathbb{R}}_+^n$ is given by $f(\xi) = [f_1(\xi_1),\ldots,f_n(\xi_n)]^T, f(0) = 0, f_{d_i}: \bar{\mathbb{R}}_+^n \to \bar{\mathbb{R}}_+^n, i = 1,\ldots,n_d$ and $f_d(0) = 0$

Furthermore, $f_i(\bullet), i = 1,\ldots,n$ and $f_i(0) = 0$.

3. Iterative identification algorithm and its convergence analysis

3.1. Iterative Identification Algorithm

In power system identification, how to evaluate the quality of the identification model is the main problem that we pay attention to. Currently, scholars believe that identification error is composed of two parts. The two components are bias error and variance, respectively. The variance is easily caused by the random noise of the observed data when performing system identification experiments. The bias error is brought about by the limited degree of model complexity[8],which can be described as follows:

$$G_0(e^{jw}) - G(e^{jw},\hat{\theta}_N) = \underbrace{G_0(e^{jw}) - G(e^{jw},\theta^*)}_{bias\ error} + \underbrace{G(e^{jw},\theta^*) - G(e^{jw},\hat{\theta}_N)}_{variance\ error} \tag{7}$$

where $G_0(e^{j\omega})$ is a real power system model, $G(e^{j\omega},\hat{\theta}_N)$ is an identification model, θ^* is the convergence point of $\hat{\theta}_N$,then $G(e^{j\omega},\theta^*)$ is the asymptotic convergence model of $G(e^{jw},\hat{\theta}_N)$.

During the research of power system iterative identification, model uncertain sets are used to analyze the robust performance and control performance. If an uncertain model set is ζ ,a real power system object is $G_0 \in \zeta$ and the power system estimation model is $\hat{G}$,then the robust controller W must satisfy two characteristics. One is that the controller W can stabilize all the models within the set ζ. The other is that the controller W can meet the control performance requirements of set ζ. So the structure of power system uncertain model set is very important during the whole design process.

To get the identification model of the power system and to implement the algorithm for iterative identification, the whole process is summarized according to[9] as follows:

① Fig.1(b) show the closed-loop power system, G₀ is the unknown model, Wi=0 is the initial controller, is chosen to satisfy the system stability;

② In order to obtain the uncertainty set nominal model Gi can be obtained (G, Wi) identification conditions.

③ Calculate $\delta_W(G_i,\zeta_i)$, and written as ρ ,and design controller W_{op};

④ Judgment condition if the condition is satisfied $W_{op} \in W(G_i,\zeta_i)$,if $W_{op} \in W(G_i,\zeta_i)$,then judge whether the performance indicator is satisfied, otherwise if find the optimal controller in W_{i+1} from $W(G_i,\zeta_i)$;

⑤ If the program requirements are not met④. Let $i = i+1$ and repeatedly do from the program②to ⑤

3.2. Convergence analysis

Based on the dynamic behaviour of iterative identification algorithm, we use the compartmentalized system model defined in equation (6) to describe the iterative identification algorithm. In this section, we will prove that this dynamic system with time lags meets the requirement of semi-stability. First, we give some strict mathematical definitions of synchronization.

Definition 3. $D \subseteq \Omega$ is a nonempty subset, if $\xi_i^0(t) \in D$, make it possible for $\xi(t) \in \Omega$ for all $t \geq 0$, and

$$\lim_{t\to\infty} \left\| \xi_i(t) - \xi_v(t) \right\| = 0, 1 \leq i \leq n \tag{8}$$

$\mathcal{G}$ s a dynamic network illustrating that network synchronization is achieved and $D \times \cdots \times D$ is called the synchronization region of the dynamic network.

Note that the stability of the zero solution needs to be slightly corrected for the normal stability definition given in [10-11] to deal with time delay non negative systems.In practice, A subset of equilibrium solutions of the system is defined in the literature [12]. The following definitions generalize non negative concepts to vector fields.

Definition 4[12]. f is nonnegative if $f_i(\xi) \geq 0$ for all $i = 1,...,n$ and $\xi \in \overline{\mathbb{R}}_+^n$ such that $\xi_i(t) = 0$, where $f = [f_1,...,f_n]^{\mathrm{T}} : \mathcal{H} \to \mathbb{R}^n$.

Definition 5[13]. According to the nonlinear delay system in (4). If $f(\bullet)$ is nonnegative , $f_d(\bullet)$ is nonnegative, so we get $\phi(\bullet) \in C_+, C_+ \triangleq \{\varphi(\bullet) \in C : \varphi(\theta) \geq 0, \theta \in [-\tau^*, 0]\}$, and solve $\xi_i(t)$ with $t \geq 0$ and $\mathcal{G}$ are non-negative.

Lemma[13]. If $\mathbf{e}^{\mathrm{T}}(f(\xi) + \sum_{i=1}^{n_d} f_{d_i}(\xi)) = 0$, there exist nonnegative diagonal matrices $Q_i \in \overline{\mathbb{R}}_+^{n\times n}$, such that $Q \triangleq \sum_{i=1}^{n_d} Q_i > 0$

$$Q_i^D Q_i f_{d_i}(\xi) = f_{d_i}(\xi), \xi \in \overline{\mathbb{R}}_+^n, i = 1,\ldots,n_d \tag{9}$$

$$\sum_{i=1}^{n_d} f_{d_i}^{\mathrm{T}}(\xi) Q_i f_{d_i}(\xi) \leq f^{\mathrm{T}}(\xi) Q f(\xi), \xi \in \overline{\mathbb{R}}_+^n \tag{10}$$

Proof. the Lyapunov functional $V : C_+ \to \mathbb{R}$ is given by

$$V(\zeta(\bullet)) = -\sum_{i=1}^{n} \int_0^{\zeta_i(0)} Q_{(i,i)} f_i(\vartheta) d\vartheta + \sum_{i=1}^{n_d} \int_{-\tau_i}^{0} f_{d_i}^{\mathrm{T}}(\zeta(\theta)) Q_i f_{d_i}(\zeta(\theta)) d\theta \tag{11}$$

It follows that $V(\zeta) \geq \sum_{i=1}^{n} Q_{i,i}[-f_i(\delta_i \zeta_i(0))]\zeta_i(0) > 0$ for all $\zeta(0) \neq 0$,where $0 < \delta_i < 1$. Due to exists a class $\mathcal{K}$ function $M(\bullet)$, $f_i(\bullet)$ is a strictly decreasing function, so there such that $V(\zeta) > M(\|\zeta(0)\|)$ .The derivative of $V(\xi_t)$ along the trajectories of (6) is

$$\begin{aligned}
\dot{V}(\xi(t)) &= -f^{\mathrm{T}}(\xi(t) Q \dot{\xi}(t)) + \sum_{i=1}^{n_d} f_{d_i}^{\mathrm{T}}(\xi(t)) Q_i f_{d_i}(\xi(t)) - \sum_{i=1}^{n_d} f_{d_i}^{\mathrm{T}}(\xi(t-\tau_i)) Q_i f_{d_i}(\xi(t-\tau_i)) \\
&= -f^{\mathrm{T}}(\xi(t)) Q f(\xi(t)) - \sum_{i=1}^{n_d} f^{\mathrm{T}}(\xi(t)) Q f_{d_i}(\xi(t-\tau_i)) + \sum_{i=1}^{n_d} f_{d_i}^{\mathrm{T}}(\xi(t)) Q_i f_{d_i}(\xi(t)) - \sum_{i=1}^{n_d} f_{d_i}^{\mathrm{T}}(\xi(t-\tau_i)) Q_i f_{d_i}(\xi(t-\tau_i)) \\
&\leq -f^{\mathrm{T}}(\xi(t)) Q f(\xi(t)) - \sum_{i=1}^{n_d} f^{\mathrm{T}}(\xi(t)) Q Q_i^D Q_i f_{d_i}(\xi(t-\tau_i)) - \sum_{i=1}^{n_d} f_{d_i}^{\mathrm{T}}(\xi(t-\tau_i)) Q_i Q_i^D Q_i f_{d_i}(\xi(t-\tau_i)) \\
&\quad -\sum_{i=1}^{n_d} [Q f(\xi(t)) + Q_i f_{d_i}(\xi(t-\tau_i))]^{\mathrm{T}} \times Q_i^D [Q f(\xi(t)) + Q_i f_{d_i}(\xi(t-\tau_i))] \leq 0, \quad t \geq 0
\end{aligned} \tag{12}$$

from $f^{\mathrm{T}}(\xi) Q f(\xi) = \sum_{i=1}^{n_d} f^{\mathrm{T}}(\xi) Q Q_i^D Q f(\xi), \xi \in \overline{\mathbb{R}}_+^n$ and since the positive orbit $\rho^+(\phi(\theta))$ of (6)is delimited,where $\rho^+(\phi(\theta)) \in C_+$, we can get (11). According to the [10], we know that $\lim_{t\to\infty} \xi(t) = \mathcal{M}$,where $\mathcal{M}$ indicate the largest invariant set contained in $\mathcal{P}$. Since $\mathbf{e}^{\mathrm{T}}(f(\xi) + \sum_{i=1}^{n_d} f_{d_i}(\xi)) = 0, \xi \in \overline{\mathbb{R}}_+^n$ and

$\mathcal{P}\subset\hat{\mathcal{P}}\triangleq\{\zeta(\bullet)\in C_+ : f(\zeta(0))+\sum_{i=1}^{n_d} f_{d_i}(\xi(-\tau_i))=0\}=\{\zeta(\bullet)\in C_+ : \zeta(\theta)=M\mathbf{e}, \theta\in[-\tau^*,0], M\geq 0\}$, which implies that $\lim_{t\to\infty}\xi(t)=\hat{\mathcal{P}}$. Then, we consider the following function $W: C_+\to\mathbb{R}$ by $W(\zeta(\bullet))=\mathbf{e}^{\mathrm{T}}\zeta(0)+\sum_{i=1}^{n_d}\int_{-\tau_i}^{0}\mathbf{e}^{\mathrm{T}} f_{d_i}(\zeta(\theta))d\theta$ .Thus, for all $t\geq 0$, along the trajectory of (6) is:

$$W(\xi(t))=W(\phi(\bullet))=\mathbf{e}^{\mathrm{T}}\phi(0)+\sum_{i=1}^{n_d}\int_{-\tau_i}^{0}\mathbf{e}^{\mathrm{T}} f_{d_i}(\phi(\theta))d\theta \tag{13}$$

which the implie that $\xi(t)\to\hat{\mathcal{P}}\cap\mathcal{T}$,where $\mathcal{T}\triangleq\{\zeta(\bullet)\in C_+ : W(\zeta(\bullet))=W(\phi(\bullet))\}$. Hence, $\hat{\mathcal{P}}\cap\mathcal{T}=\{M^*\cdot\mathbf{e}\}$ and $\lim_{t\to\infty}\xi(t)=M^*\cdot\mathbf{e}$, where M^* satisfie (11). Finally, we use the Lyapunov functional as follows

$$V(\zeta(\bullet))=-\sum_{i=1}^{n}\int_{M}^{\zeta_i(0)} Q_{(i,i)}(f_i(\vartheta)-f_i(M))d\vartheta+\sum_{i=1}^{n_d}\int_{-\tau_i}^{0}[f_{d_i}(\zeta(\theta))-f_{d_i}(M\mathbf{e})]^{\mathrm{T}} Q_i[f_{d_i}(\zeta(\theta))-f_{d_i}(M\mathbf{e})]d\theta$$

and noting that $V(\zeta)\geq\sum_{i=1}^{n_d} Q_{i,i}[f_i(M)-f_i(M+\delta_i(\zeta_i(0)-M))]\bullet(\zeta_i(0)-M)>0$, for all $\zeta_i(0)\neq M$, where $0<\delta_i<1$. So $M\cdot\mathbf{e}, M\geq 0$ is a semistable equilibrium point of (6)

4. Results and Discussion

We can verify the rationality of our algorithm through the previous sections, and the following simulation is verified through four motors with two regions:

$$\begin{aligned} G(q)&=\frac{q^4+0.708q^3+0.46q^2+0.248q+0.16}{q^5-0.83q^4-0.66q^3+0.285q^2+0.216q-0.34} \\ H(q)&=\frac{q^4-1.782q^3+1.423q^2-0.593q+0.08348}{q^4-1.8q^3+0.35q^2+1.016q-0.685} \\ W(q)&=\frac{-0.24+0.58q}{-0.6+q} \end{aligned} \tag{14}$$

We can see that fig 2 represents the interference curve and output curve of the fourth identification. The system sampling time of $T=5s$ was chosen so that an iterative identification procedure could be designed and a model for identification could be obtained, and the correctness of the method was verified in the simulation.

$$\begin{aligned} G_2(q)&=\frac{1.04q^4+0.652q^3-0.23q^2+0.148q+0.23}{0.805q^5-1.932q^4+0.36q^3+0.175q^2+0.6q-0.1805} \\ W_3(q)&=\frac{0.675q(q-0.4)}{(q+0.78)(q-0.1)} \end{aligned} \tag{15}$$

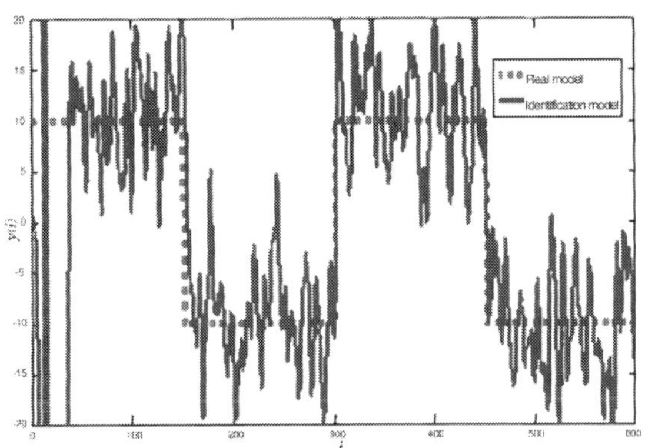

Fig.2 The fourth identification output in the power system

The performance of the real system, and the step responses curve of the closed-loop power system are represented in Fig 3. From figure, it can be obtained that after the fourth iteration of identification, $[G, W_3]$ is closed to the ture power system.

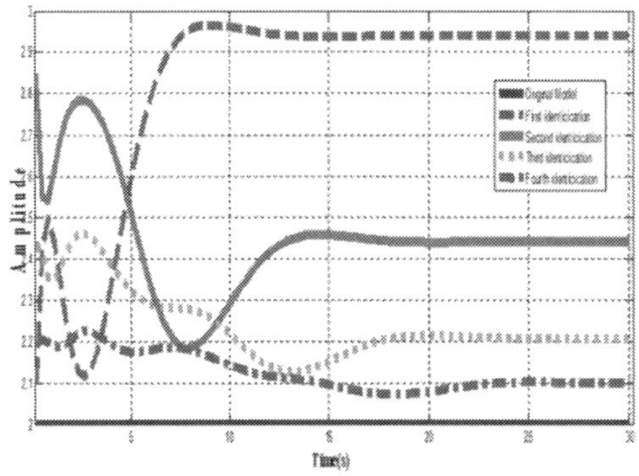

Fig.3 Closed loops step response curves of different types of power systems

The simulation result in Fig.3 tell us that the iterative identification mechanism is very superior to other methods and can make the power system stability improve and remain stable.

5. Conclusions

In this paper, we propose an iterative identification algorithm based on environmental signals, which can identify closed-loop power systems and solve the research work of setting and modeling the controller of the power system. This paper starts with an uncertain power system model and gives specific conditions for stability proof, from which Lyapunov asymptotic stability is found as the equilibrium of the dynamic network, proving that the force iterative identification algorithm is in a convergent state. The validity and reasonableness of the method proposed in this paper is verified by simulating the four-stage two-drive system. In the next step of the study, the time delay problem can be added and the effect of time lag on the power system controller.

Acknowledgments

This work is supported by State Key Laboratory of Pyramid Talent Training Project of Beijing University of Civil Engineering and Architecture (No.JDYC20200324); Graduate Education and Teaching Quality Improvement Project of Beijing University of Civil Engineering and Architecture (No.J2021016); the BUCEA Post Graduate Innovation Project(PG2021090); Key Project of Educational Science Research of Beijing University of Civil Engineering and Architecture (No.Y19-12); Security Control and Simulation for Power System and Large Power Generation Equipment (No.SKLD20M17); National innovation and entrepreneurship training program for College Students(No.202110016052, No.S202110016122, No.S202110016123, No.X202110016177, No.X202110016178, No.X202110016179); Project of Beijing Association of Higher Education(NO. YB2021131); Social Practice and Innovation and Entrepreneurship Curriculum Project of Beijing University of Civil Engineering and Architecture (No.SJSC1913); National Natural Science Foundation of China (No. 51407201)

References

[1] Ninness B, Goodwin G C. (1995) Estimation of model quality.Automatica. J ,31(12),1771–1797.
[2] Gevers M. (2007) A personal view of the development of system identification: A 30-year journey through an exciting field. IEEE Control Systems Magazine, 26(6), 93- 105..

[3] Yonghui N, Hantong X, Guowei C, et. al. (2020) Design of Wide-area Damping Controller Based on Immune System. J Power System Technology, 44(12): 4713-4721.
[4] Peiping L, Lingkang Z, Jianfeng W, et. al. (2020) Adaptive Wide-area Coordinated Damping Control Considering Bidirectional Random Communication Time-delay Compensation. J Power System Technology, 44(10): 56-65.
[5] Yan Z, Meng Z, Han G. (2021) Damping Control for Grid Connected DFIG System Based on Dissipated Energy Coefficient. J Power System Technology, 45(07): 2781-2795.
[6] Jing Y, Tong W, Jingtian B, et. al. (2020) Robust Damping Control of Subsynchronous Oscillation in Power System with Direct-drive Wind Turbines. J Automation of Electric Power Systems, 44(03): 56-65.
[7] Miao Y, Chao L. (2012) A new optimized method of excitation signal for closed-loop identification of power system based on ambient signals. J Applied Mechanics and Materials, 157-158,277-285.
[8] Hjalmarsson H. (2015) Experiment design and identification for control.Springer London; pp 396-405.
[9] Miao Y, Chao L. (2015) A new iterative identification and control method of closed-loop power system based on ambient signals. J Applied Mechanics and Materials, 798, 261-265.
[10] Hale J K, Lunel S V. (1993) Introduction to functional differential equations, Springer-Verlag,New York.
[11] Ebadat A,Valenzuela P, et. al. (2014) Applications oriented input design for closed-loop system identification: A Graph-Theory Approach.IEEE conference on Decision and Control;pp.15-17.
[12] Haddad W M,Chellaboina V. (2005) Stability and dissipativity theory for nonnegative dynamical systems: A unified analysis framework for biological and physiological systems. J Nonlinear Analysis:Real World Applications, 6(1), 35-65.
[13] Haddad W M,Chellaboina V. (2004) Stability theory for nonnegative and compartmental dynamical systems with time delay.System and Control Letters, 51(5), 355-361.

Wireless charging system load characteristic analysis and optimal load condition of of autonomous Underwater robot

Chong Li[1*], Bin Chen[1], Laifu Wei, Hong Yuan[1], and Haibo Liu[1]

[1]State Grid Intelligence Technology Co., LTD, Jinan, Shandong, 250000, China

*Corresponding author's e-mail: cse_pengf@ujn.edu.cn

Abstract. Autonomous underwater robots mainly obtains energy supplement through Wireless Power Transmission (WPT). Magnetic coupled resonant wireless charging is achieved through magnetic field coupling, with no direct connection on electricity, and can achieve intelligent control, autonomous charging, flexible and safe, frictionless wear. Based on the underwater robot wireless charging system, this paper proposes a high efficiency energy conversion technology for receiving end of wireless charging system. Though analysis of load change to effect the transfer characteristic of wireless charging system and wireless charging system optimal load conditions, this paper lays the foundation of the optimal load for subsequent research real-time tracking control, ensures the reliability of the charging system of underwater robot and improves the efficiency of wireless charging system.

1. Introduction

The safe and efficient operation of the charging system is the premise of ensuring the work of the robot. The traditional contact charging caused by frequent plugging and the sealing problem of underwater application all affect the safety of its underwater application. Due to the special working environment, wireless charging has a good applicability to underwater robots. Numerous and extensive research have been done on wireless charging in both military and civilian fields [1], The US Navy developed wireless charging systems for submarines, and electric vehicle companies developed wireless charging systems for electric vehicles.

Radio energy transmission can be divided into three types: microwave type, electromagnetic induction type, magnetic coupled resonant type [2]. The basic principle of the microwave type is electromagnetic radiation, with a long transmission distance but a low transmission efficiency. Electromagnetic induction transmission efficiency is high, but close transmission distance and high charging location. The magnetic coupled resonant has the advantages of transmission distance and transmission efficiency[3].

For magnetically coupled resonant wireless charging, the transmission distance and the load conditions are different under different working conditions[4]. Since the transmission characteristics of wireless charging system are greatly affected by transmission distance and load conditions, it is necessary to study the energy conversion technology at the receiving end of the robot charging system[5].

1.1. Wireless charging system modeling

The wireless charging system stimulates the resonant magnetic field, and the receiving terminal receives the energy from the power supply terminal, but it generally receives high-frequency AC power and a low voltage, while what is required for the inspection robot power battery charging is DC power, so it cannot be directly charged. The function of the receiving terminal is to convert the high-frequency AC power into the DC current for the robot charging. In this paper, the full-bridge rectifier circuit is selected to achieve larger average output current and smaller current pulsation. In addition, buck-Boost voltage lifting DC converter circuit is adopted in this paper, and its structure is shown in the figure1.

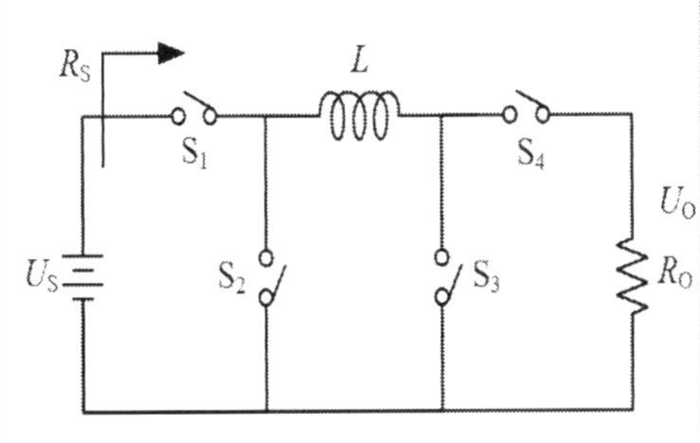

Figure 1 Schematic diagram of a four-switch buck-Boost circuit

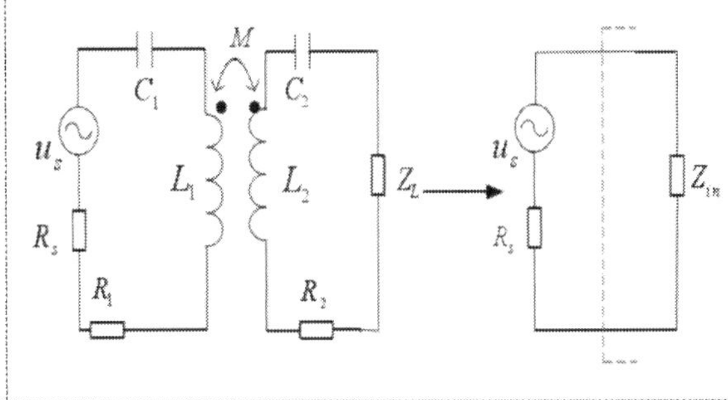

Figure 2 equivalent transformation plot of the two coil

Assuming the circuit is ideal circuit and each component is ideal component, the inspection robot load is equivalent to resistance R_O to simplify the analysis, The charging voltage at both ends is U_O. The coil includes the rectified filter circuit, BUCK-BOOST circuit and load equivalent to resistance R_S. The rectified output voltage is U_S. The resonant coil accepts voltage is U_S. The duty cycle of the MOS tube is d, cycle is T. According from the voltage relationship of BUCK-BOOST circuit in BUCK mode $U_S=(1/d)U_O$. Let $k=U_S/U_O$, according to the relationship between input and output voltages of BUCK-BOOST circuit in different modes, we can know $U_S=kU_O$, which can be obtained from the energy conservation law: $\frac{{U_S}^2}{R_S}=\frac{{U_O}^2}{R_O}$. Thus, the impedance transformation relation $R_S=k^2R_o$ is obtained. The charging load of inspection robot can be transformed from R_O to R_S by impedance transformation, so the optimal load can be obtained by impedance transformation for different charging requirements in this way.

2. Load characteristic analysis and the optimal load condition

In the case of resistive load, inductive load and capacitive load, the transformation relationship between the input impedance of WPT system and the equivalent impedance of the inspection robot charging is analyzed. Based on this, the influence of load change on the energy transmission characteristics of the system is studied. Finally, the optimal load condition of the system is obtained. It lays a foundation for the research of optimal load real-time tracking control and dynamic adjustment strategy of inspection robot charging load.

2.1 Analysis of resistance load characteristics

For the two-coil structure, when the load is pure resistive impedance, i.e. $Z_L=R_L$, and the two coils are resonant. The function of transmission efficiency and load can be derived to obtain the following formula:

$$R_L=\sqrt{R_2^2+\frac{R_2\omega^2M^2}{R_S+R_1}} \tag{1}$$

From formula. (1), the optimal load value of the efficiency is related to the resonant frequency and the coupling coefficient. The coupling coefficient is generally related to the transmission distance, coupling area, coil parameters, so for a specific coupling coefficient, at the resonant frequency, the transmission efficiency of the system will reach the maximum value.

Give parameters conventional assignments respectively, the relationship between load resistance and transmission efficiency under different coupling coefficients by MATLAB simulation as shown in Figure3.

From the figure 3, the figure shows that the transmission efficiency of the system increases with the load; when the coupling coefficient increases rapidly and changes gradually; each particular coupling coefficient has an optimal load value that increases with the coupling coefficient, which decreases with the transmission distance (the coupling coefficient decreases).

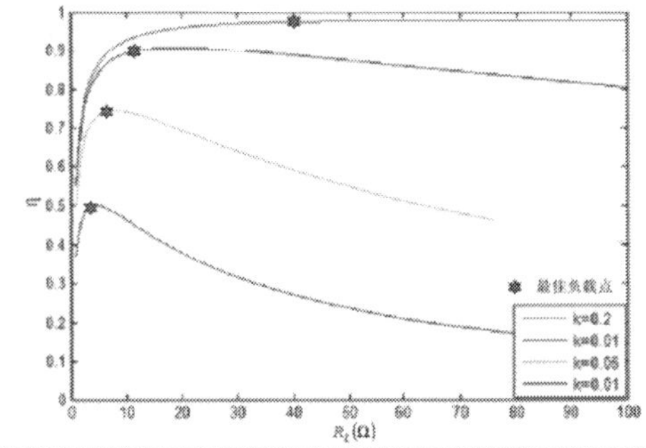

Figure3 Relationship between load resistance and transmission efficiency

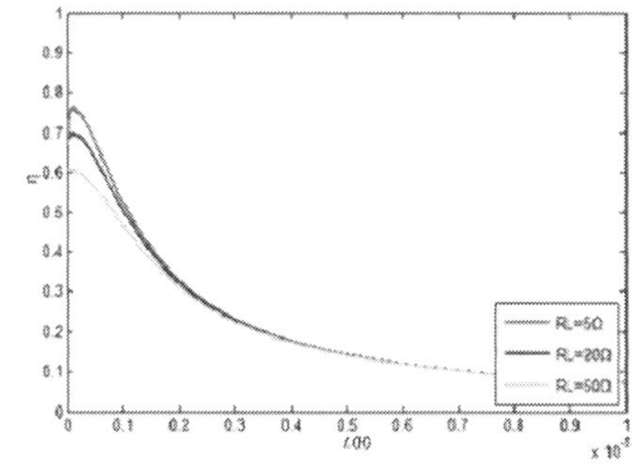

Figure4 Effect of inductive load inductive changes in both coils on efficiency

2.2 Analysis of non-resistive load characteristics

2.2.1 Analysis of the perceptual load characteristics

For the two-coil structure, when the load is a perceptual load, when the $Z_L=R_L+j\omega L$.

$$\eta = \frac{(\omega M)^2 (R_L + j\omega L)}{(R_2 + R_L + j\omega L)\left[(R_S + R_1)(R_2 + R_L + j\omega L) + (\omega M)^2\right]} \tag{2}$$

The figure 4 is the effect of load receptor change on the transmission efficiency of several fixed load resistance values based on the two-coil structure. It can be seen that when the electric inductance value is very small, the load resistance value has an obvious impact on the system transmission efficiency. With the increase of the inductance value, the three curves gradually tend to coincide. At this time, the main factor affecting the size of the transmission efficiency is the load inductance value.

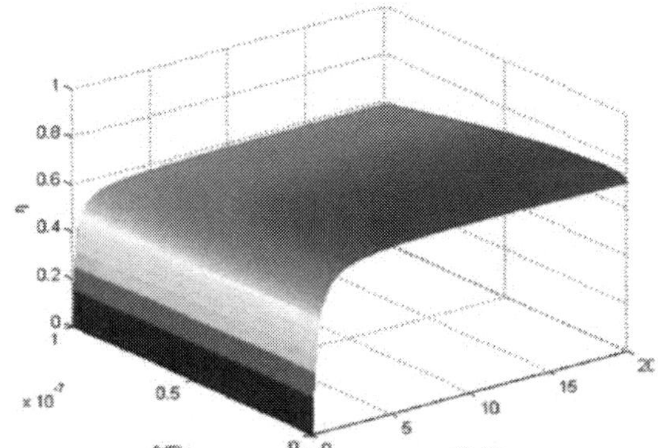

Figure5 Effect of variable tolerance load impedance parameters on efficiency

2.2.2 Analysis of tolerance load characteristics

For two-coil structures, when the load is a tolerance load, $Z_L=R_L+ 1/ j\omega C$. In the same way, give parameters conventional assignments respectively, It can be seen in figure5 that the transmission efficiency of the system decreases with the load impedance capacitance value and decreases first with the load impedance value.

2.3 Optimum load characteristics at different frequencies

In order to obtain the optimal load changes with the system operating frequency under different load properties, this section is analyzed by finite element simulations under the two-coil structure. When selecting parameters, parameters other than frequency can be fixed and the effect of the three loads on the system transmission efficiency and output power can study the frequency changes alone. The simulation results are shown in the figure 6.

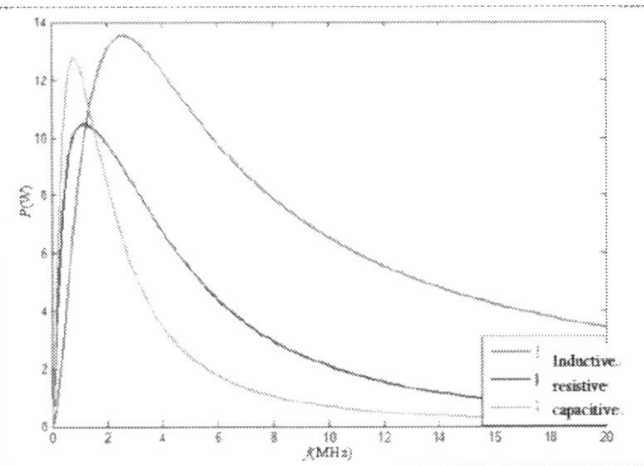

Figureure6 The power changes with frequency in the three load cases

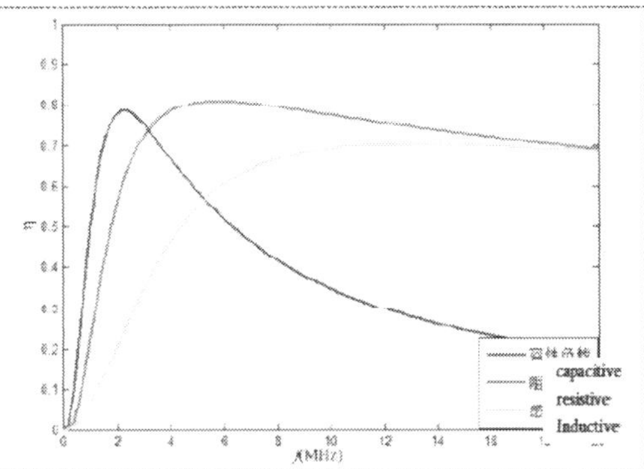

Figure7 The efficiency changes with frequency in the three load cases

It can be concluded from figure 6 that the operating frequency has a large power near the resonant frequency of the system, the pure resistance load adaptability to frequency change is relatively weak, the perceptual load and tolerance load are more adaptable to frequency change, so the perceptual load is more suitable for higher frequency.

It can be obtained from the figure 7 that the system efficiency fluctuates greatly with the frequency for the tolerance load, and the system transmission efficiency is low when the frequency is large.For resistive and perceptual load, the system transmission efficiency increases first and then tends to gentle.At large frequencies, the resistive load and the perceptual load can easily obtain greater transmission efficiency.

2.4 Optimum load condition

As the transmission characteristics of wireless charging system are greatly affected by transmission distance and load conditions, it is necessary to use two-port network theory to summarize the derivation method of optimal load conditions for analyzing the transmission characteristics of WPT system and further system optimization.

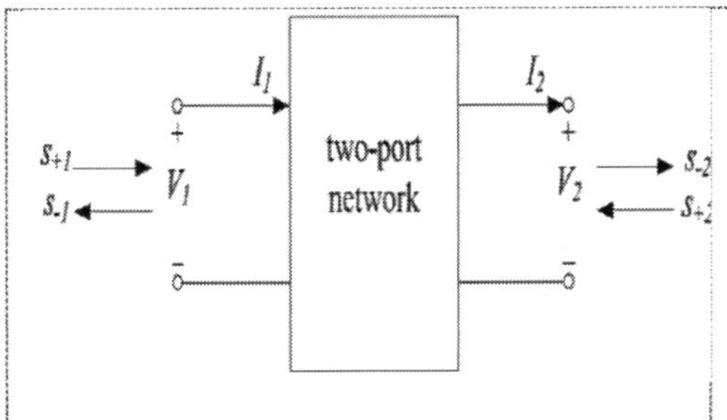

Figure 8 Two-port network

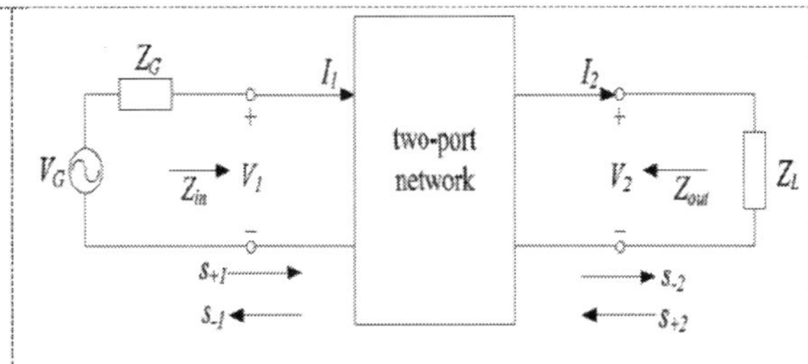

Figure 9 A two-port network that considers source and load

In this section, different equivalent circuit parameters are characterized in combination with the two-port network as shown in the figure 8.

Transmission and impedance matrix are widely used in circuit analysis, but at high frequency cases the corresponding parameter measurements become difficult. Conversely, at high frequency, the scattering matrix becomes preferred due to the presence of a network analyzer, which can measure the scattering matrix parameters in a wider frequency range. Figure 9 shows a two-port network with source and load. The power transmission efficiency of the entire system can then be expressed as:

$$\eta_1 = \frac{(1-|\Gamma_G|^2)|S_{21}|^2(1-|\Gamma_L|^2)}{|(1-S_{11}\Gamma_G)(1-S_{22}\Gamma_L)-S_{12}S_{21}\Gamma_G\Gamma_L|^2} \quad (3)$$

If both the source and the load match, yes, the reflection coefficient can be expressed as: $Z_G = Z_L = Z_0$

$$\Gamma_L = \Gamma_G = 0, \Gamma_{in} = S_{11}, \Gamma_{out} = S_{22} \quad (4)$$

Thus the relation between the transmission efficiency and the S parameter can be simplified as:

$$\eta_1 = |S_{21}|^2 \quad (5)$$

3.Conclusion

This paper first briefly introduces the underwater robot wireless charging system, constructs the mathematical model of the wireless charging system, and further studies the efficient energy efficiency conversion technology of the wireless charging system. This paper analyzes the relationship between the input impedance and the equivalent impedance of the underwater robot and studies the influence of the load change on the energy transmission characteristics and the system to provide a theoretical basis for an efficient and stable real-time tracking control and underwater robot charging load.

Acknowledgements:

This work was supported by the Science and Technology Project of State Grid Corporation of China "Research on wireless charging technology of underwater cable inspection robot based on new magnetic device"(5500-202016256A-0-0-00).

References

[1] Chen Guodong. Research on key Technologies of Wireless Charging for Underwater Cable Inspection Robot [D]. Chongqing University,2019.

[2] Xue M, Et al. (2021) Application status and Key Issues of Radio Energy Transmission Technology. Transactions of China Electrotechnical Society 36.08(2021):1547-1568.

[3] Shi Huangxia & Fu Tao.(2019). Research on wireless charging of substation inspection robot based on electromagnetic coupling Resonance. Electronic measurement techniques (20), 40-43.

[4] Bin Chen,Rong Li,Wanjian Bai,Jianxiang Li,Yue Zhou,Rui Guo.(2018) Application Analysis of Autonomous Underwater Vehicle in Submarine Cable Detection Operation[P]. Robotics, Control and Automation Engineering.

[5] Yang Yong.(2020). Research on transmission method of maximum efficiency of Wireless Charging for Inspection Robot. Automation instrument (12), 56-59.

A Multilayer Path Planning Method for High Voltage Distribution Network Based on the Floyd-Warshall Algorithm

Zhaolong Quan[1], Jie Xing[1*] and Ruilin Cao[1]

[1] School of Information Science and Technology, Donghua University, Shanghai, 201620, China

[*]Corresponding author's e-mail: xingj@dhu.edu.cn

Abstract. With the development of the city, a huge number of distribution networks are waiting for planning. A reasonable planning scheme can meet the power demand and reduce the investment cost. In this paper, a life cycle cost model including the investments of substation and wiring is established with the constraints about load flow calculation and maxi-mum load of wiring. Additionally, a multilayer planning method based on the Floyd-Warshall algorithm has been proposed to solve the model. The area of the city containing substations is divided based on the position of load through the hybrid clusters algorithm in the method. Then, using the divided result of power supply area, the cost matrix for the multilayer path planning method can be constructed through the principle of the method. Lastly, with the cost matrix, the planning scheme in each area will be provided by the Floyd-Warshall algorithm. The result on the actual planning area between the two algorithms shows the total cost of the investment is decreased through using the planning method in this paper.

1. Introduction

The distribution network connects between the power transmission network and users as a bridge, which was essential in the power grid to ensure power transmit from the grid to the low voltage network [1]. With the pressure of development, plenty of new urban planning caused a huge number of distribution networks were waiting to be planned. The planning problem of distribution was a highly complex system planning problem that including forecast and calculates the load, determines the capacity and location of the sub-station, and design the wiring connection under the constrained condition [2,3]. A reasonable distribution network planning scheme can satisfy the demand for power in a region and de-crease the cost of investment of society for a long time [4].

To solve the problem of distribution network planning, various innovation models and methods had been proposed. Geographic and load information were the first condition that needs to be considered in the planning model. Usually, the geographic information such as route in city or barrier in the wilder-ness can be abstracted by graph theory [5], while the weight of the path in the graph should consider the real situation in the urban area like reusable corridors and existing conduit structuring [6].

The results of power supply area division can influence the planning situation to some extent, and therefore, the principle of dividing supply area should not base on average distribution, national policy or administrative division but aim to minimize the investment [7,8]. The load situation also should be considered when divided the supply area, and some methods that complemented the load characteristic to decrease the investment of equipment in the distribution network were introduced [9,10]. It can im-prove the economy of the distribution network, but reliability was more important than economic in realistic planning. Once the load had changed, it may affect the stability and safety of the power grid.

Cluster algorithm had been used to make the decision for choosing the site with considered load distribution in MV distribution network [11], but it is not satisfied the demand of high voltage (HV) distribution network. Geographic information can provide a significant criterion for planning problems that make problem can be solved with various algorithms after simplification, simultaneously [5,12-14]. The planning result may divorce from reality without geographic information [15].

Some intelligent algorithms were used to solve complex problems in all walks of life including the planning problem in distribution network unexceptionally [5,15,16,17]. Different from dynamic programming algorithms such as the Floyd-Warshall algorithm [18], it was easily made the result fell into the local optimal value in the meantime. To solve this problem, a hybrid clusters algorithm is proposed in this paper. According to this algorithm, some adverse effects that come from the disadvantage of the clustering algorithm can be reduced and the algorithm can describe the data in the planning area with more accuracy. Subsequently, a method for multilayer planning based on the Floyd-Warshall algorithm, which can plan the different typical wiring connection schemes with a multilayer map, will be established.

2. Methodology

2.1. Framework of the HV distribution network planning

The framework of the HV distribution network planning was illustrated in Figure 1. It started with spatial load forecasting, which was what most planning problems need to deal with. Then the power supply area should be divided for subsequent work. After that, the framework was divided into two parts, those were the substation planning part and the wiring planning part, which needed to be discussed in this methodology. The substation planning part proceeded from the substation classification to model calculation, and the wiring planning part aimed to generate the path wight, which could express the correlation between the different substations correctly, through calculating the wiring model. Subsequently, a life cycle cost model was established with information of two parts that could be used to generate the matrix. Finally, the model could be solved by the Floyd-Warshall algorithm in each area.

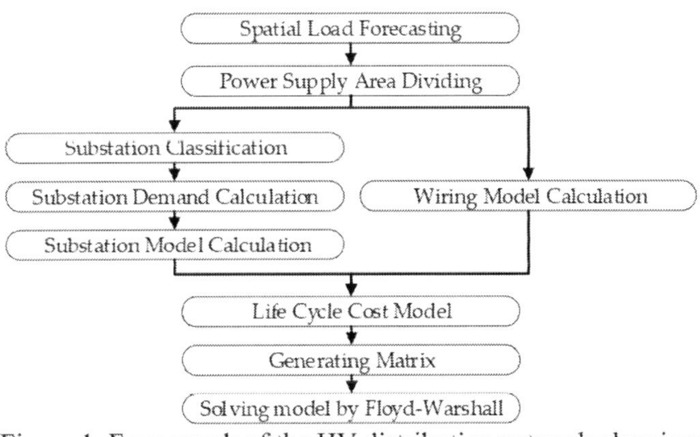

Figure 1. Framework of the HV distribution network planning.

2.2. Planning model of HV distribution network

The planning model of the HV distribution network given by Equation (1) had been mainly divided into two parts.

$$\min C^{Co} = \sum_{i \in N_b} (C_i^{TS} + C_i^{CW}) \tag{1}$$

where C_{Co} represented the total cost of the planning model; N_b was the number of supply areas; C_i^{TS} represented the cost of wiring and C_i^{CW} represented the substation in each supply areas.

Considering the real conditions, there was still a large proportion of other costs when the construction cost had been subtracted. To reflect the economy more accurately, a life cycle cost model should be taken into consideration to reduce the deviation between planning cost and actual cost. In the substation part of planning, the main objective was to establish the cost model of substation as realistically as possible. For that reason, the substation part of the planning model which was incorporated with construction cost, operation cost and recovery cost was expressed as Equation (2).

$$C_i^{\mathrm{TS}} = I_i^{\mathrm{TS}} + \sum\nolimits_{t\in T} \sum\nolimits_{n\in N_{\mathrm{TS}}} ((1+r)^{-t}(C_{i,t,n,\mathrm{L}}^{\mathrm{TS}} + C_{i,t,n,\mathrm{O}}^{\mathrm{TS}} + C_{i,t,n,\mathrm{M}}^{\mathrm{TS}}) + (1+r)^{-\tau} C_{i,n,\mathrm{D}}^{\mathrm{TS}}) \tag{2}$$

where I_i^{TS} represented the investment of substation; T represented the set of the year of planning horizon; N_{TS} was the number of transformer substation; r represented the rate of interest; $C_{i,t,n,\mathrm{L}}^{\mathrm{TS}}$, $C_{i,t,n,\mathrm{O}}^{\mathrm{TS}}$ and $C_{i,t,n,\mathrm{M}}^{\mathrm{TS}}$ were parts of operation cost and represent loss cost of substation, overhaul cost of substation and malfunction cost of substation respectively. τ was a constant value of years that depends on designed life from construction to demolition. $C_{i,n,\mathrm{D}}^{\mathrm{TS}}$ represented the recovery cost.

Similar to the substation part, a life cycle cost model had been established as Equation (3) to obtain a more accurate result, especially in the distribution network planning problem.

$$C_i^{\mathrm{W}} = I_i^{\mathrm{W}} + \sum\nolimits_{t\in T} \sum\nolimits_{m\in N_{\mathrm{W}}} (1+r)^{-t}(C_{i,t,m,\mathrm{L}}^{\mathrm{W}} + C_{i,t,m,\mathrm{O}}^{\mathrm{W}} + C_{i,t,m,\mathrm{M}}^{\mathrm{W}}) \tag{3}$$

The structural difference between the wiring part and substation part of the planning model was reflected in the wiring was not worthy of recycling like transformers in the substation, when it was broken or reach the end of life. That was the reason why Equation (3) has not the item about recovery cost. In this equation, I_i^{W} represents the investment of wiring; N_{W} was the number of sets of typical wiring connection; $C_{i,t,n,\mathrm{L}}^{\mathrm{W}}$, $C_{i,t,n,\mathrm{O}}^{\mathrm{W}}$ and $C_{i,t,n,\mathrm{M}}^{\mathrm{W}}$ were also both part of operation cost and represent loss cost of wiring, overhaul cost of wiring and malfunction cost of wiring respectively.

In addition, the planning model of HV distribution network should be constrained with Equation (4).

$$\begin{aligned} & f(TYP') = 0 \\ & g(TYP') \le 0 \\ & P_m \le P_{m,\max} \qquad m \in N_{\mathrm{W}} \end{aligned} \tag{4}$$

where the function $f(TYP')$ was the equality constraints of load flow calculation; function $g(TYP')$ was the inequality constraints of load flow calculation; P_m and $P_{m,\max}$ were represent the load of wiring m and the maximum load of wiring m respectively.

3. Multilayer path planning method for HV distribution network

3.1. Hybrid clusters algorithm for HV distribution network

The hybrid clusters algorithm was a combination of two clustering algorithms, which were the DBSCAN and the K-means algorithm, that could describe the density distribution of samples and abided by the minimum error criterion to assign data respectively. The algorithm of DBSCAN could use a set of neighborhood parameters ε and *MinPts* to describe the density distribution. ε represented the radius of neighborhoods; *MinPts* represented the minimum size of the sample in the neighborhood.

The algorithm of K-means that could try to get the minimum error criterion works as follows.

$$d = \sum\nolimits_{i=1}^{k} \sum\nolimits_{\mathbf{x}\in CS_i} \left\| \mathbf{x} - \frac{1}{|CS_i|} \sum\nolimits_{\mathbf{x}\in CS_i} \mathbf{x} \right\|_2^2 \tag{5}$$

where k represented the number of clusters; the CS_i represented the set of samples. Equation (5) could be used to minimizes the sum of the distance between each sample and average vector with the quantitative constraint of parameter k by iterative method. Different values of k would cause different results, for that reason, the initialization of k and the average vector should be constrained by DBSCAN, in the hybrid clusters algorithm. The parameters k and the average vectors depended on the number and the center coordinates which calculated in Equation (6) respectively of the power supply area generated by DBSCAN.

$$\mathbf{v} = \sum_{i=1}^{k} \frac{\sum_{\mathbf{x}\in CS_i} \mathbf{x}P_{\mathbf{x}}}{\sum_{\mathbf{x}\in CS_i} P_{\mathbf{x}}} \tag{6}$$

where **v** represented the average vector; P_x represented the load of each sample. The advantage of the method, which used the load torque to calculate the average vectors, was the uneven distribution of heavy loads had been considered to increase the reasonable substation classification in each power supply area. By the change of average vectors, equation (5) should be rewritten as follows:

$$d = \sum_{i=1}^{k} \sum_{\mathbf{x}\in CS_i} \left\| \mathbf{x} - \frac{\sum_{\mathbf{x}\in CS_i} \mathbf{x}P_{\mathbf{x}}}{\sum_{\mathbf{x}\in CS_i} P_{\mathbf{x}}} \right\|_2^2 \tag{7}$$

In the urban area, the distribution of load had strong density characteristics that could be used for the DBSCAN algorithm to obtain the supply area of the city. This algorithm could get a good result in load clustering, but could not classify the substation site that waiting to be selected. Through hybridize the DBSCAN and *K*-means, two algorithms could accomplish the objective of substation part with the complementarity of each other.

3.2. Multilayer path planning method based Floyd-Warshall algorithm

Floyd-Warshall was an algorithm based on dynamic programming to solve the all-pairs shortest paths (APSP) problem, that is, found the shortest path between multiple points with the weighted graph. The cost of civil engineering usually accounted for a large proportion of the wiring part when considering the real situation, and the planning method should consider the minimum of the cost rather than the shortest length of the path. Thus, reused the cable conduit that had been already constructed was a major method to reduce the cost.

Multilayer path planning method that based Floyd-Warshall algorithm, in this paper, was an extension of Floyd-Warshall algorithm, which illustrated in Figure 2, to satisfy the requirements of HV distribution network planning.

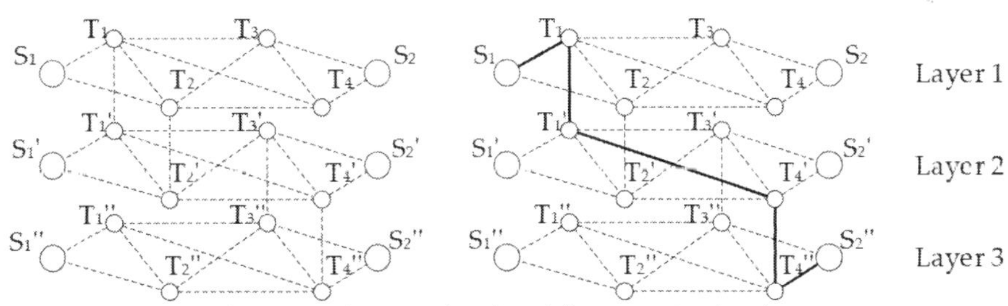

Figure 2. The sample of multilayer path planning.

In Figure 2, there was a three-layer structure as an example to explain the method. The left part was entire structure with connection path between different layer, the right part was same structure with shortest path. All the layers described a same situation of the planning area, once any path weight had been changed, all of layers would been change at same time. According to this structure, some points such as T_1 and T_4 as necessary points had been added to the solving process to calculate the shortest path between S_1 and S_4. The principle of method was shown in Equation (8)-(11).

$$\boldsymbol{M}_{\text{Cost}} = \boldsymbol{M}_{\text{R}} \otimes \boldsymbol{M}_{\text{P}} + \boldsymbol{I} \otimes (\boldsymbol{M}_{\text{P}} * \boldsymbol{M}_{\text{CV}}) + \boldsymbol{M}_{\text{CT}} \tag{8}$$

$$\boldsymbol{M}_{\text{R}} = \begin{bmatrix} m_1 & 0 & \text{inf} \\ 0 & m_2 & 0 \\ \text{inf} & 0 & m_3 \end{bmatrix} \tag{9}$$

$$\boldsymbol{M}_{\text{P}} * \boldsymbol{M}_{\text{CV}} = [l_{(i,j)}]_{n\times n} * [p_{\text{CV}(i,j)}]_{n\times n} \tag{10}$$

$$\boldsymbol{M}_{\mathrm{CT}}=\begin{bmatrix}0 & C_{12} & 0\\ C_{21} & 0 & C_{23}\\ 0 & C_{32} & 0\end{bmatrix} \tag{11}$$

where $\boldsymbol{M}_{\mathrm{cost}}$ represented the cost matrix that to be used by Floyd-Warshall algorithm; $\boldsymbol{M}_{\mathrm{R}}$ represented the matrix with the relationship of multilayer and price of different cable types; $\boldsymbol{I}$ represented the identity matrix; $\boldsymbol{M}_{\mathrm{CV}}$ was a matrix that represented cost of civil engineering, which would change the value after finished one wiring connection; $\boldsymbol{M}_{\mathrm{CT}}$ represented cost of transformer substation and the elements C_{ij} in matrix represented the connection situation between two layers; Operational character $\otimes$ and * represented the Kronecker Product and Hadamard Product respectively. Through the Kronecker Product, the matrix $\boldsymbol{M}_{\mathrm{R}}$, which included the information of price in each layer, would be expanded to a 3n orders of matrix. With the matrix $\boldsymbol{M}_{\mathrm{Cost}}$, in which all the elements in the planning model of HV distribution network had been included, that is, the minimum cost of planning problem could be solved through the Floyd-Warshall algorithm.

4. Case Study

In this section, a real urban area in China to be planned was used to be an example of the demonstration of the planning method in this paper. The map of the load area was illustrated in Figure 3.

Figure 3. The map of the area to be planned with load points.

Figure 4. The clustering of the load by using the hybrid clusters algorithm.

The overall load of the area was 1130.042 MVA. With the clustered of the load by used the hybrid clusters algorithm in Figure 4, the load in clusters was 1121.391 MVA that accounts for 99.23% because of the feature of DBSCAN. So, the load which had not cluster could be considered to have no influence on the results. To satisfy the demand of each supply area, the optimal solution of substation for the area was shown in Table 1.

Table 1. The optimal solution of substation for each area.

Area	Load (MVA)	Demand (MAV)	Result (MVA)	Average load rate
1	147.322	294.644	3×(50×2)	49.11%
2	210.837	421.674	2×(63+50)+2×(50×2)	49.49%
3	202.234	404.268	1×(63+50)+3×(50×2)	48.97%
4	380.161	760.322	5×(63+50)+2×(50×2)	49.69%
5	74.562	149.124	2×(50×2)	37.28%
6	106.374	212.748	1×(63+50)+1×(50×2)	49.94%

To ensure the reliability of the power supply, a scheme of two-sided power was preferred. And remaining substations adopt the single-side scheme.

The contrast between two connective results of planning in this area with the method in this paper and method of shortest path respectively was illustrated in Figure 5 and Figure 6, and the results of data shown in Table 2.

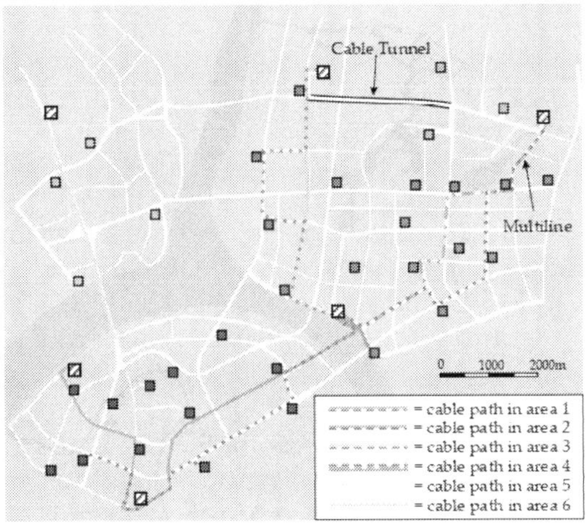

Figure 5. A multilayer path planning method that based Floyd-Warshall algorithm

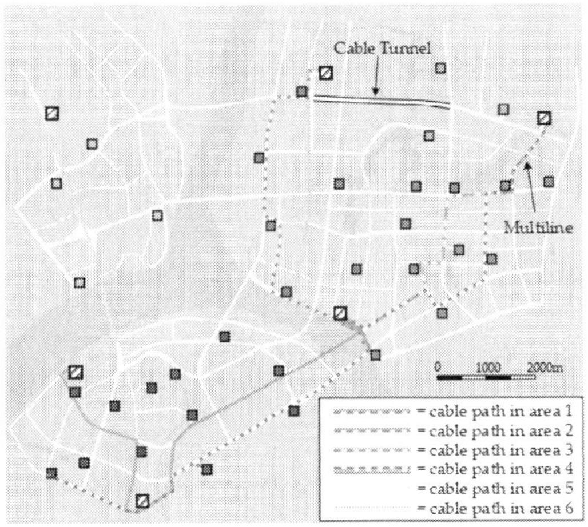

Figure 6. The method that based shortest path.

Table 2. The contrast results between two planning method.

Area	Multilayer path planning method		Area	Shortest path planning method		Cost variance (million CNY)
	Length (km)	Cost (million CNY)		Length (km)	Cost (million CNY)	
1	4.19	146.21	1	4.19	146.21	2.12
	2.04	70.15		1.95	72.27	
2	6.82	205.61	2	6.82	205.61	7.94
	7.39	196.57		7.33	204.51	
3	5.86	182.56	3	5.86	182.56	6.76
	7.47	194.02		7.10	200.78	
4	7.44	264.70	4	7.44	264.70	-2.08
	7.64	248.91		7.64	248.91	
	7.76	250.91		7.76	250.91	
	1.30	83.11		1.30	81.03	
5	8.70	167.01	5	8.31	176.52	9.51
6	6.67	170.45	6	6.67	170.45	0

From the comparison result of both methods, the shortest path planning method aimed to choose the shortest path for each wiring to be planed as usual, but the multilayer path planning method chose the path considering the total cost of the investment rather than chose the shortest path as planning scheme. It would tend to use the tunnel, which had been already established for the pipeline. By using the method in this paper, the total cost of the investment could be reduced 24.26 million CNY.

5. Conclusions

In this paper, the power supply area and substation were divided by the hybrid clustering algorithm. And the multi-layer path planning method based on dynamic programming was used to solve the life cycle cost model. It provided a new way for distribution network planning under complex urban road conditions. The method was verified in an actual area. The comparisons between the proposed method and the traditional shortest path planning method show that the proposed method was more economic. This method still had some imperfections, such as the medium voltage distribution network and new energy resources system had not been considered. Those would be solved in the following research.

References

[1] Kazmi, S.A.A.; Shahzad, M.K.; Shin, D.R. (2017) Multi-Objective Planning Techniques in Distribution Networks: A Composite Review. Energies, 10:208.
[2] Vahidinasab, V.; Tabarzadi, M.; Arasteh, H.; Alizadeh, M.I.; Beigi, M.M.; Sheikhzadeh, H.R.; Mehran, K.; Sepasian, M.S. (2020) Overview of Electric Energy Distribution Networks Expansion Planning. IEEE Access, 8:34750-34769.
[3] Nahman, J.M.; Peric, D.M. (2008) Optimal Planning of Radial Distribution Networks by Simulated Annealing Technique. IEEE Transactions on Power Systems. 23:790-795.
[4] Farhangi, H. (2010) The path of the smart grid. IEEE Power Energy Mag. 18–28.
[5] Wang, Z.; Lin, D.; Zeng, G.; Yu, T. (2020) A Practical Large-Scale Distribution Network Planning Model Based on Elite Ant-Q. IEEE Access. 8:58912-58922.
[6] Li, Z.; Wu, W.; Zhang, B.; Tai, X. (2020) Feeder-corridor-based distribution network planning model with explicit reliability constraints. IET Gener. Transm. Distrib. 14:5310-5318.
[7] Li, L.; Xie, Y.; He, L.; Ge, H.; Wu, D. (2018) Research on substation planning of distribution network with microgrid. In: IOP Conference Series-Earth and Environmental Science, 2018 4th International Conference on Environmental Science and Material Application. Xian, PEOPLES R CHINA. 032025, 252.
[8] Sun, K.; He, D.; Wang, L.; Wang, H.; Pan, H.; Sun, Z. (2018) A Grid-Based Method for Distribution Network Planning in Urban Areas. In: IOP Conference Series-Earth and Environmental Science, 2018 First International Conference on Environment Prevention and Pollution Control Technology, Tokyo Univ Sci. Tokyo, JAPAN. 052053, 199.
[9] Liu, W.; Liu, H.; Wang, F.; Wang, C. Zhao, L.; Wang, H. (2020) Practical Automatic Planning for MV Distribution Network Considering Complementation of Load Characteristic and Power Supply Unit Partitioning. IEEE Access. 8:91807-91817.
[10] Alarcon, J.A.; Santamaria, F.; Al-Sumaiti, A.S.; Rivera, S. (2020) Low-Capacity Exploitation of Distribution Networks and Its Effect on the Planning of Distribution Networks. Energies. 13:1920.
[11] Li, Y.; Du, M.; Xie, W.; Yang, B.; Fang, C.; Zhang, Y.; Wang, S. (2018) Method for division of urban load power supply district based on cluster analysis. IET Gener. Transm. Distrib. 12:4577-4581.
[12] Valenzuela, A.; Inga, E.; Simani, S. (2019) Planning of a Resilient Underground Distribution Network Using Georeferenced Data. Energies. 12:644.
[13] Valenzuela, A.; Montalvo, I.; Inga, E. (2019) A Decision-Making Tool for Electric Distribution Network Planning Based on Heuristics and Georeferenced Data. Energies. 12:4065.
[14] Li, Z.; Wu, W.; Zhang, B.; Tai, X. (2020) Hexagon raster-based method for distribution network planning considering line routes and pole locations. IET Gener. Transm. Distrib. 14:1420-1429.
[15] Wu, K.; Li, K.; Liang, R.; Ma, R.; Zhao, Y.; Wang, J.; Qi, L.; Liu, S.; Han, C.; Yang, L.; Huang, M. (2018) A Joint Planning Method for Substations and Lines in Distribution Systems Based on the Parallel Bird Swarm Algorithm. Energies. 11:2669.
[16] Ahmadian, A.; Elkamel, A.; Mazouz, A. (2019) An Improved Hybrid Particle Swarm Optimization and Tabu Search Algorithm for Expansion Planning of Large Dimension Electric Distribution Network. Energies. 12:3052.
[17] Nahman, J.; Peric, D. (2008) Optimal planning of radial distribution networks by simulated annealing technique. IEEE Transactions on Power Systems 23:790-795.
[18] Risald.; Mirino, A.; Suyoto. (2017) Best Routes Selection Using Dijkstra And Floyd-Warshall Algorithm. International Conference on Information & Communication Technology and Systems, 11th International Conference on In-formation and Communication Technology and System. Surabaya, INDONESIA. 155-158.

An optimized demand-response operation method of regional integrated energy system considering 5G base station energy storage

Yue Yunli[1*], Sun Beibei[2], Yue Hao[1], Chen Dan[1], Zhou Yi[1], Lu Jun[3]

[1] Economic and Technological Research Institute, State Grid Jibei Electric Power Co., Ltd, Beijing 100052, China

[2] State Grid Jibei Electric Power Co., Ltd, Beijing 100052, China

[3] North China Electric Power University, Beijing 102208, China

*Corresponding author's e-mail: yueyunli999@163.com

Abstract—The scheduling technology of regional integrated energy system is one of the key technologies to realize carbon neutralization by utilizing wind-power. Aiming at the optimal scheduling problem of regional electrothermal integrated energy system considering wind-power utilization and load side energy consumption, this paper proposes an optimized demand-response operation method of regional integrated energy system considering 5G base station energy storage. The regional integrated energy system of load side demand response is constructed based on the comprehensive consideration of technical and economic factors such as wind-power utilization and economic costs and load side peak valley difference. Finally, a two-layer particle swarm optimization method is proposed to solve the model. The experimental results show that the proposed method can effectively achieve wind-power utilization, economic dispatch and reduce the peak valley difference through load side demand response, which can improve the economic efficiency, environmental protection and low-carbon operation of regional integrated energy system.

1. Introduction

With the proposal of the national strategic goal of "dual carbon", the construction and development of energy Internet has faced new technical challenges [1]. How to ensure the safe and stable operation of a high proportion of clean energy power system with the help of advanced technologies such as solar storage technology, demand side response technology and large grid interconnection while promoting clean replacement and electric energy replacement has become one of the keys to achieve carbon peak and carbon neutrality in the power industry [2]. Regional Integrated Energy Systems (RIES) have the technical advantages of multiple Energy forms, absorbing renewable Energy, supporting demand side response and providing multiple demands for multiple loads [3]. RIES coordinates and optimizes the distribution, conversion, storage and consumption of various types of energy, including electricity, heat and gas, to break the original independent operation mode, improve system flexibility, make full use of renewable energy and effectively improve energy efficiency [4].

At present, scholars at home and abroad have carried out a lot of work on regional integrated energy system from two aspects of demand response and optimal scheduling [3][4][5] [6][7][8][9].Therefore, on the basis of comprehensive consideration of wind power consumption, economic and environmental costs, and load side peak-valley differential and other technical and economic factors, this paper constructs a multi-objective optimization operation model of regional integrated energy system RIES for economic cost optimization and peak-valley differential optimization.

2. Mathematical Description of Integrated Energy System Model

RIES systems may include power systems, thermal systems, renewable energy generation systems, energy storage systems and integrated load users. Typical centralized RIES containing wind power generation system (hereinafter referred to as wind power system) and 5G base station energy storage system are shown in Figure 1.

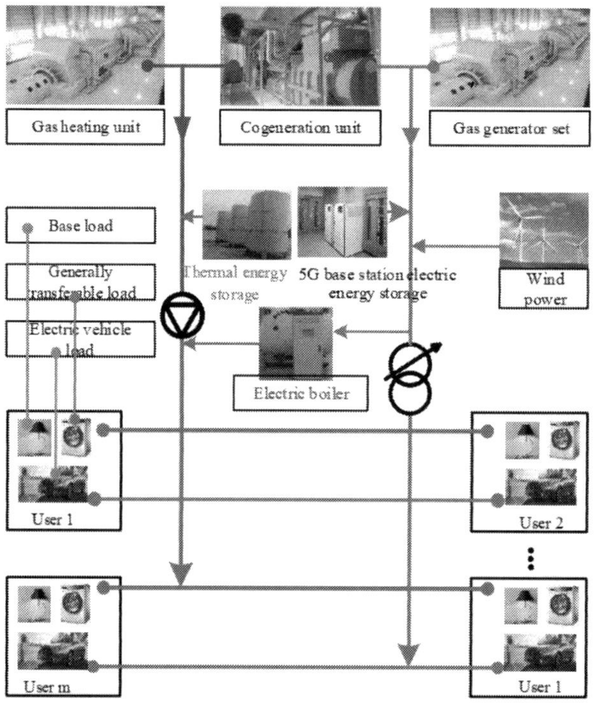

Figure 1. Operation architecture of typical centralized RIES with wind-power and energy storage

2.1. Energy supply model of cogeneration unit

The CHP, as the core of RIES 'coordinated dispatching, is powered by coal or natural gas (natural gas in this case) to generate both electricity and heat. The energy supply model of unit cogeneration unit in the time period is shown in Formula (1) .

$$\begin{pmatrix} T_{CHP}(t) \\ P_{CHP}(t) \end{pmatrix} = (\eta_T, \eta_P) \cdot \begin{pmatrix} u_c \cdot M_T(t) \\ u_c \cdot M_P(t) \end{pmatrix} \tag{1}$$

Where, $T_{CHP}(t)$ and $P_{CHP}(t)$ are thermal power output and electric power of cogeneration unit at the time, η_T and η_P are heating efficiency and power generation efficiency of the unit respectively, u_c is gas burning energy per unit (kJ/ 1000Nm3), $M_T(t)$ and $M_P(t)$ are gas consumption of cogeneration unit at the time (Nm3). $T_{CHP}(t)$ and $P_{CHP}(t)$ can be converted to each other with the help of the ratio of electric output and thermal output of the hot spot cogeneration unit. In the normal mode of "fixed electricity by heat", $T_{CHP}(t) = \eta_{conversion} \cdot P_{CHP}(t)$.

2.2. Energy storage model
Energy Storage system is an important part of RIES. The energy Storage model consists of Electric Storage (ES) of 5G base station and Thermal Storage (TS). This model uses storage battery and heat storage tank to store heat.

2.2.1. Energy storage model of 5G base station.
The dynamic model of 5G base station energy storage device is expressed by the following formula.

$$E(t)=E(t-1)+\frac{P_c(t)\eta_c}{N}\Delta t \qquad E(t)=E(t-1)+\frac{P_d(t)/\eta_d}{N}\Delta t \tag{2}$$

Where, $E(t)$ is the electric quantity stored during the period t, $P_c(t)$ and $P_d(t)$ are respectively the charge and discharge power, η_c and η_d are respectively the charge and discharge efficiency of the battery, and N is the rated capacity of the energy storage battery.

2.2.2. Thermal storage model.
The thermal storage TS model (in this paper, the heat storage tank is used as an example) participates in the dynamic heat storage characteristics of RIES, including capacity, output and input characteristics, and thermal conversion efficiency, as shown in Equation (3) [9].

$$H_{TS}(t)=(1-k_r)H_{TS}(t-1)+[Q_{st}(t)k_{st}-Q_{es}(t)k_{es}]\Delta t \tag{3}$$

Where $H_{HS}(t)$ is thermal energy storage capacity of time period t; k_r is the heat loss rate of thermal energy storage, $Q_{st}(t)$、$Q_{es}(t)$ and k_{st}、k_{es}, are respectively the heat absorption and heat release power efficiency.

3. Energy Optimization Modeling for Integrated Energy Systems
The constraint conditions and optimization objective function of the model are introduced in turn.

3.1. The constraint
RIES constraints are mainly equipment involved in scheduling optimization and user-side DR load constraints.

3.1.1. CHP constraints on cogeneration units.
CHP constraints of CHP units include upper and lower limits of output and climbing rate constraints, as shown in Equations (4) [3].

$$P_{CHP}^{min}\le P_{CHP}(t)\le P_{CHP}^{max} \qquad Q_{CHP}^{min}\le Q_{CHP}(t)\le Q_{CHP}^{max} \tag{4}$$

Where $P_w(t)$ is the generation power of the unit during the period t, P_{CHP}^{min}、P_{CHT}^{max}、Q_{CHP}^{min} and Q_{CHT}^{max} are respectively the maximum output power and minimum power of generation and heating of CHP.

3.1.2. Energy storage constraint of 5G base station.
Electric energy storage ES of 5G base station are shown in Equation (5).

$$S_{ES}^{min}\le SOC(t)\le S_{ES}^{max} \tag{5}$$

Where, S_{ES}^{min} and S_{ES}^{max} are respectively the upper and lower limits of the charged state of the electric energy storage ES, and $SOC(t)$ is the value of the charged state of the electric energy storage ES in the current period t.

For other constraints, please refer to [3][4][5] [6][7][8][9].

3.2. The objective function

RIES system economic cost is calculated according to Equation (6).

$$C_{\mathrm{run}}(t) = C^{o}_{CHP}(\mathrm{t}) + C_{WT}(\mathrm{t}) + C_{ES}(\mathrm{t}) + C_{TS}(\mathrm{t}) \tag{6}$$

Where C_{run} is the operating cost of RIES, including DR compensation cost, $C_{WT}(\mathrm{t})$、$C_{ES}(\mathrm{t})$、$C_{HS}(\mathrm{t})$ and $C_{EB}(\mathrm{t})$ and are respectively WT wind turbine abandon cost, ES charge and discharge cost of electric energy storage, HS operating cost of thermal energy storage and EB operating cost of electric boiler. For specific calculation, please refer to reference [9].

RIES system technical cost is calculated according to Equation (7).

$$\delta_{\mathrm{L}}^{\mathrm{DR}}(t) = \frac{1}{\Delta t}\sum_{\Delta t}[P_{\mathrm{EL}}(t) - \overline{P_{\mathrm{EL}}(t)}]^2 \tag{7}$$

Where Δt is the scheduling time in unit time period, and $\overline{P_{\mathrm{EL}}(t)}$ is the mathematical expectation within the time period t.

4. The Example Analysis

In order to verify the effectiveness of the model proposed in this paper, an urban area in northern China was selected, and the regional power supply and heating was provided jointly by CHP, WT, ES, TS and EB. The other parameters of the model are set as follows: The average gas cost of CHP is 3.6 yuan/Nm3 (commercial), the heating efficiency η_{T} is 0.8924, the power generation efficiency η_{P} is 0.3539, and the electric heating ratio $\eta_{\mathrm{conversion}}$ is 1. Unit wind power cost is calculated according to [9]. The technical and economic parameters of operation and maintenance of RIES equipment [9] are set in Table 1.

Table 1. Operational parameters of different units

Unit type	P_N / MW	P_N / kW · h	P_{min} MW	P_{max} MW	K_i $yuan \cdot (kW \cdot h)^{-1}$
Cogeneration system	200	—	100	200	0.0946
TW	40	—	0	2	0.0296
EB	10	—	0	10	0.0500
ES	—	200×96	0.024	0.078	0.0832
TS	—	4×5000	1.5	4	0.0450

In experimental case 1, only CHP, TW and heat load were involved in dispatching. On the basis of case 1, thermal energy storage, electric energy storage, electric boiler and DR electric load are added into case 2 to participate in scheduling.Figure 2 shows the comparison of wind power consumption results between case 1 and case 2. Case 1 wind power consumption rate is 47.68%; In case 2, the wind power consumption rate was increased to 90.43%, which greatly improved and reduced the waste of wind power resources.

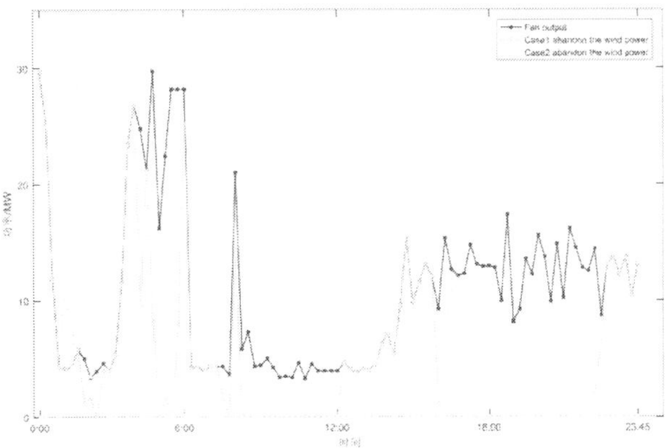

Figure 2. Wind power utilization comparison.

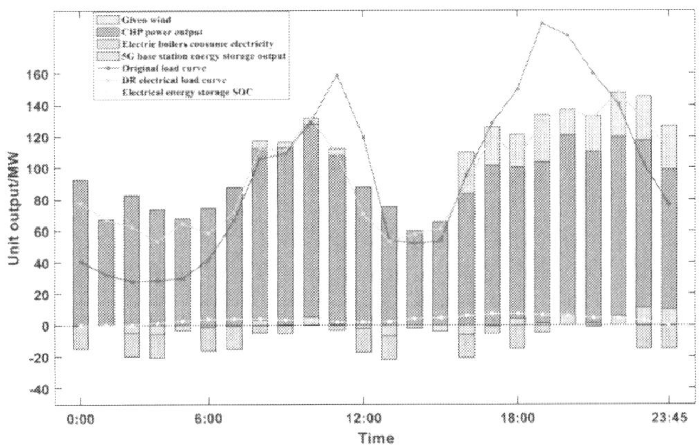

Figure 3. Power load balance and SOC curve of ES in Case 2.

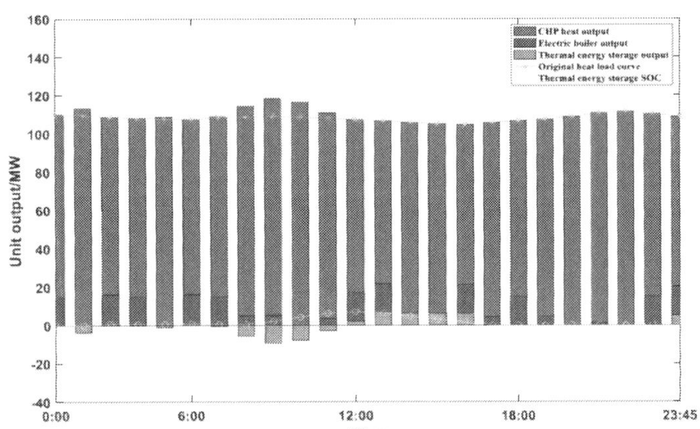

Figure 4. Power load balance and SOC curve of ES in Case 2.

Figure 3 shows that fan output is vigorous during 18:00-24:00. At 00:00-06:00 and 13:00-18:00, the electric energy storage output, electric boiler output and DR output cooperate with each other to improve the electric load in the changing period and provide space for wind power consumption. Figure 4 shows that from 15:00 to 17:00, thermal energy storage increases to meet the thermal load, reducing fuel cost, absorbing a certain amount of wind power, and greatly reducing fuel cost and wind abandoning cost.

Table 2. Economic cost comparison

	case1（Ten thousand yuan）	case2（Ten thousand yuan）
The total cost	80.4193	72.0350
CFU	51.7899	43.5208
CWT	6.8904	3.2317
CTS	/	0.0070
CES	/	0.4173
CME	26.7390	24.8582

The economic expenditure pairs of case 1 and case 2 are shown in Table 2. In case 1, only CHP, TW and thermal load participate in scheduling, and the overall total cost is relatively high. On the basis of example 1, thermal energy storage, 5G base station electrical energy storage, electric boiler and DR electrical load are added to the scheduling in example 2. Through DR electrical load dispatching, the wind power is greatly absorbed, the waste of wind power resources is reduced, and the cost of abandoning wind is greatly reduced.

5. Conclusion

This paper takes the optimization operation of regional integrated energy system that supports the carbon-neutral regional 5G base station energy storage to participate in the demand response as the entry point. The proposed method can effectively realize wind power consumption and economic dispatching, so as to improve the economic efficiency, environmental protection and low-carbon operation of regional integrated energy system. The next step is to further study the optimization operation mechanism and scheduling of regional integrated energy system of 5G base station energy storage and distribution grid.

Acknowledgments

This work is supported by State Grid Jibei Electric Power Company Science and Technology Project (Research on Deepening demand side resource development and utilization and comprehensive energy services to adapt to energy transformation).

References

[1] Ma Zhao. "Dual carbon" target forces the construction of new power system to accelerate [N]. China Energy News,2021-03-22(004).
[2] Qu B, LIU C, LI D Z, et al. Research on energy substitution development strategy under the goal of "carbon neutral" [J]. Power Demand Side Management, 201,23(02):1-3+9.
[3] Zici Wang, Bingyin Lei, Lingyi Yang, Yuqing Su, Weizhe Sun. Multi-objective bi-level optimal scheduling of regional integrated energy system considering combined demand response of electric heating [J]. Journal of electric power system and its automation: 1-10 [2021-04-10]. HTTP: / / https://doi.org/10.19635/j.cnki.csu-epsa.000750.
[4] Li Zilin, LIU Ronghui. Operation optimization of regional integrated energy system with energy storage considering demand side response [J]. Modern electric power,2019,36(06):61-67.
[5] Bai Hongkun, Zhang Peng, Yin Shuo, Yang Meng, Yang Qinchen. Journal of Henan Polytechnic University (Natural Science Edition), 201,40(02):127-134.
[6] Cao Huiqiu. Day-Ahead Economic Dispatch of Wind Integrated Power System Considering Demand Response[D]. Wuhan： Wuhan University, 2018.
[7] Wang Yali, Liu Hong, ZhaoJuan, et al. Distribution network economic dispatch based on network reconfiguration and power-heat combined demand response[J]. Electric Power Construction,

2019, 40(4): 90-97.
[8] Fang Shaofeng, Zhou Renjun, Xu Fulu, et al. Optimal operation of integrated energy system for park micro-grid considering comprehensive demand response of power and thermal loads[J]. Proceedings of the CSU-EPSA, 2020,32(1):50-57.
[9] Liu Hong, Chen Xingyi, Li Jifeng, Xu Ke. Electric power automation equipment, 2017,37(6):193-200.
[10] Jing Weizhe, Liu Yang, Xiang Yue, et al. Power construction,2017,38(12):68-76.
[11] Lu Jun, PENG Wenhao, Zhu Yanping, et al. Power grid technology,2017,41(07):2370-2377.
[12] Zhengmao, ZHANG Feng,LIANG Jun,et al. Optimization on microgrid with combined heat and power system[J].Proceedings of the CSEE, 2015, 35(14): 3569-3576.
[13] Lu Jun, Zhu Yanping, PENG Wenhao, et al. Automation of electric power systems,2017,41(17):113-120.
[14] Zhenbo Wei, Xiaolin Ren, Yuhan Huang. Multi-objective optimal scheduling of regional integrated energy system considering comprehensive demand side response [J]. Electric power construction,2020,41(07):92-99.
[15] Naveed A. K.,Guftaar A. S.S., Ahmed B. A. etc. Modeling and operation optimization of RE integrated microgrids considering economic, energy, and environmental aspects[J].International Journal of Energy Research. 2019,43(13),pp:6721-39.
[16] Hu Xiaotong, Liu Tianqi, He Chuan, et al. Proceedings of the csee,2016,36(10):2674-2681.

Research and Design of Active Equalization System for Multi-stage Series Li-ion Battery

Jinli Xu[1], Haiqing Chang[1*], Hehui Xun[1], Jiaowu Wang[1]

[1]School of Opto-Electronic and Communications Engineering, Xiamen University of Technology, Xiamen, Fujian, 361024, China

*Corresponding author's e-mail:changhaiqing@xmut.edu.cn

Abstract: In view of the obvious phenomenon of the difference between the single cells of the series battery pack after multiple cycles of recycling, it is necessary to adopt balanced control to extend the working life of the battery pack, improve the energy utilization rate, and ensure the safety and stability of the battery pack. This paper proposes a single battery voltage as the control variable, adopts staged equalization, charge-discharge two-level equalization protection and isolated flyback converter circuit to realize the active equalization of battery pack energy, and designs the corresponding equalization control circuit. Through the comparison of a variety of control methods, the application of this system can achieve the final voltage balance. Tests show that the proposed equilibrium system can make full use of the larger capacity monomers, quickly improve the energy imbalance in the battery pack, and can effectively extend the working life of the battery pack and improve energy utilization.

1. Introduction

As one of the energy storage methods widely used by humans, Li-ion batteries have the advantages of low self-discharge ratio, fast cycle charge and discharge speed, and no memory effect. In addition, Li-ion batteries mostly exist in series in actual high-voltage and high-capacity applications. However, the series connection of batteries will be affected by many factors such as the capacity of single cells, internal resistance, and electrochemical characteristics, which can cause imbalance in the battery pack. If the imbalance in the battery pack is not dealt with in time, it will lead to overcharge and overdischarge of the battery pack, accelerated polarization, low energy utilization, shorten the life of the battery pack, and cause safety hazards, etc.[1], so the research on the balance control of the battery pack is particularly important.

At present, and domestic foreign companies and related scholars have studied the balance control of battery packs, and most of them use capacity, voltage, state of charge, etc. as control variables [2]. In the literature [3], a switch resistor topology is used for capacity balance control, which is relatively simple and low cost, but a large amount of power waste will be generated during the balance process and the actual working time will be shortened. In literature [4], a parallel non-isolated conversion circuit is used to balance the voltage at the single battery end. In industrial applications, the battery voltage is often used as the criterion. Compared with the former, this method uses an active balance method to achieve efficient use of electrical energy, But it can only complete the energy transfer between adjacent cells and is not suitable for multi-level battery series. The literature [5-6] regards the state of charge as the research control object. The state of charge is an important reference for measuring the real-time status of the single battery, but it cannot be directly measured. It is mostly used in the simulation stage. The estimation accuracy depends on the accuracy of the algorithm, and easily affected by the external environment.

Based on the analysis of the above balance control, and taking into account the system loss, as well as the safety and service life of the battery, it is necessary to ensure that the voltage of the single battery cannot be greater than 4.25V and less than 2.75V. Therefore, this paper adopts the directly measurable voltage value as the control variable, combined with staged balance, charge-discharge two-level balance protection and isolated flyback converter circuit to achieve the purpose of fast and active balance and full utilization of battery pack power. First of all, build and analyze the principle of the overall architecture of the active balance control system of the battery pack. Secondly, research and design the equalization process control circuit and algorithm. Finally, verify and analyze the adopted control system by building a balanced experimental platform.

2. The overall architecture and principle of the active balance control system

2.1. The overall architecture of the active balance system

On the one hand, the active balance control method adopted in this paper avoids the energy loss caused by passive balance using energy-consuming components. On the other hand, it uses staged balance, charge-discharge two-level balance protection and isolated flyback converter circuit to ensure the safe and fast balance of the control method. The overall architecture of the battery pack active balance control system is shown in Fig. 1. The battery pack consists of multiple single ternary Li-ion batteries with a nominal voltage of 3.7V and different power levels. The main controller uses its own internal analog-to-digital conversion (ADC) interface to realize the collection of the voltage of each cell in the battery pack and the first-level equalization protection control. The Li-ion battery charge and discharge protection IC implements secondary protection for the charge and discharge and equalization process according to the preset overcharge voltage threshold and overdischarge voltage threshold. The PWM control IC directly drives the NMOS switching element on the flyback converter side by outputting a PWM signal of a certain frequency to realize the energy conversion in the battery pack. Among them, the flyback converter is actually a Buck/Boost circuit that combines electrical isolation, voltage matching, and energy storage. This article uses a 100KHz PWM signal to drive the flyback converter. Therefore, when the PWM control IC outputs a high level, the flyback converter stores energy, and when the PWM control IC outputs a low level, the flyback converter releases energy.

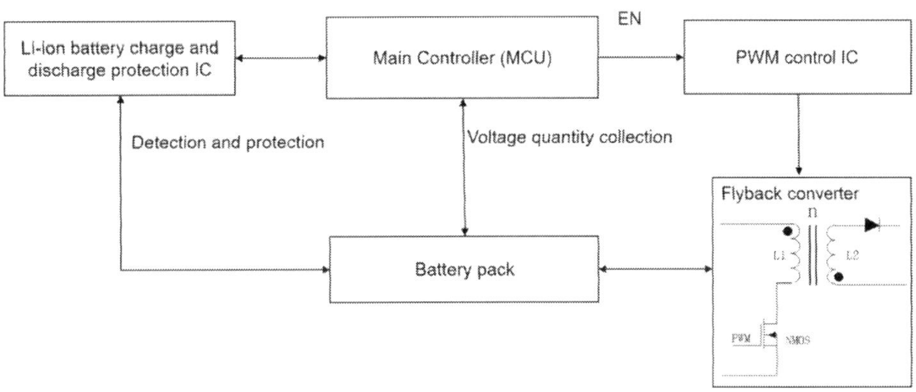

Fig. 1. Overall architecture of active balance control system

2.2. Principles of Active Balance Control

Since the battery pack is recycled for many times, based on the manufacturing process and the difference in electrolyte, there are obvious differences in the single cells in the pack, showing a phenomenon of different capacity values. During the charging and discharging process, the battery voltage is positively correlated with the remaining power. During the period, the battery with a smaller capacity value usually takes the least time to reach the cut-off voltage (4.25V, 2.75V) [7]. However, whether the charging and discharging process is terminated often depends on the monomer with a smaller capacity value, that is, the "barrel effect". This results in a monomer with a larger capacity value that can never be filled or discharged. Therefore, it is of great significance to prevent the cells with small capacity values from reaching the cut-off state and ensure the balance of power in the group, which is of great significance for extending the working life of the battery pack and improving the energy utilization rate. The working principle of the flyback converter is: when the NMOS switching element is in the on state, the two sides of the coils L1 and L2 have opposite polarities, so the diode on the secondary side of the converter is reversely cut off, and the converter stores energy at this time. When the NMOS switching element is in the off state, the diode on the secondary side of the converter is turned on, and the converter releases energy.

The specific control principle is: in the charging and discharging state, when the voltage of the single battery collected by the main controller presents an extremely unbalanced state. And a single battery voltage is higher than 4.1V or lower than 3V, the main controller immediately triggers the first level of equalization protection, and sends out the equalization protection high-level enable signal to the enable terminal of the PWM control IC. Since the enable end of the PWM control IC is high-level effective, it triggers the flyback converter to execute the high-voltage single battery power flow to the low-voltage single battery. Until the voltage of the single cell in the battery pack reaches the overcharge cut-off voltage of 4.25V or the overdischarge cut-off voltage of the Li-ion battery charge-discharge protection IC, the Li-ion battery charge-discharge protection IC starts to implement secondary protection. When the battery pack is in a static state and the voltage of the single cells is in an unbalanced state, the main controller directly sends a balanced enable high-level signal to the enable terminal of the PWM control IC. Then trigger the flyback converter to execute the higher-voltage single battery power flow to the low-voltage single battery until the voltage difference between the maximum single battery voltage value and the minimum single battery voltage value in the battery pack is lower than 0.05V, which is considered to be balanced.

3. Balanced circuit design and control process

In order to reduce the complexity of circuit design and verify the feasibility of the active balance control system used in this article, the circuit design and control are mainly aimed at four-string, 3000mAh rated capacity ternary Li-ion Battery with nominal voltage 3.7V, respectively. Variable acquisition circuit, charge and discharge protection circuit, energy balance topology circuit, balance control algorithm design.

3.1. Control variable acquisition circuit

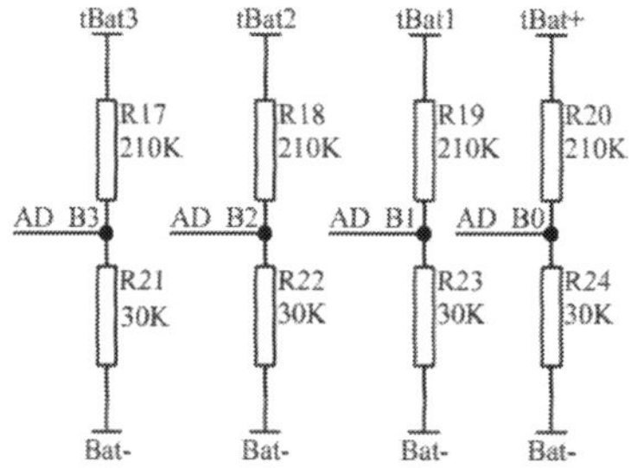

Fig. 2. Voltage acquisition circuit

For the collection of the voltage of the control variable, the internal ADC of the main controller is used. This article uses the CMS8S5885 chip as the main controller. It has multiple built-in AD conversion channels and the conversion digits are 10 bits, including 1.2V, 2.0V, 2.4V, 3.0V 4 kinds of reference voltages (V_{ref}) are available, based on the maximum voltage value of the four-string Li-ion battery pack is about 17V, V_{ref}=2.4V is selected as the reference voltage (the minimum resolvable input voltage change is 2.3mV), and use the voltage divider resistance to measure the voltage to collect the voltage, the voltage collection circuit is shown in Fig. 2. According to the 30KΩ voltage divider resistance:

$$V_{Batx} = V_{By} \times \frac{30K + 210K}{30K} \tag{1}$$

$$V_{CELL1} = V_{Bat+} - V_{Bat1} \tag{2}$$

$$V_{CELL2} = V_{Bat1} - V_{Bat2} \tag{3}$$

$$V_{CELL3} = V_{Bat2} - V_{Bat3} \tag{4}$$

$$V_{CELL4} = V_{Bat3} - 0 \tag{5}$$

In the Eq. (1), V_{Batx} and V_{By} respectively represent V_{Bat+}, V_{B0}; V_{Bat1}, V_{B1}; V_{Bat2}, V_{B2}; V_{B3}, V_{Bat3}. In the Eq. (2), Eq. (3), Eq. (4), Eq. (5), V_{CELL1}, V_{CELL2}, V_{CELL3}, and V_{CELL4} represent the actual voltages of the single battery 1, single battery 2, single battery 3, and single battery 4, respectively.

3.2. Charge and discharge protection circuit

The voltage value of the Li-ion Battery must be monitored at all times during the charging and discharging process. According to the battery industry standard, the maximum battery voltage can be about 1.2 times the nominal voltage value. This article uses a Li-ion Battery with a nominal voltage of 3.7V. Therefore, when the voltage is higher than 4.25V, it is regarded as overcharging. Overcharging will cause a strong chemical reaction in the electrolyte inside, and the release of gas will cause the battery to bulge, reduce the capacity, or cause an explosion. When the battery voltage is lower than 2.75V, it is regarded as over-discharge, and continuous discharge will cause its internal material structure to dissolve, causing permanent damage to the battery [8]. Therefore, the design of the charge-discharge protection circuit can effectively prevent overcharge or over-discharge during the balancing process of the battery pack, avoid the overcharge and over-discharge of the single cells from affecting the performance of the entire battery pack, and ensure the safety and stability of the battery pack. The design of the charge and discharge protection circuit in this article is based on the R5432V404BA chip, which has the functions of protection, protection release, and battery equalization. The charging and discharging protection circuit is shown in Fig. 3. The circuit schematic diagram of this paper is drawn and simulated by Altium Designer software. In order to prevent excessive current from damaging the switching devices during the charging and discharging process, four groups of power NMOS transistors are used in the circuit design. During system operation, when overcharge or overdischarge is detected, corresponding protection actions will be executed. Until the voltage of the single cells in the battery pack is lower than 4.1V or higher than 3V, the charging and discharging protection restrictions are lifted.

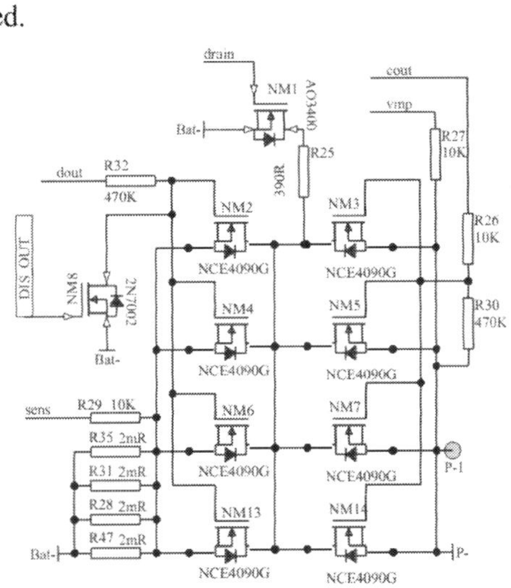

Fig. 3. Charge and discharge protection circuit

3.3. Energy balance topology circuit

In order to avoid the impact of hysteresis characteristics in the charging and discharging process of the battery pack [9], a phased equilibrium method is adopted, that is, the energy transfer balance in the charging and discharging phase and the energy transfer in the resting phase are balanced. Based on the state of the battery pack and the charging and discharging protection circuit, the main controller executes the corresponding energy balance algorithm to perform energy balance conversion. The energy balance topology circuit is shown in Fig. 4. Design four single battery sides to connect PWM control IC chips in parallel respectively. In the working state, the gate pin of the chip can output a PWM signal with a frequency range of 50KHz~1MHz and a pulse width of 0~85% according to the load current. In order to reduce the switching loss, this article uses a 100KHz switching frequency signal to drive the gate drive NMOS tube , And then control the flyback converter to perform energy conversion, C25, C26, C27, C28 used to eliminate the ripple voltage generated in the switching process. This balanced topology circuit has a simple principle and strong controllability. The use of staged balance can also avoid excessive magnetic loss and switch tube loss during the energy transfer process, and improve the balance efficiency.

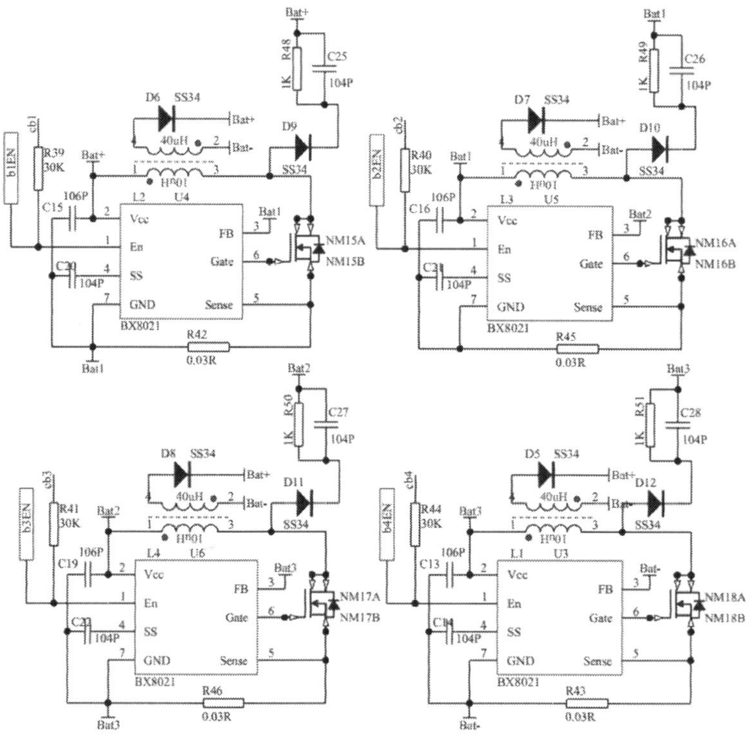

Fig. 4. Energy balance topology circuit

In the charging (discharging) state, if the voltage of the single battery 1 is the highest and higher than 4.1V (if the voltage of the single battery 2 is the lowest and lower than 3V, and the voltage of the single battery 1 in the battery pack is the highest). The main controller controls b1EN to be high. At this time, the PWM control IC chip in parallel with the single battery 1 starts to work, the NM15 switching element starts to be turned on and off periodically, and the series flyback converter starts to perform the transfer of the single battery 1 power to the battery pack at the same time, that is, the battery 1 discharges, Battery 2, Battery 3, Battery 4 are charged. Until the voltage of single cell 1 is lower than the voltage of other single cells, b1EN returns to a low potential. The main controller continues to detect the battery voltage and loops this control until it detects that the single battery voltage is higher than 4.25V (lower than 2.75V), and the charge (discharge) equalization ends.

In the static state, if the V_{CELL4} voltage in the battery pack is the highest, the main controller controls b4EN to be high, and the flyback converter starts to transfer the power of the single battery 4 to the battery pack. After a period of time, if the highest voltage of V_{CELL2} is detected, the main controller controls b2EN to be high, b4EN returns to low, and the flyback converter starts to transfer the power of the single battery 2 to the battery pack. According to this equalization algorithm, the main controller continues to detect and control until the voltage of the single cells in the battery pack reaches equilibrium, and the equalization ends.

3.4. Balance control algorithm

The balance control algorithm is to perform corresponding balance control according to the state of the battery pack and the collected voltage. The process of the equalization algorithm used in this paper is shown in Fig. 5. The state of the battery pack is determined according to the detection value of the Hall current detection device MT9223 in series with the battery pack. When the MT9223 detects a current of about 1A, it is judged to be the charging and discharging stage, otherwise it is the resting stage.

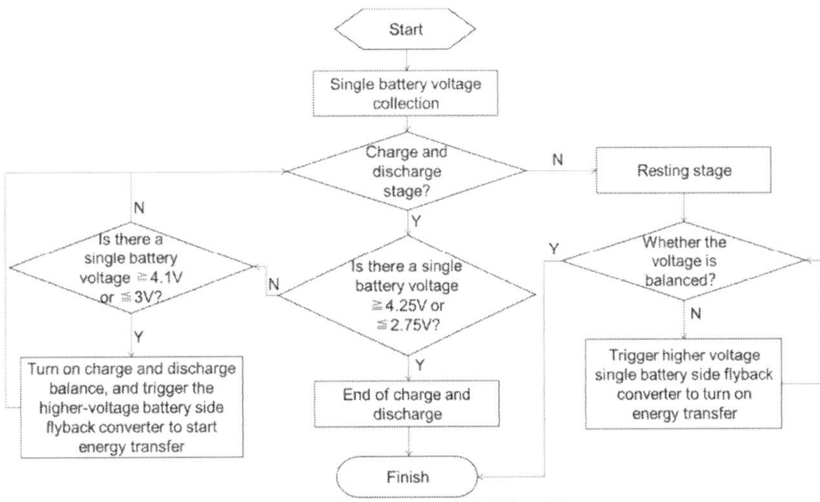

Fig. 5. Balanced algorithm flow

4. Test verification

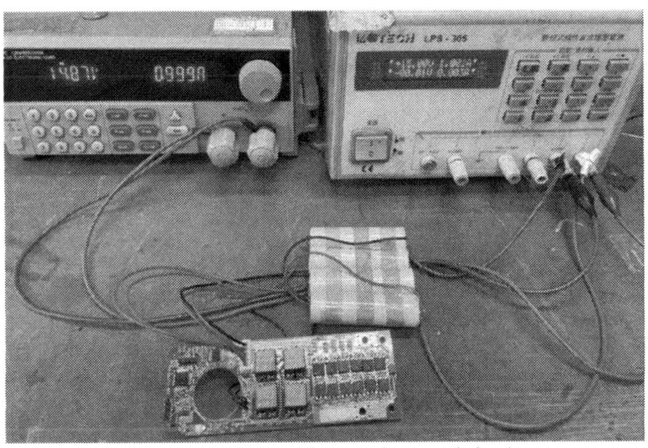

Fig. 6. Balanced experimental device

The balance experiment is carried out for a battery pack with four ternary Li-ion cells connected in series. The experimental platform includes electronic load meter, DC stabilized power supply, high-precision multimeter, four-channel oscilloscope, etc. The equilibrium experiment device is shown in Fig. 6.

4.1. Balance test in resting state

For the balance test of the battery pack in the static state, the battery pack is first discharged using an electronic load machine at a rate of 0.2C for 1 hour. Due to the correlation between the terminal voltage and the state of charge of the battery in the resting state [10], the terminal voltage balance is the state of charge balance. Therefore, connect the equalization circuit after standing for a period of time, and use an oscilloscope and a high-precision multimeter to track the voltage of the battery pack in real time. The real-time voltage of the equalization process is shown in Fig. 7.

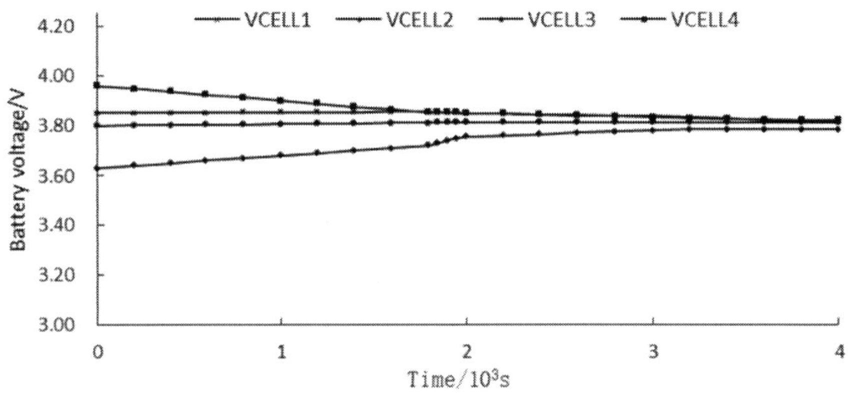

Fig. 7. Real-time voltage during equalization

In Figure 7, V_{CELL1}, V_{CELL2}, V_{CELL3}, and V_{CELL4} represent the real-time voltages of cells 1, 2, 3, and 4, respectively. It can be seen from the curve that although the voltage after the equalization of the control system fluctuates, the voltage difference between the maximum single cell voltage value and the minimum single cell voltage value starts to be lower than 0.05V after 3000s, that is, it tends to balance. It shows that the equilibrium system used in this paper has good performance and the overall equilibrium speed is relatively fast.

4.2. Battery pack working time comparison test

The battery pack working time test is divided into two groups. Discharge each single battery first, and then use a DC stabilized power supply at the same room temperature. The first group uses the equilibrium system of this paper to charge the battery pack to saturation at a rate of 17V/0.3C. The second group does not use the equalization system of this paper to charge at the same rate until a certain cell voltage is greater than 4.25V. After that, use an electronic load meter to discharge at a constant current of 0.2C to cut-off. The statistics of charging time and discharging working time adopts high-precision timer and manual monitoring method, and its accuracy can reach 99%. The test data is shown in TABLE 1

TABLE 1. Battery pack charge and discharge test

	Charging time（h）	**Discharge working time（h）**
The first group (using this equalization system)	3.28	4.26
The second group (not use this equalization system)	3.05	3.97

Tests show that the use of this balance control system is 0.29h longer than the unused working time. This data shows that the equilibrium control system in this paper can make full use of the larger capacity monomers, extend the working life of the battery pack, and improve energy utilization.

5. Conclusion

This article is based on a comprehensive analysis of various equilibrium controls. This paper proposes a single battery voltage as the control variable, adopts a staged equalization, charge-discharge two-level equalization protection strategy and an isolated flyback converter circuit to achieve active equalization control of battery pack energy. The circuit and algorithm required for equalization are analyzed and designed. In order to verify the feasibility of the active equalization control system used in this article, a test was conducted on a four-string ternary Li-ion battery pack. The test results show that the entire battery pack can be balanced for about 3000s in a static state. Compared with the unused balance control system, the working time of the balance control system in this article can be extended by 0.29h. The results show that the equilibrium system in this paper can make full use of the larger capacity monomers, quickly improve the energy imbalance in the battery pack, and can effectively extend the working life of the battery pack and improve the energy utilization rate.

Acknowledgments
Xiamen University of Technology High-level Talent Project (YKJ17020R)

References

[1] Zhu Lingzhen, Sun Xiujuan, Wang Chuanjiang, Jia Wang. (2020) Design of Lithium Battery Management System Based on Active Balance. Power Technology, 44(12): 1788-1791+1799.
[2] Wang Yixiu, Wei Xuezhe, Fang Qiaohua, Zhu Jiangong, Dai Haifeng. (2020) Electric vehicle battery consistency change law and equilibrium strategy for maximizing the life span of the whole group. Chinese Journal of Mechanical Engineering, 56(22): 176-183.
[3] XU Jun, MEI Xuesong, WANG Junping. (2019) A high power low-cost balancing system for battery strings. Energy Procedia, 158: 2948-2953.
[4] Dong Liang, Lei Liangyu, Li Xueyuan, Liu Bing, Zheng Huifeng. (2018) Design of Active Equalization Circuit for Lithium Battery Based on Ancient Card Transformation. Power Supply Technology, 42(12): 1835-1836+1840.
[5] Hannan M A, Hoque M M, Peng S E, et al. (2017) Lithium-ion battery charge equalization algorithm for electric vehicle applications. IEEE Transactions on Industry Applications, 2017, 53(3):2541-2549.
[6] Meng J, Ricco M, Luo G, et al. (2017) An Overview and Comparison of Online Implementable SOC Estimation Methods for Lithium-ion Battery. IEEE Transactions on Industry Applications, 2017: 1-1.
[7] Fan Zhiwei, Qiao Dan, Cui Haigang. (2020) Research on the influence of lithium-ion battery charge-discharge rate on capacity attenuation. Power Technology, 44(03): 325-329.
[8] Tuo Anping, Wei Zhiming, Li Jialiang, Duan Fuwei. (2019) Research on lithium battery protection technology for space portable equipment. Power Technology, 43(06): 1017-1019.
[9] Sun Jinlei,Lu Rengui,Wei Guo,Guo Yao,Zhu Chunbo. (2015) Design of a two-way full-bridge SOC equalization control system for series battery packs.Journal of Electrical Machines and Control,,19(03):76-81.
[10] Xu Ying, Shen Ying. (2014) Battery SOC estimation based on improved Kalman filter. Journal of Beijing University of Aeronautics and Astronautics, 40(06): 855-860.

Research on tomography algorithm of furnace flame temperature field based on acoustic method in thermal power station

Zhe Kan[1], Shuqiang Ding[1], Xiaolei Wang[2]*

[1] School of Information and Control Engineering, Liaoning Petrochemical University,Fushun,China

[2] School of Computer and Communication Engineering,Liaoning Petrochemical University,Fushun,China

*Corresponding author email:kanzhe @lnpu.edu.cn

Abstract. An improved Tikhonov regularization algorithm MCTR based on Markov radial basis function is proposed to obtain the feedback of closed-loop control for pulverized coal injection process in thermal power plant, so as to solve the ill posed problem in temperature field reconstruction in furnace. The algorithm constructs a new regularization matrix to replace the standard Tikhonov regularization element matrix, so as to achieve the correction effect of the new algorithm. MCTR algorithm determines the boundary singular value of small singular value defined in this paper by determining the proportion between the standard deviation component corresponding to small singular value and the sum of standard deviation components corresponding to all singular values after singular value decomposition of coefficient matrix. After determining the boundary, a new regularization matrix is constructed according to the eigenvector corresponding to small singular value. Compared with the MTR algorithm using the identity matrix as the regularization matrix, MCTR algorithm only selectively modifies the parameters corresponding to small singular values after determining the regularization parameters, which improves the stability of the parameter solution and the reconstruction accuracy of the temperature field, and is helpful to the automatic control of thermal power plants and the efficiency of coal combustion.

1. Introduction

Temperature is of great significance in industrial automation control and daily life. At present, most of the temperature measurements are indirect. By measuring the changes of sensor parameters with specific relationship with temperature, the change of temperature can be deduced[1]. According to whether the measured object is in contact with the corresponding sensor during measurement, the temperature measurement technology can be divided into two categories: contact and non-contact. Contact temperature measurement includes thermocouple temperature measurement, thermal resistance temperature measurement, optical fiber temperature measurement, etc. The non-contact temperature measurement technology does not require the sensor to directly contact the object to be measured, and has the characteristics of non-invasive and non-interference[2]. At the same time, the non-contact temperature measurement technology has small thermal inertia, small time delay and can dynamically reflect the temperature field change of the object to be measured. Common non-contact temperature measurement technologies mainly include radiation temperature measurement technology, thermal imaging temperature measurement technology, acoustic temperature measurement technology, etc.

Many scholars are exploring acoustic reconstruction of temperature field through various methods (Mayer, s.f.green, M. bramanti, Barth et al). Their research results enable other scholars to better apply acoustic temperature field tomography algorithm. However, in the process of temperature field reconstruction algorithm, there are still some unsolved problems, such as improving the stability of parameter solution, so as to improve the reconstruction accuracy of temperature field and achieve better anti noise ability. Therefore, many scholars still need to carry out some simulations and experiments to study and overcome the problems in acoustic reconstruction of temperature field.

Based on the MTR algorithm using identity matrix as regularization matrix, a new algorithm MCTR is proposed and discussed in this paper. At the same time, through the calculation of correcting the square difference of the regularization matrix, the errors of the two are compared, and the advantages of the new reconstruction algorithm are obtained through MATLAB simulation analysis. The part 2 is the principle of measuring temperature field by acoustic method. The part 3 is the regularization matrix correction process to determine the boundary of small singular values defined in this paper. The part 4 is the simulation analysis. The advantages of the new reconstruction algorithm are obtained through the simulation effect. The part 5 is the conclusion.

2. The Principle of Acoustic Temperature Field Reconstruction

When the sound wave passes through the measurement plane of the furnace flame temperature field, its propagation speed has a specific functional relationship with the absolute temperature of the flue gas in the furnace[3]. Therefore, in the process of acoustic method temperature field reconstruction, the temperature distribution in the furnace can be obtained according to the measured sound wave path in the furnace temperature field propagation velocity[4-6]. The sound velocity c in a gaseous medium at an absolute temperature T is given by

$$c=\sqrt{\frac{\gamma RT}{m}}=Z\sqrt{T} \qquad (1)$$

Where Z is a constant decided by gas composition. The value of Z for air is 20.05.

The reconstruction of acoustic temperature field is mainly divided into two steps: the first step is to establish the forward problem model, obtain the coefficient matrix describing the acoustic system according to the layout of the temperature acoustic transceiver and the division of grid pixels, and use the measured acoustic flight time to establish a linear relationship to model the problem to be solved. In the second step, the mathematical model describing the problem is solved, and the appropriate temperature field reconstruction algorithm is used to reconstruct the temperature distribution of the furnace flame. When reconstructing the actual temperature field, whether it is the simulation verification of the computer in the laboratory or the actual measurement of the utility boiler, the reconstruction matrix describing the acoustic system must be obtained and solved first.

The typical two-dimensional temperature field reconstruction and three-dimensional temperature field reconstruction acoustic path are shown in the figure below. In this paper, the two-dimensional temperature field is mainly reconstructed.

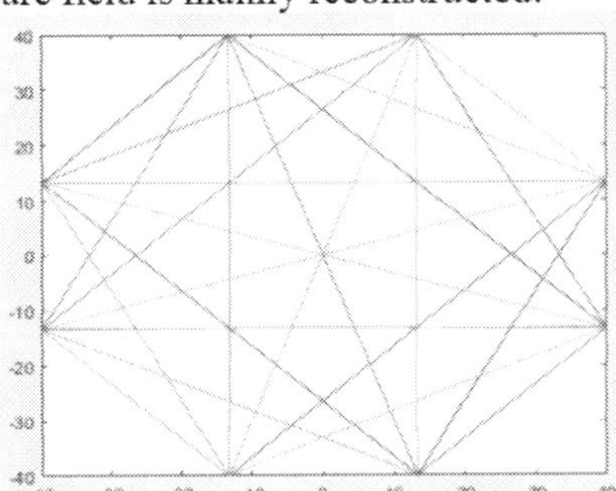

Fig. 1.Two-dimensional acoustic path distribution

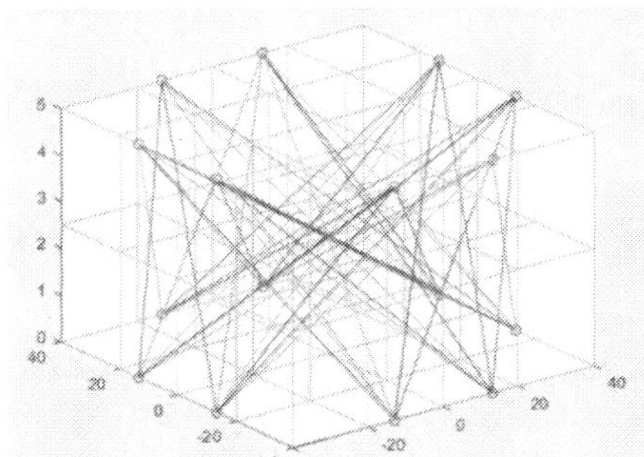

Fig. 2.Three-dimensional acoustic path distribution

3. Construction of new regularization matrix

The standard Tikhonov method takes the identity matrix as the regularization matrix, which can be used to a great extentreduce the variance of parameter estimation and improve the stability of parameter solution. According to the transmission law of covariance, we can know the standardTikhonov's covariance formula is[7]:

$$\mathrm{cov}(\hat{\varepsilon})=\sigma_0^2\left(A^TA+\alpha I\right)^{-1}A^TA\left(A^TA+\alpha I\right)^{-1} \quad (2)$$

According to the covariance formula, the trace of the coefficient matrix is:

$$D(\hat{\varepsilon})=\mathrm{tr}\left[\sigma_0^2\left(A^TA+\alpha I\right)^{-1}A^TA\left(A^TA+\alpha I\right)^{-1}\right] \quad (3)$$

The eigenvalues of square array ATA are decomposed as follows:

$$A^TA=G\wedge G^T=\sum_{i=1}^{n}G_i\wedge_i G_i^T \quad (4)$$

Where eigenvalue $\wedge_i=\lambda_i^2$, the eigenvalue of the square matrix ATA is the square of the singular value of the matrix A, and the eigenvalue of $\wedge_i$ is the right singular vector of the singular value λ. Substitute Formula (4) into formula (3) to obtain:

$$D(\hat{\varepsilon})=\mathrm{tr}\left[\sigma_0^2\left(\left((A^TA)^{-1}\right)\left(\sum_{i=1}^{n}G_i\wedge_i G_i^T+\alpha\sum_{i=1}^{n}G_iG_i^T\right)\right)^{-1}\left(A^TA+\alpha I\right)^{-1}\right]$$

$$=\sigma_0^2\sum_{i=1}^{n}\frac{\wedge_i}{(\wedge_i+\alpha)^2} \quad (5)$$

The standard Tikhonov is a biased estimation and the deviation formula is:

$$\left(\mathrm{bias}(\hat{\varepsilon_\alpha})\right)^2=\alpha^2\left(A^TA+\alpha I\right)^{-1}\varepsilon\varepsilon^T\left(A^TA+\alpha I\right)^{-1} \quad (6)$$

The trace can be obtained by decomposing the eigenvalues in the above formula:

$$\mathrm{tr}\left[\left(\mathrm{bias}(\hat{\varepsilon_\alpha})\right)^2\right]=\sum_{i=1}^{n}G_i\frac{\alpha^2}{(\wedge_i+\alpha)^2}G_i^T\varepsilon \quad (7)$$

It can be seen from formula (5) that the standard Tikhonov obtains a stable parameter solution after reducing the parameter variance by modifying the eigenvalue of the square matrix ATA. Because the unit matrix is used, after the regularization parameters are determined, each eigenvalue is modified to the same extent[8]. According to formula (7), standard Tikhonov regularization introduces deviation when reducing variance estimation[9]. The size of the introduced deviation is related to the selected regularization matrix. Different regularization matrices will correct the eigenvalues to varying degrees and the regularization method can be further modified.

Formula 5 shows that the influence caused by the serious ill condition of the equation is fully reflected in the amplification of the variance of parameter estimation by the eigenvalue with small value. According to $\wedge_i=\lambda_i^2$, the influence of the eigenvalues on variance can be transformed into the influence of singular values on standard deviation[10-11].

$$\mathrm{sta}=\left\{\sigma_0\frac{1}{\lambda_1},\sigma_0\frac{1}{\lambda_2},L,\sigma_0\frac{1}{\lambda_n}\right\} \quad (8)$$

According to the above formula, the smaller the singular value, the greater the impact on the standard deviation, according to the characteristics of several orders of magnitude difference between large singular values and small singular values of ill posed matrix, when the sum of standard deviation components of small singular values accounts for more than 95% of the standard deviation, the boundaries of these singular values are determined and regularized to reduce their impact on the

IOP Publishing
2121 (2021) 012009 doi:10.1088/1742-6596/2121/1/012009

standard deviation[12]. The judgment conditions for singular value boundaries are expressed as follows:

$$\sum_{i=k}^{m}\frac{\sigma_0}{\lambda_i}\geq 95\%\sum_{i=k}^{m}\frac{\sigma_0}{\lambda_i} \quad (9)$$

In the matrix Scomposed of singular values,$\lambda_1 > \lambda_2 > \cdots > \lambda_k > \cdots > \lambda_m$,$\lambda_k$is the boundary value of the small singular value, and the eigenvector corresponding to the small singular value is selected to construct the regularization matrix:

$$R=\sum_{i=k}^{m} G_i G_i^T \quad (10)$$

The regularization of the newly constructed regularization matrix can be expressed as:

$$\arg\min_{\varepsilon}=\|A\hat{\varepsilon}-g\|^2+\alpha\hat{\varepsilon}R\hat{\varepsilon} \quad (11)$$

$$\hat{\varepsilon}=\left(A^TA+\alpha\sum_{i=k}^{m} G_iG_i^T\right)^{-1}A^Tg \quad (12)$$

By substituting the regularization matrix R into formula (5), the trace of the variance matrix is:

$$D(\hat{\varepsilon})=tr\left[\sigma_0^2\left(\left((A^TA)^{-1}\right)\left(\sum_{i=1}^{n} G_i\Lambda_iG_i^T+\alpha\sum_{i=1}^{n} G_iG_i^T\right)\right)^{-1}(A^TA+\alpha I)^{-1}\right]$$

$$=\sigma_0^2\left[\sum_{i=1}^{k-1}\frac{1}{\Lambda_i}+\sum_{i=k}^{m}\frac{\Lambda_i}{(\Lambda_i+\alpha)^2}\right] \quad (13)$$

By substituting the newly constructed regularization matrix into equation (7), the deviation formula of regularization can be obtained as follows:

$$tr\left[\left(bias(\hat{\varepsilon}_\alpha)\right)^2\right]=\sum_{i=k}^{n}\varepsilon^TG_i\frac{\alpha^2}{(\Lambda_i+\alpha)^2}G_i^T\varepsilon \quad (14)$$

Comparing formula (5) with formula (14), it can be seen that the deviation value corresponding to formula (14) is always less than that corresponding to formula (5). It is proved that the regularization method using the newly constructed regularization matrix can reduce the variance of parameter estimation and reduce the introduction of deviation compared with the standard Tikhonov regularization method, so as to improve the stability of the undetermined parameter solution.

4. Simulation Research on the New Reconstruction Algorithm

In order to verify the reconstruction effect and accuracy of the two algorithms, simulation needs to be carried out in different temperature field modelsverification:

Single peak symmetric temperature model:

$$T(x,y)=a+b\times\cos(k_1\pi x)\times\cos(k_2\pi y) \quad (15)$$

Single peak bias temperature model：

$$T(x,y)=a+b\times\cos\left(k_1\pi(x-x_1)\right)\times\cos\left(k_2\pi(y-y_1)\right) \quad (16)$$

Bimodal temperature model:

$$T(x,y)=a_i\left[(x-x_1)^2+\left(\frac{y}{c}-y_1\right)^2\right]^{\frac{i}{2}}+b_i\left[(x-x_2)^2+\left(\frac{y}{d}-y_2\right)^2\right]^{\frac{i}{2}} \quad (17)$$

Where a and b are constants, and a represents the average temperature distribution of the temperature field; b represents the temperature gradient of the temperature field; k_1 and k_2 are constants, indicating the peak value of the temperature field; (x_1, y_1) represents the peak position of the single peak bias temperature field; (x_1, cy_1), (x_2, dy_2) respectively represent the positions of the two peaks; $a_i$and$b_i$represents the temperature gradient of the temperature field.

In order to analyze the reconstruction accuracy of MTR and MCTR algorithms, simulation verification is carried out in three temperature field models respectively. To evaluate the quality of the reconstructions, the root-mean-squared percent error of the reconstructed field is used, which is defined as follows.

$$E_{rms}=\frac{\sqrt{\frac{1}{N}\sum_{j=1}^{N}[T(j)-\hat{T}(j)]^2}}{T_{ave}}\times 100\% \qquad (18)$$

Where T(j) represents the model temperature of the first network, $\hat{T}(j)$ represents the reconstruction temperature of the second grid, and T_{ave} represents the average temperature of the model temperature field, set the mesh division to 8*8 and the noise standard to 5E-4.

Table 1.Error analysis of reconstruction results

Model	Reconstruction algorithm	Meshing	Noise standard	E_{rms} (%)
Single peak symmetrical				2.87 2.86
Single peak offset	MTR MCTR	8*8	5E-4	6.63 6.62
Double peak				3.29 3.27

From the above simulation studies, the simulation results of MTR and MCTR are made respectively. The simulation results of MTR and MCTR under the same grid division and the same noise standard under the conditions of Single peak symmetry, Single peak offset and Double peak are given.The results show that the new algorithm is more suitable for the reconstruction of complex temperature field, and the new algorithm can also reconstruct complex field well.

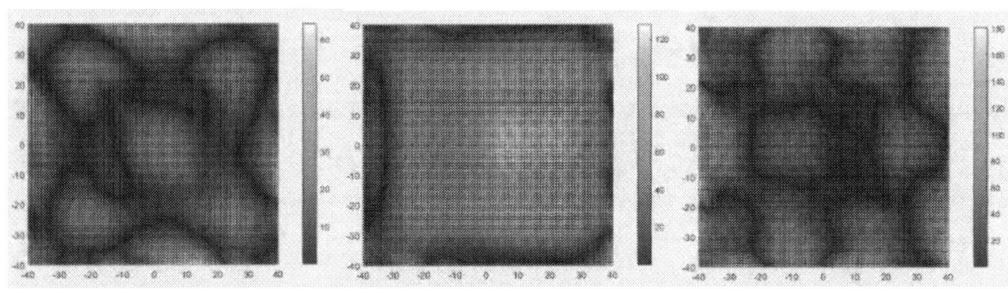

Fig. 3.Reconstruction error of MTR

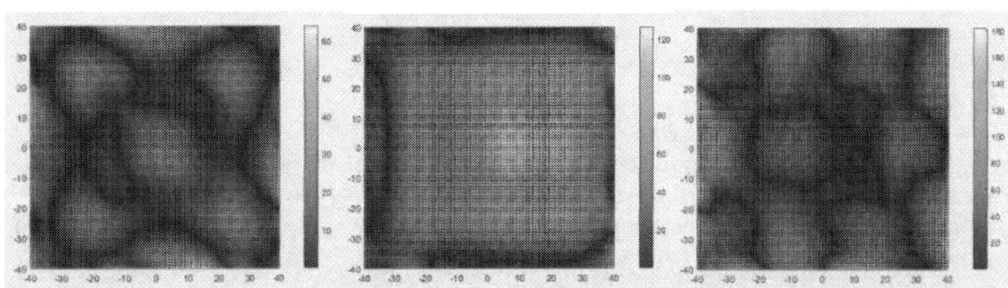

Fig. 4.Reconstruction error of MCTR

5. Conclusion

According to the simulation results of root mean square error of three temperature field models in the above table, MCTR effectively improves the accuracy of temperature field reconstruction. Compared with MTR algorithm, MCTR algorithm has further improved the reconstruction accuracy under different models and different noise levels, which shows that this algorithm has better anti noise ability, and proves the reconstruction advantages of the new algorithm compared with MTR algorithm.

In order to realize the large-scale commercial application of acoustic temperature measurement technology in furnace flame temperature measurement of thermal power station, there are still many problems to be studied. The accurate measurement of acoustic flight time is directly related to the reconstruction accuracy of temperature field. At present, the acoustic line bending effect based on variational principle will be considered in the measurement of acoustic flight time. However, in the actual measurement process, the improvement is not significant when the temperature gradient is small. How to accurately measure and reasonably correct the acoustic flight time in order to improve the applicability of different temperature field reconstruction is also the key research content.

Acknowledgments

This work was supported by the General Project of theLiaoning Province Education Department (L2020019), China.

References

[1] Hua Yan, Guannan Chen, Yang Qi, Lijun Liu. Research on complicated temperature field reconstruction based on acoustic CT[J].Chinese Journal of Acoustics,2013,32(01):51–62.
[2] Dapeng Yan, Feng Liu, Zhendong Wang, Anzhi He, Wu Ding. An improved algebraic iterative reconstruction technique and its application in 3D temperature field reconstruction [J]. ActaopticaSinica, 1996 (09): 82-86
[3] RongBao Chen, CiBian Jing,Fanghui Meng. Research on Closed-Loop Control System Based on Image-Signal Processing of Furnace Flame[J]. Advanced Materials Research, 2013, 16–21
[4] Liwei Zhang, Mingji Wang, Wen Cao. Experimental Measurement of Two-Dimensional Circular Boundary Temperature Field by Acoustic Method[J]. Advanced Materials Research, 2014, 31–49
[5] Lina Jia,Yan Gao,Decai Lu. Two-Dimension Temperature Field Reconstruction in Furnace by Acoustic Method[J]. Applied Mechanics and Materials,2013,21–42:
[6] Hongbo Lu,Weihua Sun,Zhihe Li. Research on Reconstruction Algorithm of Temperature Field by Acoustic Measurement in Boiler[J]. Advanced Materials Research, 2012,11–16
[7] Yunzhong Shen, Peiliang Xu, Bofeng Li. Bias-corrected Regularized Solution to Inverse Ill-posed Models [J]. Journal of Geodesy, 2012, 86(8): 597- 608.
[8] Qian Kong, Genshan Jiang,Yuechao Liu, Jianhao Sun,FaustoArpino. 3D Temperature Distribution Reconstruction in Furnace Based on Acoustic Tomography[J]. Mathematical Problems in Engineering,2019,19–21
[9] Yanqiu Li, Shi Liu, Schlaberg H. Inaki. Dynamic Reconstruction Algorithm of Three-Dimensional Temperature Field Measurement by Acoustic Tomography[J]. Sensors,2017,17(9)
[10] R.X. Wang. Investigations on Acoustic Detecting Techniques and Simulation of 3-D Temperature Field. Daqing Petroleum institute (2007)
[11] M. Barth, A. Raabe. Acoustic tomographic imaging of temperature and flow fields in air[J]. Measurement Science and Technology, 2011, 22(3):1-13.
[12] Kunpu Ji, Yunzhong Shen. Unbiased estimation of unit weight variance of TSVD canonical solution method [J]. Journal of Wuhan University (Information Science Edition), 2020,45 (04): 626-632

Research on the efficiency of single-wire power transmission

Fei Zhang[1],Jing Guo[1*],Jingming Wang[1], Xiaoying Ren[1],Lu Gao[1], Chen Tian[1], Qiuyu Duan[1], Yanning Gao[1]

[1]Inner Mongolia University of Science and Technology, Inner Mongolia Autonomous Region, Baotou City, 014010, China

*Corresponding author's e-mail: 2020022092@stu.imust.edu.cn

Abstract. With the development of related theoretical research on Single Wire Power Transfer(SWPT), the SWPT system has demonstrated its advantages in a variety of special environments. In this paper, the influence of capacitance and inductance on transmission efficiency is studied in order to improve the transmission efficiency of swpt system. This article simplifies and analyzes the circuit model of the SWPT system. It can be found that the system is a second-order circuit, and its response distribution trend conforms to the Gaussian function distribution. There should be an optimal ratio between the capacitor and the inductance to maximize the circuit response. Through the influence of different capacitance, inductance, and transmission distance on transmission efficiency, multiple sets of experiments and simulations are carried out. The experiment found that the capacitance, inductance, and transmission efficiency are in accordance with the Gaussian non-normal distribution, and the acquired data is fitted by the Gaussian function. The average relative error of the fitting is less than 1% under different transmission distances, which can be better reflect the relationship between the three. The study of single-wire power transmission system will provide a key reference for wireless power transmission.

1. Introduction

Wireless power transmission (WPT) is one of the current hot research directions in the field of electrical engineering. The United States, Japan, Germany, France and other countries in this area started early and they occupy an important position in the fields of smart electronics and electric vehicles. China Association for Science and Technology also listed it as ten science and technology leading the future[1-2], this technology has shown advantages in smart grid, aerospace, military and other fields.

In daily life, WPT system has also be widely used, literature [3-6] describes the application of WPT system in the context of electric vehicles, drones, mobile phones, human implantable medical devices, and household appliances. The power supply mode of the WPT system allows people to get rid of the space constraints caused by the wired power supply network.

In the current power network system, the power transmission method is still based on multiple wires[7]. However, in addition to the dangers and shortcomings of such a power network, it is also greatly restricted in space, if the WPT system can be applied to the power network system, it can get rid of many restrictions caused by wired power transmission.

But the WPT system is limited in the transmission distance and transmission power[8].In order to break through the barriers of wireless technology, people have turned their attention to single-conductor power transmission technology (SWPT). If a single wire can be replaced by natural conductors such as the earth, wireless power transmission in a broad sense can be realized[9].

Compared with the WPT system, the SWPT system uses a wire, so the SWPT system can achieve greater transmission power and longer transmission distance and the SWPT system has advantages that the WPT system does not have in some special environments, for example, the power supply problem of isolated islands or the power supply problem of wireless sensor network sensors[10], SWPT system has proposed solutions for these problems with its advantages of low cost and easy laying.

Literature [11] briefly described the impact of capacitance on transmission efficiency, but did not carry out a detailed description and systematic research on it; Literature [12] conducted related experiments and analysis on the equivalent method of the capacitance between the two Tesla coils in the SWPT system, and further analyzed the circuit model of the system, but there was still no systematic discussion on the relationship of transmission efficiency; In [13], a wireless power supply platform based on the SWPT system was constructed by using the capacitance to ground generated by the spiral coil, and the transmission effect of capacitance and electric energy was theoretically analyzed and experimentally verified.

Aiming at the problem of the transmission efficiency of the SWPT system, in this article, we study the influence of capacitance and inductance on the transmission efficiency, which will provide a new way to improve the transmission efficiency of the SWPT system. This article is based on the previous research. Through theoretical derivation, experiment and simulation, the relationship between the two and the transmission efficiency was briefly studied.

2. The influence of capacitance and inductance on transmission efficiency

2.1 Theoretical Derivation of Single-Wire Power Transmission System

In order to realizes the high-power electric energy transmission of the SWPT system, not only need for two Tesla coils to be coupled at the same frequency, but also closely related to the inductance and capacitance at the resonance frequency. Therefore, there should be a certain relationship between inductance, capacitance and transmission efficiency, so that the transmission efficiency of inductance and capacitance can reach the maximum value only at a specific value.

To verify the above relationship, first analyze the circuit model of the SWPT system. Literature [11] proposed a circuit model of the SWPT system, as shown in Figure 1.

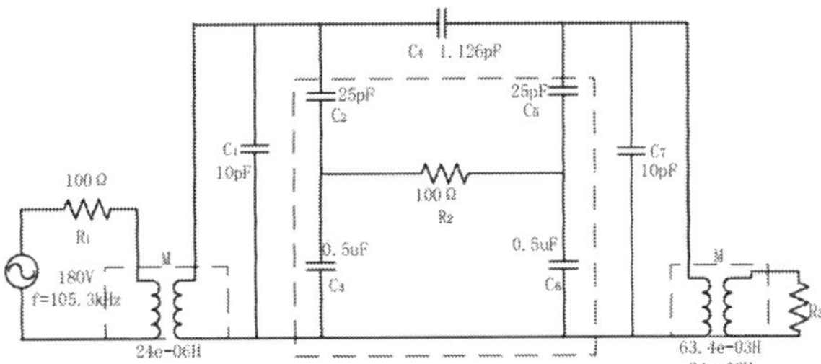

Figure 1. SWPT system equivalent circuit diagram

Literature [11] pointed out that the resistance R_2 will increase due to the transmission distance and skin effect, which can be equivalent to infinity; at this time, the capacitors C_2 and C_3 are connected in series and then connected in parallel with C_1, which is recorded as C_{12}. After C_5 and C_6 are connected in series, they are connected in parallel with C_7 and recorded as C_{57}. Since the wire is close to the ground, the capacitors connected in parallel on both sides are respectively: $C_{12} \approx C_1 + C_2$ and $C_{57} \approx C_5 + C_7$, As the transmission distance increases, the capacitance C_4 will decrease rapidly. When C_4 is small enough, it can be equivalent to a wire. At this time, the equivalent capacitance of C_{12} and C_{57} in parallel is recorded as C_8, and the Tesla coil is regarded as a transformer. Its decoupling is equivalent to an inductance, and the two inductances are connected in parallel, denoted as L_1, and the circuit is simplified to get the circuit diagram shown in Figure 2.

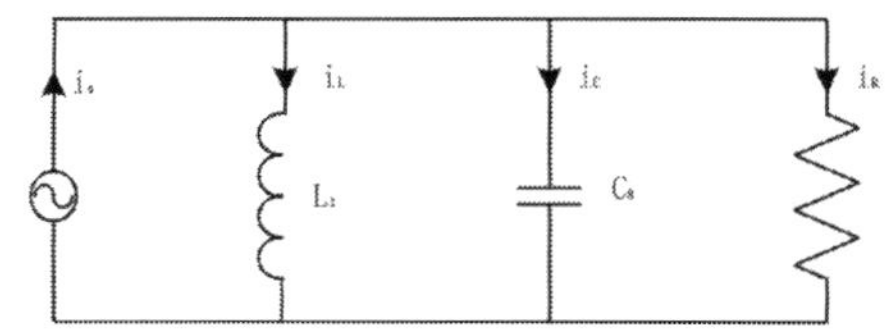

Figure 2. Simplified equivalent circuit diagram of SWPT system

By obtaining the current response i_L flowing through the inductor in the above figure 2, the KCL law can be obtained:

$$LC\frac{d^2i_L}{dt^2}+\frac{L}{R}\frac{di_L}{dt}+i_L=i_s \quad (1)$$

It can be seen from formula (1) that the model is a second-order LC resonant circuit, so the characteristic equation of formula (1) is shown in formula (2):

$$\lambda^2+\frac{1}{RC}\lambda+\frac{1}{LC}=0 \quad (2)$$

After substituting the corresponding parameters into equation (2), it can be seen that the characteristic equation has two different solutions, then the general solution of the corresponding homogeneous equation is shown in equation (3):

$$i_L=C_1e^{\lambda_1x}+C_2e^{\lambda_2x} \quad (3)$$

Therefore, the transformation trend of the voltage across the load with the capacitance and inductance should conform to the distribution trend of the Gaussian function, then when the capacitance and inductance are at a certain value, the voltage across the load will reach a peak value.

2.2 Single-wire power transmission system experiment

2.2.1 Experimental conditions and test equipment

Experimental conditions: 0.4mm enameled wire, PVC pipe with an outer diameter of 25mm, a ball with a diameter of 35mm, aluminum foil, wire, universal board, and several resistors.

Experimental instruments: signal generator, Tektronix oscilloscope, inductance tester. Among them, the maximum output voltage of the signal generator is 10V, and the maximum operating frequency is 60MHz.

2.2.2 Single-wire power transmission experiment

In order to verify the assumption that there is a certain relationship between inductance, capacitance and transmission efficiency, a SWPT system as shown in Figure 3 [14] is built.

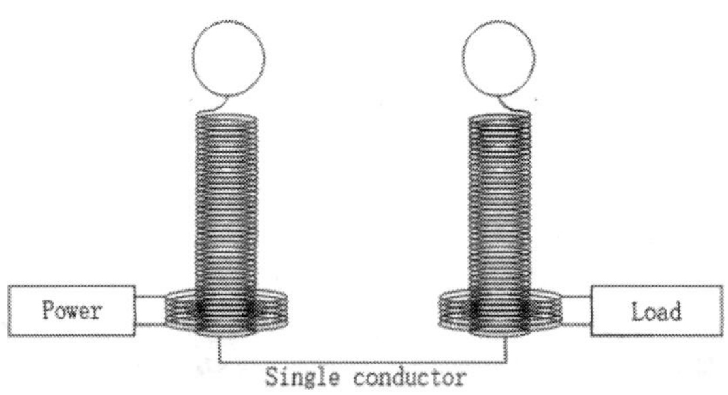

Figure 3. The single-wire power transmission structure used in this article

With 10V alternating current as the input voltage of the system, the working frequency of the signal generator is constantly adjusted, and the coil is gradually increased from 30 turns to 180 turns, and the experiment is carried out in steps of 10 turns. The distance between the receiving end and the transmitting end is fixed at 3m, 10m, and 40m respectively. With a 50Ω resistor as the load, the peak-to-peak voltage at both ends of the load is detected by an oscilloscope. When the peak-to-peak voltage reaches the maximum, the system is considered to have reached the resonance frequency. After the resonant frequency is obtained, the capacitance to ground of each coil is obtained by formula (4), and the peak-to-peak voltage on both sides of the load at the receiving end under different turns is recorded.

$$f=\frac{1}{2\pi\sqrt{LC}} \quad (4)$$

Adjust the output voltage of the signal generator to 10V and the operating frequency to 17MHz, and gradually reduce it. Calculate the capacitance of the coil to ground. The values are shown in Table 1.

Table 1. Receiving terminal voltage and resonance frequency under different turns

Turns	Inductance (μH)	Capacitance (pF)	Resonance Frequency (MHz)	Turns	Inductance (μH)	Capacitance (pF)	Resonance Frequency (MHz)
30	36	2.5376	16.66	110	129.5	3.1081	7.937
40	48	2.544	14.45	120	143	3.1448	7.576
50	61	2.6472	12.85	130	154	3.1814	7.194
60	74	2.7545	11.57	140	166.5	3.2131	6.978
70	84.5	2.8617	10.24	150	176	3.2447	6.8
80	94.5	2.9355	9.398	160	188.5	3.3067	6.378
90	105.5	3.0092	8.937	170	201	3.3269	6.203
100	123	2.9881	8.306	180	212	3.3471	6.039

Considering that the SWPT system realizes the transmission of electric energy through the resonance of the Tesla coils at both ends, the resonance frequency determines the transmission efficiency to a large extent. It can be seen from equation (4) that when the inductance and capacitance reach the appropriate When matching, the transmission efficiency is the largest, so the relationship between the three is explored by taking the product of inductance and capacitance.

Under the transmission distance of 3m, 10m and 40m, the relationship between the product of the inductance L and the capacitance C and the received voltage is shown in Figure 4, where the abscissa is the product of the inductance and the capacitance, and the ordinate is the voltage across the load.

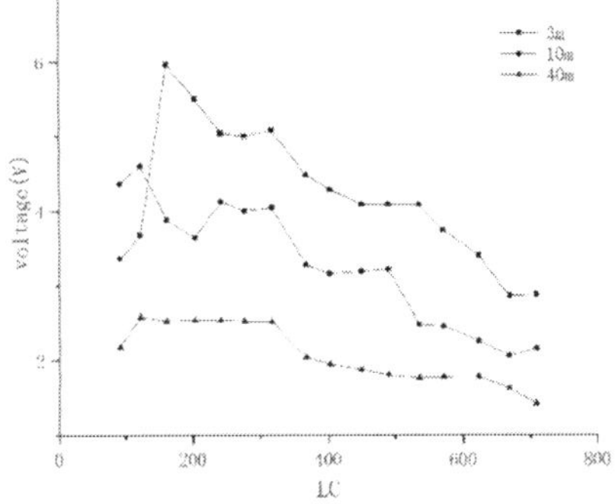

Figure 4. Waveforms of the voltage at the receiving terminal and the capacitance and inductance under different line lengths

 2121 (2021) 012010 doi:10.1088/1742-6596/2121/1/012010

2.3 Experimental simulation of single-wire power transmission

In order to further verify the influence of capacitance and inductance on transmission efficiency, the experiment process was simulated through Simulink. Set up the simulation model shown in Figure 2 in Simulink, the signal input amplitude is a sine wave of 10V, the inductance and capacitance are respectively expanded by 1000 times on the basis of Table 1, and the frequency is correspondingly reduced by 1000 times (because Simulink will have an error when calculating the reciprocal of the too small number or matrix or matrix), use a 50KΩ resistor as the load to detect the peak-to-peak voltage across the load, and then use the product of the inductance L and the capacitance C as the abscissa, and the peak-to-peak voltage across the load as the On the ordinate, plot the experimental and simulated waveforms, as shown in Figure 5.

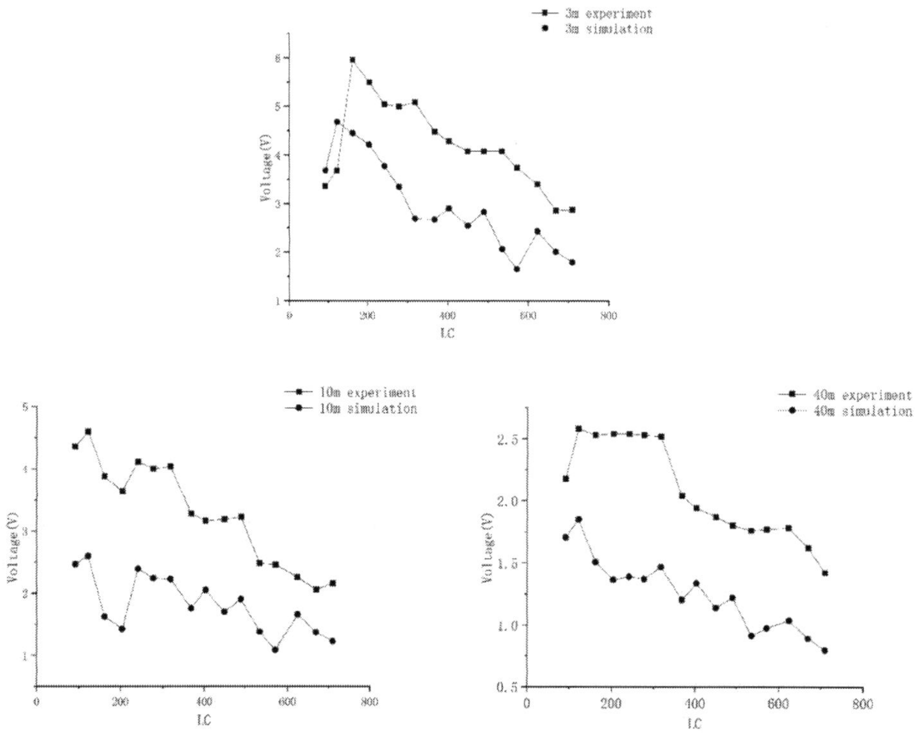

Figure 5. Waveform diagram of the peak-to-peak voltage of the load at different transmission distances as a function of the product of capacitance and inductance

Observe Figure 5, it can be seen that the overall change trend is relatively similar, but the overall amplitude of Figure 5 has a certain degree of decline compared to Figure 4. Comprehensive considerations, it is believed that the following reasons are mainly caused: (1) Due to the circuit The model is a lumped parameter model, while the experiment is a distributed parameter, which causes a certain change in the waveform; (2) The load during simulation is larger than the load in the experiment.

3. Analysis of results

It can be seen from Figure 4 and Figure 5 that the increase in line length leads to a rapid drop in the voltage across the load, which is consistent with the conclusion drawn in the literature [11]. Secondly, it can be found that although the transmission efficiency is significantly reduced, the change trend of the voltage at both ends of the load with the capacitance and inductance under different line lengths is still roughly the same, and the maximum value of the transmission voltage is obtained at 40 turns to 50 turns, and then As the inductance and capacitance increase to decrease rapidly, this is consistent with the previous description, so there is a certain specific relationship between capacitance, inductance and receiving voltage.

In the second section of this article analyzed the circuit model and determined that the circuit model is a second-order circuit. Its waveform should be consistent with the Gaussian function distribution trend. From Figure 4 and Figure 5, we can see the trend of that the voltage across the load varies with inductance and capacitance,which changed waveform is similar to the Gaussian function.This is consistent with the conclusion that in the second section of this article when after obtaining the response of the circuit model . Therefore, use the data of 3m, 10m and 40m are respectively fitted by the Gaussian function through MATLAB to obtain the formula (5):

$$V_L=\alpha_1*e^{(-(\frac{LC-\beta_1}{\gamma_1})^2)}+\alpha_2*e^{(-(\frac{LC-\beta_2}{\gamma_2})^2)}+\alpha_3*e^{(-(\frac{LC-\beta_3}{\gamma_3})^2)} \quad (5)$$

Among them, V represents the voltage across the load at the receiving end, in V, L represents the coil inductance, in μH, and C represents the coil capacitance, in pF. Among them, the parameter values are shown in Table 2.

Table 2. The parameter values of the fitting equation

L	3m	10m	40m
α_1	2.469	0.994	1.525
β_1	175.2	114	115.9
γ_1	30.42	39.03	83.05
α_2	1.005	0.6787	1.253
β_2	268.6	284.4	263.8
γ_2	88.12	63.4	111.1
α_3	4.242	3.649	1.8
β_3	365.	104.5	524.7
γ_3	558.7	795.5	440.7

Based on the above-fitted model and the measured data, the comparison is shown in Table 3

Table 3. Comparison of model calculation data and measured data

	3m line length model			10m line length model			40m line length model		
Turns	Measured data	Model calculation	Relative error (%)	Measured data	Model calculation	Relative error (%)	Measured data	Model calculation	Relative error (%)
30	3.36	3.33	0.96	4.36	4.36	0.02	2.18	2.19	0.46
40	3.68	3.7	0.55	4.6	4.60	0.03	2.58	2.54	1.55
50	5.96	5.97	0.2	3.88	3.87	0.19	2.53	2.58	1.98
60	5.5	5.48	0.35	3.64	3.73	2.54	2.54	2.49	1.97
70	5.04	5.03	0.28	4.12	3.97	3.54	2.54	2.55	0.39
80	5	5.1	2.01	4	4.15	3.78	2.53	2.83	11.86
90	5.08	4.85	4.62	4.04	3.91	3.13	2.52	2.44	3.17
100	4.48	4.49	0.3	3.28	3.39	3.44	2.04	2.11	3.43
110	4.28	4.36	1.98	3.16	3.19	1.03	3.16	1.93	0.52
120	4.08	4.24	3.9	3.19	3.02	5.22	3.19	1.82	2.67
130	4.08	4.1	0.61	3.22	2.89	10.38	1.98	1.81	0.55
140	4.08	3.91	4.25	2.48	2.72	9.79	2.48	1.8	2.27
150	3.74	3.71	0.69	2.46	2.59	5.16	2.46	1.78	0.56
160	3.4	3.39	0.16	2.26	2.38	5.52	2.26	1.71	3.93
170	2.86	3.09	7.97	2.06	2.21	7.11	2.06	1.62	0
180	2.88	2.8	2.81	2.16	2.05	5.28	2.16	1.51	6.34

It can be seen from Table 3 that the average relative error of the model with a transmission distance of 3m is 0.8123%, SSE is 0.2259, R-square is 0.9840, Adj R-sq is 0.9679, and RMSE is 0.1680; the model with a transmission distance of 10m The average relative error is 0.664%, SSE is 0.3681, R-square is 0.9687, Adj R-sq is 0.9375, RMSE is 0.2145; the average relative error of the model with a transmission distance of 40m is 0.8771%, SSE is 0.0396, R-square It is 0.9855, Adj R-sq is 0.9710, and RMSE is 0.0704.

From the above indicators, the fitting accuracy of the three models is acceptable, and the average relative error with the measured data is less than 1%, which is relatively accurate. It is confirmed that there is indeed a certain relationship between capacitance, inductance and transmission efficiency. As the transmission distance increases, the transmission efficiency decreases, and the coefficients of the fitted mathematical model also change accordingly, but there is no obvious change in the overall change trend.

Based on the above model, we can consider adjusting the parameters when the capacitance or inductance is known to control the quality factor in a more appropriate range, so that the receiving voltage at the receiving end can achieve the desired receiving effect.

4. Conclusion

In this paper, three sets of SWPT systems with different transmission distances are built through 18 sets of Tesla coils. The hypothesis that capacitance and inductance have nonlinear effects on transmission efficiency is deduced and verified by theory, experiment and simulation, it is confirmed that there is a correlation between the three. When the capacitance and inductance are at the right value, the transmission efficiency can be maximized. After obtaining the experimental data, the Gaussian function is used to fit the relationship between the three, and the error is analyzed. Various error indicators show that the fitting relationship can accurately reflect the relationship of capacitance, inductance and transmission efficiency. It provides new ideas and directions for improving the transmission efficiency of the SWPT system.

Acknowledgments

Thanks to Professor Xiyou Chen of Dalian University of Technology

This work is supported by Key Special Sub-projects of International Cooperation in Science and Technology Innovation between Governments,No. 2017YFE0109000、Inner Mongolia Autonomous Region Science and Technology Plan Project,No. 2021GG0046、Inner Mongolia Autonomous Region Science and Technology Plan Project,No. 2021GG0048.

References

[1] Shuoyi Chen, Li Zhang, Mingsheng Tang, et al. Development status and trend of radio energy transmission technology [J]science and Technology China, 2018 (07): 7-10.

[2] Ming Xue, Qingxin Yang, Pengcheng Zhang, et al. Application research status and key problems of radio energy transmission technology [J]Journal of engineering and technology 2021,36 (08) 1547-1568 .

[3] Mohamed Ahmed A. S.; Marim A.A.; Mohammed O. A . Magnetic Design Considerations of Bidirectional Inductive Wireless Power Transfer System for EV Applications [J]. IEEE Transactions on Magnetics, 2017,53(6):1-5.

[4] Milan Simic; Cees Bil; Vuk Vojisavljevic. Investigation in Wireless Power Transmission for UAV Charging [J]. Procedia Computer Science, 2015, 60:1846-1855.

[5] N. Saranya; T. Kesavamurthy.Review on next generation wireless power transmission technology for implantable biomedical devices [J]. International Journal of Biomedical Engineering and Technology, 2021, 35(3):207-222.

[6] Chunyang Mou,Shizhong Li, Guoqiang Liang. Wireless power transmission system based on Tesla coil [J]. Science and technology and Cheng, 2015,15 (13): 77-81 + 86

[7] Liu G , Zhang B . Analytical Model of a 25-50m Robust Single-wire Electric-field Coupling Power Transfer System Using a Limiter[J]. IEEE Transactions on Circuits & Systems II Express Briefs, 2018.66(6):1-1.
[8] Susanna, Vital, de, et al. A Novel Method for Data and Power Transmission Through Metallic Structures[J]. IEEE Transactions on Industrial Electronics, 2017,64(5):4027-4036.
[9] Zeran Zhang , Xiyou Chen, Yuxiang Zhou, et al. Experimental study on single line power transmission [J]. New electrical energy technology, 2016,35 (10) 69.
[10] Yang Li, Rui Wang, Yu-Jie Zhai et al. A Novel Single-Wire Power Transfer Method for Wireless Sensor Networks[J]. Nutrients,2020,13(19):5182-5182.
[11] Xiyou Chen, Jianhui Chen, Xianmin Mou, et al. Simulation, understanding and construction of space electric field coupled single line power transmission [J]. Chinese Journal of electrical engineering, 2017,37 (19) 5749-5758 + 5852
[12] Ziqi Zhao, Zifang Zhang, Jiazhi Zhao, et al. New research practice of Tesla coil power transmission [J]. University physics experiment ,2021,34 (01): 47-51.
[13] Huangui Qin. Relationship between circuit power factor and vibration [J]. Communication power supply technology, 2020,37 (09). 87-88
[14] Ke Huo. Single line power transmission characteristics and experimental research [D] Tianjin University of technology, 2020

Journal of Physics: Conference Series **2121** (2021) 012011 doi:10.1088/1742-6596/2121/1/012011

Research on Classification of Daily Load Curve of Distribution Network Based on Improved SSA-FCM

Haoran Shi[1], Rong Cao[2], Wenbo Hao[2], Mingyu Xu[2], Heng Hu[2], Peng Jiang[2], and Feng Zhou[1]*

[1] College of Electrical Engineering and Automation, Shandong University of Science and Technology, Qingdao, Shangdong Province, 266590, P.R.China

[2] State Grid Heilongjiang Electric Power Company Limited, Heilongjiang Province, 150090, P.R.China

*Corresponding author's e-mail: zhoufeng@sdust.edu.cn

Abstract. In the analysis of three-phase unbalance in distribution network, the accuracy of daily load curve classification results determines the size of three-phase unbalance.Aiming at the shortcomings of Fuzzy C-Means (FCM), a fuzzy C-Means clustering algorithm (SSA-FCM) optimized based on Sparrow Search Algorithm (SSA) is proposed. The cluster validity evaluation index is introduced to get the optimal quantity of clusters, and the SSA is used to search for the initial cluster center, which solves the problem that the FCM algorithm relies on the initial value and is easy to converge to local optimal solution. The simulation results show that, compared with the FCM algorithm, the load curves classified into the same category by SSA-FCM are closer together.

1. Introduction

With the progress of my country's technological level and economic development, the scale of the power system **continues** to expand and the structure becomes more complex. At present, in our country's low-voltage power supply system distribution network, the user's load types are complex, and the difference in daily load characteristics is becoming more and more huge[1]. It is important to study a fast and accurate classification method based on user's actual daily load curve[2].

At present, in the daily load characteristic curve classification, the main methods used are: K-means algorithm[3],FCM[4], DBSCAN algorithm[5], hierarchical clustering algorithm[6], Gaussian mixture model (GMM) algorithm[7], support vector machine (SVM) algorithm[8] and extreme learning machine (ELM) algorithm[9], etc.

Among these methods, the FCM algorithm is one of the most widely used power load characteristic classification algorithms. However, there are three shortcomings in applying FCM algorithm to classify users' daily load curves:

- The fuzzy weighted index m can only be selected based on experience, and there is still a lack of theoretical guidance.
- The number of clusters directly affects the final result, but the best number of clusters cannot be obtained in advance
- FCM is easy to converge to the local optimal solution, but can not get the global optimal solution. The starts impact the effectiveness of FCM greatly .

Journal of Physics: Conference Series **2121** (2021) 012011 doi:10.1088/1742-6596/2121/1/012011

In view of the current problems of the FCM ,the sparrow search algorithm is used to improve FCM, and SSA is used to quickly seek the optimal initial cluster centroid for FCM, so as to reduce the possibility of FCM falling into local optimization and obtain an approximate global optimal cluster partition.

2. Fuzzy C-means clustering algorithm based on SSA

2.1. Sparrow search algorithm

The sparrow search algorithm is a swarm intelligence optimization algorithm proposed based on the behavior of sparrows foraging and evading predators[10]. In the sparrow search algorithm, individuals are divided into discoverers, followers, and guards, and each individual position corresponds to a solution. According to the algorithm setting, the population proportion of the guards is 10%~20%.

The following is the location update of the discoverer:

$$X_{i,j}^{t+1}=\begin{cases} X_{i,j}.\exp(-\frac{i}{\alpha . iter_{\max}}), if\ R_2<ST \\ X_{i,j}+QL, if\ R_2\geq ST \end{cases} \tag{1}$$

Among them, t represents the current number of iterations; $iter_{max}$ represents the maximum number of iterations; Xi,j represents the position information of the i-th sparrow in the j-th dimension. $\alpha\in(0,1]$ is a random number. R2($R2\in[0,1]$) and ST($ST\in[0.5,1]$) represent the warning value and safety value respectively; Q is a random number that obeys a normal distribution Number. .L is a 1 by d matrix.

The location update description of the joiner is:

$$X_{i,j}^{t+1}=\begin{cases} Q.\exp\left(\frac{X_{worst}-X_{i,j}^{t}}{i^2}\right), if\ i>n/2 \\ X_P^{t+1}+\left|X_{i,j}-X_P^{t+1}\right|.A^{+}.L, otherwise \end{cases} \tag{2}$$

Among them, X_P is the optimal position of the discoverer in the population; X_{worse} represents the worst position of the sparrow in the current population. A represents a 1×k matrix, where each element is randomly assigned a value of 1 or -1, and $A^{+}=A^{T}(AA^{T})^{-1}$.

The initial position of the vigilant is randomly generated in the population. The expression is as follows:

$$X_{i,j}^{t+1}=\begin{cases} X_{best}^{t}+\beta.|X_{i,j}^{t}-X_{best}^{t}|, if\ f_i>f_g \\ X_{i,j}^{t}+K\left(\frac{|X_{i,j}^{t}-X_{worst}^{t}|}{(f_i-f_w)+\varepsilon}\right), if\ f_i=f_g \end{cases} \tag{3}$$

Among them, X_{best} is the current global optimal position; β is a random number subject to a normal distribution with a mean of 0 and a variance of 1; $K\in[-1,1]$ is a random number, and f_i is the adaptation of the current sparrow individual Degree value; f_g and f_w are the current global best and worst fitness values respectively; ε is a constant.

2.2. SSA-FCM algorithm

The SSA algorithm has good global search characteristics. Aiming at the FCM algorithm's dependence on the initial value and the shortcomings of converging to local extremum easily, the sparrow search algorithm is used to optimize the original algorithm, and the global search of the sparrow algorithm is used to replace the gradient descent of the FCM algorithm. Process to avoid local extreme values and reduce dependence on initial values. In the search process, for the evaluation of each individual sparrow, the fitness function defined is consistent with the objective function of the FCM algorithm. The specific steps for SSA-FCM are:

- Determine the range of the number of clusters $C=[2,C_{max}]$.
- Initialize the population. Determine the number of iteration terminations, the population size, initialize the discoverer-to-participant ratio, randomly select c cluster centers as sparrow

individuals from the sample, and repeatedly generate multiple sparrow individuals to generate the initial population.

- Execute FCM clustering algorithm. Calculate the membership degree of each individual sparrow, then update the cluster center, and traverse each particle to achieve the effect of compactness and separation between clusters.
- Calculate the fitness function value, calculate the fitness value of each individual, and record the individual extreme value and the global extreme value of the entire group.
- Check termination condition. If the maximum number of iterations is reached or if the fractional positive epsilon or the fuzzy matrix less than specified is not changed, the calculation ends. Otherwise, according to formula (1) and formula (2) And formula (3) update the location of the discoverer, joiner and guard. Form a new cluster center, go to step 2.
- Set the initial value of the FCM to the result obtained in step 5, the FCM algorithm is executed for classification, and output the result.

3. Daily load curve classification based on SSA-FCM algorithm

3.1. Construction of the input data set

The user load data obtained by the smart collection terminal in the smart grid contains bad data and abnormal data, which cannot be cleaned and eliminated during the data collection process. Therefore, it is also necessary to clean the input samples. After removing the bad data and abnormal data in the daily load data, the curve data which is missing a small amount of data is processed by smoothing formula.

3.2. Selection of fuzzy weighting index m

The previous article summarized the shortcomings of the FCM algorithm. Among them, the fuzzy weighting index m is determined based on experience. The value of m impacts the classification results greatly. The larger the value of m, the more fuzzy the clustering. Bezdek gave a range of experience; later, from the physical explanation, m=2 was the most meaningful; Chan et al. obtained from the application background of Chinese character recognition that the best value of m should be between 1.25 and 1.75; Pal et al. obtained from the experimental study on clustering effectiveness that the optimal selection interval of m was [1.5, 2.5], and generally take the median of the interval m=2 without special requirements.

3.3. Cluster validity evaluation index

The traditional clustering effectiveness indicators are PC, MPC, PE, XB, UV, FM. However, these indicators can only exert their advantages on specific data sets, and cannot be applied to all data sets. This paper chooses a new fuzzy clustering effectiveness index W[11], the definition is as follows:

$$W(c,U)=Var^{n}(c,U)+\frac{Cop^{n}(c,U)}{Sep^{n}(c,U)} \tag{4}$$

Where, $Var^n(c,U)$ is intra-class compactness; $Cop^n(c,U)$ is the overall overlap degree; $Sep^n(c,U)$ is the overall degree of separation.

The $Var^n(c,U)$ smaller, the closer the distance within the class; the $Sep^n(c,U)$ larger, the more scattered the distance between classes; the $Cop^n(c,U)$ smaller, the smallest overlap. Obviously, the $W(c,U)$ smaller the clustering effect is the best.

3.4. Algorithm flow design

The overall algorithm includes load data preprocessing and determining the optimal number of clusters in the SSA-FCM algorithm. The following is the overall process of user daily load curve classification using SSA-FCM algorithm:

- Carry out load data preprocessing on the original daily load curve data;

- Set and minimum cluster number c_{min} and maximum cluster number c_{max}, SSA-FCM related parameters;
- Execute the SSA-FCM algorithm for cluster analysis;
- Use W index to evaluate the effectiveness of clustering;
- Determine whether the current c is greater than the maximum number of clusters c_{max}, if yes, go to step 6. If not, c=c+1, execute step 3;
- Output the optimal number of clusters and index values.

4. SSA-FCM algorithm daily load curve classification simulation analysis

4.1. Daily load curve classification

In June 2021, 55 users'daily load data were collected in a city of Heilongjiang. Each user collected 96 times a day for 15 minutes. After processing, it finally contains 50 daily load curves.

In order to remove the influence of the magnitude of the load on the classification results, the user's daily load data is normalized before the classification. Here, extreme value sequence normalization is used. After normalization, all the data are in [0,1] .

Use SSA-FCM algorithm and FCM to classify the processed data respectively, and compare the classification results. The parameters of the SSA-FCM algorithm are set as follows: minimum number of clusters c_{min}=2, maximum number of clusters c_{max}=6, fuzzy weighting index m=2, population size 200, and maximum number of iterations $imter_{max}$=300. The FCM algorithm parameters are: fuzzy weighting index m=2; the maximum number of iterations is t_{max}=300, and the objective function threshold is 10^{-7}.

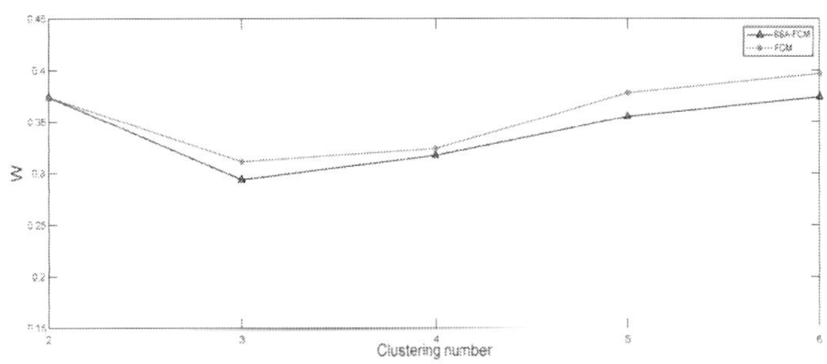

Figure 1. Trend of W indicator.

Figure 3 shows that the clustering number of the two methods is 3 when the W index reaches the minimum value, so the optimal clustering number of the two algorithms is 3. In addition, when taking the same number of clusters, the W index values of SSA-FCM algorithm are all smaller than those of FCM algorithm, so SSA-FCM has better performance.

Table 1. Number of various types of curves.

Algorithm name	Type 1	Type 2	Type3
SSA-FCM	41	4	5
FCM	45	1	4

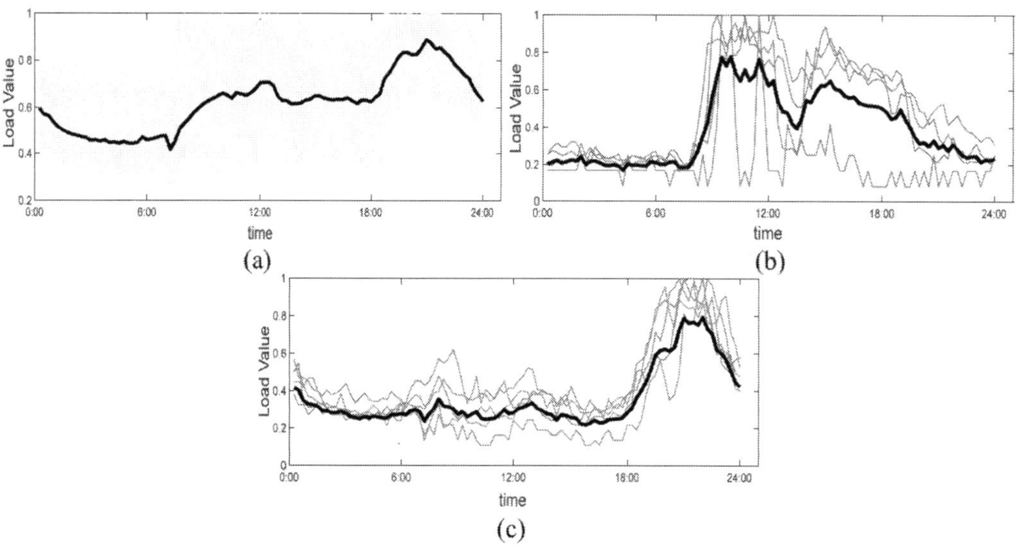

Figure 2. SSA-FCM daily load curve classification results.

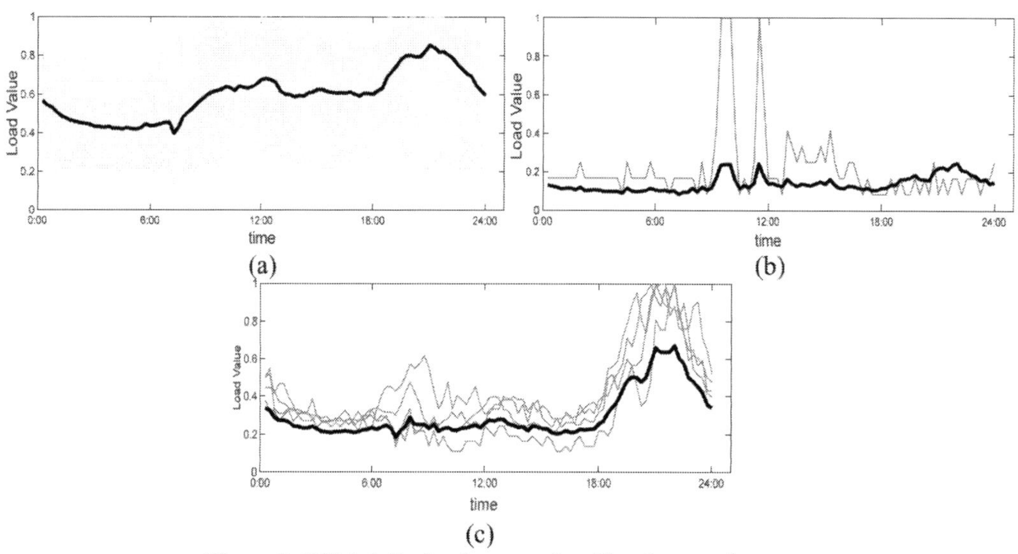

Figure 3. FCM daily load curve classification results.

Table 1 shows the number of daily load curves belonging to each type in the classification results of SSA-FCM and FCM, Figure 2 and Figure 3 compare the clustering effects of the two clustering algorithms, the bold curves in Figure 2 and Figure 3 are all kinds of clustering centers. The clustering centers obtained by SSA-FCM algorithm are different from those obtained by FCM algorithm. Combined with Table 1, Figure 2 and Figure 3, the clustering center obtained by SSA-FCM algorithm is different from that of FCM algorithm. Combined with table 1, Figure 2 and Figure 3, the classification results of the two algorithms are different. The daily load curve classification results are obtained based on SSA-FCM algorithm, the daily load curves of the same category are relatively close, while in the results of FCM algorithm, the daily load curve deviates from the cluster center more seriously. Therefore, SSA-FCM algorithm has better clustering effect.

5. Conclusion

At present, the most widely used method in daily load curve classification is the FCM algorithm. Because it is a local search algorithm, it is very sensitive to initialization and converge easily to local minimums. At the same time, when using this algorithm, the best number of clusters cannot be obtained in advance. In order to solve the problem of FCM, a FCM algorithm based on SSA algorithm is proposed.

- Adopt a new fuzzy clustering effectiveness index. This indicator is composed of three parts: compactness, separation and overlap. It can accurately find the optimal number of clusters on noise data sets and overlapping data sets.
- Use the SSA algorithm to improve the FCM clustering algorithm, and replace the FCM with SSA instead of the iterative process based on gradient descent. Simulation experiments show that, compared with the FCM, the SSA-FCM is classified into the same category of daily load curves are closer, so the improved algorithm has better clustering accuracy.

The proposed algorithm can make the power consumption characteristics of the same class of users have high similarity and there are obvious differences between different classes of users. It provides an accurate and scientific user classification method for the power department, which is of great significance to solve the three-phase imbalance problem of distribution network and reduce the network loss of distribution network.

Acknowledgments

Funded by the 2020 R&D Science and Technology Project of State Grid Heilongjiang Electric Power Company Limited: Research on loss reduction technology of three-phase unbalance distribution load in Heilongjiang Province. (523437190010)

References

[1] Chaohong, B., Xu, W., Yuan, H. (2017) Research review and prospects of energy internet planning. J. Proceedings of the Chinese Society of Electrical Engineering, 37(22): 6445-6462+6757.
[2] Xiaoxin, Z., Shuyong, C., Zongxiang, L., et al. (2018) Technical characteristics of China's new generation power system in the energy transition. J. Proceedings of the Chinese Society of Electrical Engineering, 38(7): 1893-1904.
[3] Wenqing, Z., Yaqiang, G. (2016) Load curve clustering based on Kernel K-means. J. Electric Power Automation Equipment, 36(6): 203-207.
[4] Yongguang, L., Chaoliang, S., Zhenzhen, N., et al. (2014) Research on power load characteristics classification technology of improved fuzzy C-means clustering algorithm. J. Electrical Measurement and Instrumentation, 51(18): 5-9.
[5] Guilan, W., Guoliang, Z., Hongshan, Z., et al. (2016) Fast clustering and anomaly detection technology for large-scale electricity data flow. J. Automation of Electric Power Systems, 40(24): 27-33.
[6] Xiaojuan, G., Xiaoxia, Li., Xiaoling, L. (2008) Improvement and analysis of hierarchical clustering algorithm. J. Computer Applications and Software, 25(6): 243-246.
[7] Jia, Y., Shitong, W. (2006) Research on EM Algorithm and Initialization in Gaussian Mixture Model Clustering. J. Microcomputer Information, 11(22):244-247.
[8] Xiaopu, F. (2010) Research on Power User Classification Technology Based on Actual Load Curve. J. Electric Power Science and Engineering, 26(09): 18-22.
[9] NAGI, J., YAPK, S., TIONG, S.K., et al. (2010) Nontechnical loss detection for metered customers in power utility using support vector machines. J. IEEE Transactions on Power Delivery, 25(2): 1162-1171.
[10] Jiankai, X. (2020) A novel swarm intelligence optimization approach:sparrow search algorithm. J. Systems Science & Control Engine erring An Open Access Journal, 8(1): 22-34.

[11] Jiayi, G., Xuezhong, Q., Shibing, Z. (2019) New fuzzy clustering validity index. J. Application Research of Computers, 36(4): 1001-1006.

Power system transient stability assessment based on rough set and deep residual neural network

Jian Chai[1], Xihuai Wang[2] and Jianmei Xiao[1]*

[1,2]School of Logistics Engineering, Shanghai Maritime University, Shanghai,201306, China

*Corresponding author's e-mail: jmxiao@shmtu.edu.cn

Abstract. Machine learning algorithms have been widely used in power system transient stability evaluation. The combined application of data analysis and evaluation and neural network provides a new direction for power system transient stability analysis. After the actual power grid is running, there is obviously an imbalance between stable samples and unstable samples. The current deep learning network realizes the power system transient stability assessment method with too many redundant attributes, and the characteristics will inevitably be lost during the data transmission process. This leads to serious problems with the tendency of the training of the data-driven transient stability assessment model. The rough set theory algorithm is introduced to reduce the redundant attributes of power system transient data sets, which simplifies the difficulty of data training. At the same time, as the neural network deepens, the deep residual neural network model has a higher accuracy rate and effectively avoids the "gradient explosion" and "gradient dispersion" problems. Compared with the traditional neural network, it has better Evaluate performance.

1.Introduction

With the rapid development of the power system, the trend of smart grid and regional interconnection is becoming more and more rapid. Large-scale complex and volatile new energy sources are connected to the power system, making the power system bear greater pressure. According to research[1], power angle instability caused by large disturbances in the power system is the main cause of large-scale power outages. Traditional transient stability assessment methods are time-domain simulation and direct methods, and machine learning technology based on data-driven processing is becoming more and more advanced, and this method is not affected by the complexity of the model, and it is more and more popular by current power system transient stability analysis researcher. The shallow learning method completes the mapping of the system state parameters to the transient stability by learning the input training samples, and realizes the transient evaluation of the power system[2]. Such as using artificial neural network(ANN), support vector machine(SVM), decision tree(DT), random forest(RF), etc. However, the shallow network model has limitations for the input feature processing, and the generalization ability of complex classification problems is greatly restricted. Therefore, the shallow machine learning method cannot well realize the transient stability assessment of the power system. In recent years, deep learning research has made major breakthroughs and has been gradually applied to power system transient stability assessment. However, the power system is a very complex nonlinear system, and the topological structure of its internal electrical components is also extremely complex, and there is a lot of redundancy in the coupling information between variables[3].

Therefore, to solve the above problems, a rough set and deep residual neural network power system transient stability assessment method is proposed.

2.Introduction to Rough Sets

2.1 Rough set theory

When dealing with a large amount of fuzzy and uncertain data information, using rough set theory, after constructing the initial transient feature set, the data information is analyzed and researched, and further knowledge reduction is carried out[4].The set $U(U \neq \varnothing)$ constituted by the researched objects is called the universe of discourse; The set A composed of a finite number of state attributes is called the attribute set; Attribute range set $V=\{V_1,\ldots,V_n\}$, where V_1 is the range of A_i ; Information function: $U \times A \rightarrow V, f(x_i, A_i) \in V_j$. An information system is composed of the quadruple $IS=\{U,A,V,f\}$ composed of the universe U, the attribute set $A=C \bigcup D$, the attribute value range set V and the information function f. Subsets C and D are condition attributes and decision attributes respectively.

2.2 Rough set attribute reduction algorithm

Every time the importance of all attributes is calculated, if the importance of this attribute for the current condition set is 0, it can be deleted. Then select the most important attributes to add to the current set of conditions, and proceed to the next cycle. Until the conditional information entropy of the current condition set is equal to the conditional information entropy of the original data[5]. Explain that the current collection can replace the original collection. This function is expressed as follows:

$$H(p)=-\sum_{i=1}^{n} p(x_i)\log(p(x_i)) \quad (1)$$

Let R be an equivalence relation, like $\exists r \in R$,have $ind(R)=ind(R-\{r\})$,Then r is unnecessary, call a reduction of R, the intersection of all reduction sets of R is called R's core; The importance of the reduction is $\mathrm{sig}(\alpha,R)=|ind(R \bigcup \{\alpha\})|-|ind(R)|$. The reduction of rough set is to keep the U condition of the universe unchanged, find the decisive attribute, delete the redundant attribute, and achieve the simplification of the system information.

3.Introduction to Deep Residual Neural Networks

The deep residual neural network is an improved convolutional neural network proposed by He Kaiming's team in 2015[6]. It skips the middle network layer and passes the previous activation value directly to the later network to avoid gradient explosion and the gradient disappears and a deeper network is trained. The deep structure can link neural network layers that are "closer" and "earlier" to the input layer to achieve identity mapping, so that gradient parameters are effectively transmitted, and the difficulty of training network parameters is greatly reduced, so that higher accuracy can be trained Deep network model.

3.1 Deep residual network structure

Available from Document, the overall network structure includes input layer, convolutional layer, multiple deep residual modules, batch normalization layer, Activation function, average pooling layer and output fully connected layer. From the related introduction in literature [7] The overall network structure is shown in the figure 1:

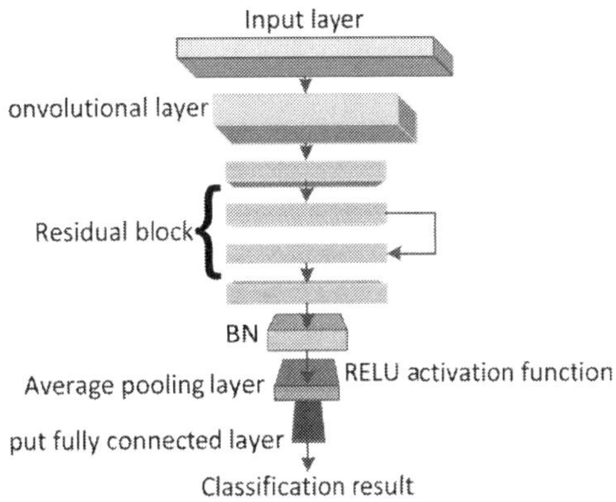

Figure 1. Overall network structure diagram

The input layer, receive the incoming transient feature set that has been processed by rough set, and the input parameter is the 32-dimensional data of the combined feature variable. In this paper, the initial processing of the data set input to the model is carried out by normalization to make the neural network converge faster. The conversion function is:

$$x^* = (x - min) / (max - min) \tag{2}$$

Convolutional layer, contains multiple convolution kernels, extracts input features, and updates and trains weights and biases. The deep residual neural network greatly reduces the amount of training parameters. The operation process of the convolutional layer is:

$$y_j = \sum\nolimits_{i \in M_j} X_i * K_{ij} + b_j \tag{3}$$

Batch standardization layer, as much as possible to reduce the difference between the data sample and the characteristics of each dimension. After traditional input data is processed with a series of weights, paranoia and activation operations, it cannot guarantee the standardization of implicit feature quantities. Preprocessing the data before the hidden layer through the batch normalization layer accelerates the optimization of the neural network [8].

In the average pooling layer, the average value is selected as the final feature in the pooling area, and the average pooling is performed in this area to reduce the dimensionality. Improve the anti-interference ability of the model. The calculation process is:

$$H_i = p(H_{i-1}) \tag{4}$$

The output fully connected layer, through the activation function, converts the incoming data of the previous layer into probability, y_{pred} defines the two-class threshold value δ, the default is 0.5, and the model output probability is:

$$y_{pred}(i) = \begin{cases} 0, & \overline{Y}(i) < \delta \\ 1, & \overline{Y}_i(i) \geq \delta \end{cases} \tag{5}$$

Among them, 1 means the system is stable, and 0 means the system is unstable.

3.2 Residual block unit

The residual network is a sub-network, which can form a deep network through stacking. Check the research theory of He Kaiming's team, refer to the literature [9], The residual neural network layer is shown in the figure 2:

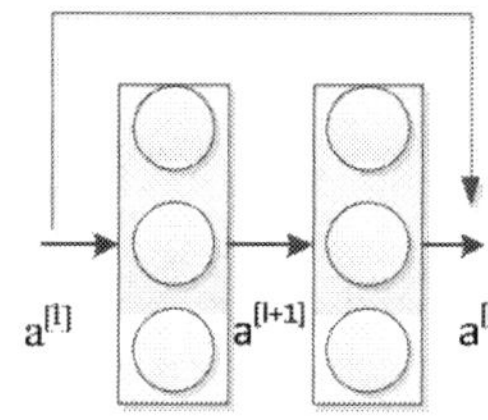

Figure 2. Residual network layer

Through jump connection, copy $a^{[l]}$ directly to the deep layer of the neural network
No longer passing along the main path, the transfer formula is

$$a^{[l+2]} = g\left(z^{[l+2]} + a^{[l]}\right) \tag{6}$$

In the back propagation, the gradient consists of two parts, where the direct mapping gradient to x is 1, and the other part is the gradient of the multi-layer ordinary neural network mapping; assuming the error is ε, the partial derivative of it is:

$$\frac{\partial \varepsilon}{\partial x_i} = \frac{\partial \varepsilon}{\partial x_L}\left(\left(\prod_{i=l}^{L-1} \lambda_i \sum_{i=l}^{L-1} F(x_i, w_i)\right)\right) \tag{7}$$

Assuming that the loss function is loss, the backpropagation formula is:

$$\frac{\partial loss}{\partial x_i} = \frac{\partial loss}{\partial x_L}\frac{\partial x_L}{\partial x_i} = \frac{\partial loss}{\partial x_L}\left(1 + \frac{\partial}{\partial x_L}\sum_{i=l}^{L-1} F(x_i, w_i)\right) \tag{8}$$

The In-depth residual infrastructure diagram is shown in the figure 3

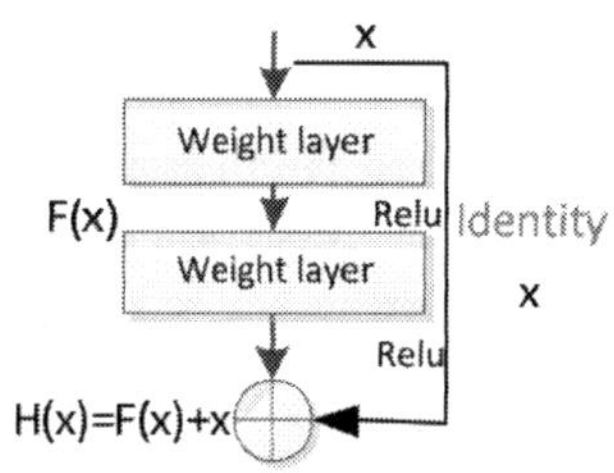

Figure 3. In-depth residual infrastructure

Through the "shortcut connections" method, the input x is directly passed to the output as the initial result, and the output result is:

$$H(x) = F(x) + x \tag{9}$$

When $F(x)=0$, then $H(x)=x$, and the identity mapping is realized.

4.Case analysis

4.1Initial feature construction

Time-domain simulation was performed on a 10-machine 39-node system in New England. The test system contained 39 buses, 46 transmission lines, and the system frequency was 60 Hz. Consider the topological structure and N-1 fault operation mode in the simulation process. With 5% as the step size, set a total of 10 different load levels from 80% to 125%, and at the same time change the generator output accordingly. After the fault is cleared, the lines overlap and the system topology remains unchanged. The simulation duration is set to 5s. Transient stability assessment is a binary classification problem.

4.2Transient stability assessment process

In [10], The confusion matrix can describe the classification performance of the model for test data with known labels. In the confusion matrix, T_S, T_{US}, F_S and F_{US} represent the number of stable samples correctly classified, the number of unstable samples correctly classified, and the prediction of unstable samples The number of misclassifications for stable samples and the number of misclassifications for predicting stable samples as unstable samples.

Table 1. Confusion matrix.

	Actual label	
Predicted label	Stablize	Instability
Stablize	T_S	F_S
Instability	T_{US}	F_{US}

4.3Model performance comparison

After inputting the transient feature set into the model training, the final classification result is shown in the figure 4 below.

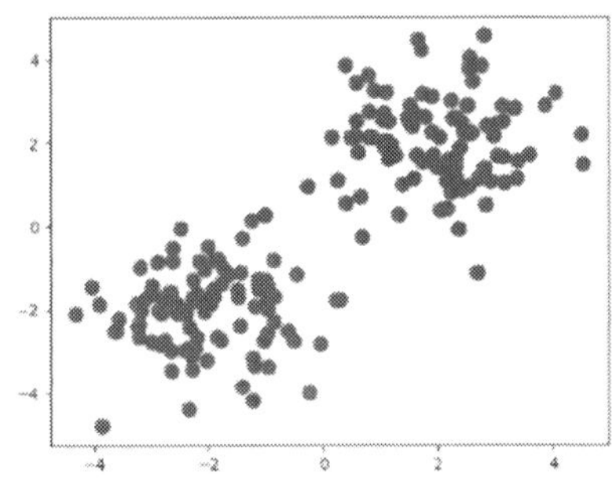

Figure 4. Schematic diagram of result classification

Table 2. Evaluation and comparison of each model.

Model	Accuracy/%	*Recall rate/%*	precisions/%	*F1 value*
DNN	96.55	90.26	96.98	95.62
DT	96.10	81.64	95.89	95.23
SVM	95.63	87.88	96.16	98.09
CNN	96.78	98.11	96.23	97.20
DRNN	98.98	98.33	98.65	98.89

5.Conclusion

1) The proposed rough set attribute reduction algorithm can not only remove redundant information data between the topological structure of the power system, and reduce the impact of redundant information on the model; at the same time, it can also strengthen and re-characterize the input features to make the provided features More sufficient.
2) Compared with deep neural networks such as CNN, DRNN can improve the model's practice feature extraction ability, and it still has a higher accuracy rate in the case of insufficient and excessive samples. The generalization ability of the model is strong.

References

[1] Liu, Y. (2018) Analysis and Enlightenment of Disconnection and Outage of Brazil's "9.13" Far Northwest Power Grid. J. Proceedings of the Chinese Society for Electrical Engineering, 38(11): 3204-3213.
[2] Yu, Y, Y. Yang, Q, W. (2007 (2018 reprint)) Power System Analysis. China Electric Power Press Publishing. Beijing:
[3] Odun Ayo, T. Crow, M, L. (2012) Structure-Preserved Power System Transient Stability Using Stochastic Energy Functions, J. IEEE Transactions on Power Systems27(3):1450-1458 .
[4] Su, F. Yang, S, H. Wang, H, Y. (2017) Research on Fast Termination Algorithm for Time Domain Simulation of Power System Transient Stability. J. Proceedings of the CSEE.37(015):4372-4378.
[5] Wang, T. Liu, J, L. Zhu, S, X. Wang, Z, P. (2020) Power system transient stability assessment and emergency control strategy based on random forest. J. Power System Technology, 44(12): 4694-4701.
[6] Song, L. (2020) An Integrated Scheme for Online Dynamic Security Assessment Based on Partial Mutual Information and Iterated Random Forest. J. IEEE Transactions on Smart Grid11(4):3606-3619 .
[7] Zhang, K. (2017) Overview of rough set theory and its applications. J. Intelligent Processing and Applications.6:93-98
[8] Li, B, Y. Xiao, J, M. Wang, X, H. (2020) Power System Transient Stability Assessment Integrating Neighborhood Rough Reduction and Deep Forest. J. Transactions of China Electrotechnical Society35(15): 3245-3257.
[9] Zhou, Y, Z, Zha, X, Lan, J. (2020) Power system transient stability prediction based on data augmentation and deep residual network. J. China Electric Power (001):22-31.
[10] He K M, Zhang X Y, Ren S Q, et al. (2016) Deep residual learning for image recognition. J. IEEE Computer Society:770-778.

Influencing Factor and Error Analysis of Operating Capacitor Voltage Transformer Harmonic Measurement

Yan Zhang[1*], Huajia Wang[1], Danwen Yu[1], Liu Yang[2], Wenjun Cao[1], Yongxin Yang[3]

[1]State Grid Shandong Electric Power Research Institute, Jinan 250003, China;

[2]Key Laboratory of Power System Intelligent Dispatch and Control of Ministry of Education, Shandong University, Jinan, 250061, China;

[3]State Grid Energy Research Institute Co., Ltd., Beijing, 100031, China;

[*]E-mail: zhangyan@yjy.sd.sgcc.com.cn

Abstract: With the increasing demands of power quality monitoring and management, the problem of harmonic measurement error caused by the widespread operation of capacitor voltage transformers (CVT) in high-voltage power grids has been unavoidable. This paper analyses the influence of environmental factors on harmonic measurement error of CVT, and proposes a new method for analysing the harmonic measurement errors of operating CVT, which explains the unstable phenomenon of CVT measurement errors based on field comparison test, and provides a progressing screening method of effective analysis data. the CVT harmonic error characteristic analysis method proposed in this paper considers the impact of field installation conditions, electromagnetic environment, and other actual operating environment, which can reflect the actual harmonic error characteristics of CVT in field operation. A CVT harmonic measurement system is established based on the proposed method. The validation and effectiveness of the proposed method is illustrated on a field measurements of 1000kV CVT in Shandong Province.

1. Introduction

Voltage transformer is an essential core equipment for power quality monitoring in power system. In high-voltage power grid, compared with traditional electromagnetic voltage transformer (PT), capacitor voltage transformer (CVT) has the advantages of small volume, light weight and low cost. At the same time, it can also avoid the problem of ferromagnetic resonance caused by iron core saturation. In addition, the capacitor voltage divider in CVT can be used as the coupling capacitor for carrier communication, which can also weaken the lightning wave head. Due to above advantages, the share of CVT in 110kV and above level substations is as high as 53.5%[1]. However, because of its inherent characteristics, CVT has large errors in measuring harmonics. Appendix D of Chinese national standard GB/T 14549-1993 clearly stipulates that CVT cannot be used for harmonic measurement[2]. In a word, the CVT has been widely used in power system now, but it cannot accurately measure harmonics, which has become a key obstacle to analysing power quality in high-voltage power system.

Actually, a lot of research has been carried out and many research results have been obtained to solve the problem of CVT harmonic voltage measurement. On the whole, these research results can be divided into two categories. One is the direct method, that is, transform the existing CVT to make it have the ability to measure harmonics. The other is the indirect method, which uses the obtained CVT harmonic transmission characteristics to correct the harmonic measurement results. These two methods have their own advantages and disadvantages, but neither completely solves the problem of harmonic measurement of CVT.

In the research of CVT harmonic measurement, the direct method was first considered, which has been studied as early as 20 years ago. [3,4] proposes a method to correct CVT harmonic measurement results by using the capacitor voltage divider composed of the end screen of current transformer. However, this method requires additional equipment and complex operation. Besides, the measured signal is small and the accuracy and reliability of the signal amplification device have a great impact on the accuracy of the measurement results. [5,6] gives a current measurement method of grounding circuit. By adding two current transformers to the grounding circuit of CVT, the voltage on the high-voltage side of CVT can be calculated according to the known capacitor value. This method is still in the verification stage now. What's more, the implementation of this method involves the accurate measurement of small current signal and noise pollution, and the accuracy of calculation results needs to be evaluated. [7] introduces a method of adding capacitor C_3 in CVT capacitor voltage divider for on-line transient overvoltage monitoring, which can also be applied to harmonic monitoring. The harmonic component in the primary voltage can be obtained by measuring the harmonic component of the voltage at both ends of C_3. This method is also in the experimental stage now and its measurement accuracy in practical applications still needs to be evaluated in depth.

Compared with the direct method, the indirect method does not need power cut to transform the CVT. It only needs to accurately obtain the harmonic transmission characteristics of CVT and then correct the harmonic measurement results, so as to realize the accurate measurement of harmonics. In [8,9], by establishing the high-frequency equivalent model of CVT, its transfer function is derived and the influence law of different parameters on harmonic voltage measurement is studied. These papers define the influencing factors of measuring harmonic voltage by CVT, but do not give the elimination method of harmonic measurement error; [10] proposed that the method of using CVT harmonic transmission characteristic curve to correct the transmission error of harmonic frequency has the disadvantage that the key parameters cannot be obtained. To solve this problem, [11] built a CVT harmonic transmission characteristics and measurement error test system to test CVT in the 35kV voltage level and corrected the error based on the test results to avoid the direct measurement of key parameters. However, without considering the randomness and contingency of the test data, the obtained CVT harmonic transmission characteristics are still not accurate enough; On the basis of experiments, [12] modifies the CVT harmonic measurement results combined with neural network algorithm. The results have certain accuracy and universality, but require a large number of experimental data and the method is time-consuming and laborious.

Based on the above, this paper studies the CVT harmonic measurement error in harmonic monitoring of high-voltage power grid. The CVT harmonic equivalent model and test platform are built and the differences between different CVT harmonic measurement errors are comprehensively analysed from the perspective of mechanism and experiment. Furthermore, an analysis method for harmonic measurement error characteristics of CVT in UHV is proposed, which provides an effective way to solve the accuracy problem of harmonic signal transmitted by CVT.

2. Harmonic equivalent model of CVT

According to the design standard of CVT, the internal intermediate transformer cannot reach saturation at 1.5 times overvoltage[13], so the iron core of the intermediate transformer will not be saturated under normal working conditions and always works in linear state. Therefore, the influence of nonlinearity of excitation branch can be ignored when establishing CVT harmonic equivalent model. Besides, according to the electromagnetic field theory, it is necessary to consider the stray capacitor between CVT internal inductance coil devices when analysing the frequency characteristics[4,5,8,14]. To sum up, the harmonic equivalent circuit model of CVT shown in figure 1 is established.

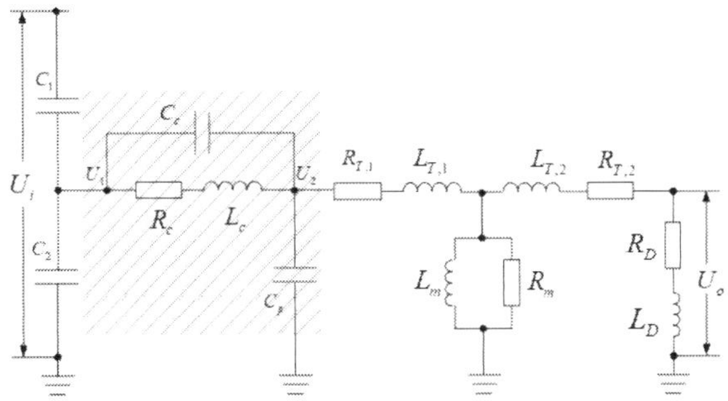

Figure 1. The harmonic equivalent circuit model of CVT.

As shown in figure 1, C_1 and C_2 are high-voltage capacitor and medium-voltage capacitor; L_c is the inductance of compensation reactor; R_c is the equivalent resistance of compensation reactor; C_c is the equivalent stray capacitance of compensation reactor; C_p is the stray capacitance to ground of primary winding of intermediate transformer; R_m and L_m are the resistance and inductance of excitation winding of intermediate transformer; R_{T1} and L_{T1} are the resistance and leakage inductance of primary winding of intermediate transformer; R_{T2} and L_{T2} are the secondary resistance and leakage inductance of intermediate transformer converted to primary side; R_D and L_D are loads.

According to the harmonic equivalent circuit shown in figure 1, the transfer function of each part is obtained by using cascade analysis theory:

$$H_1(s)=\frac{\frac{1}{sC_2}}{\frac{1}{sC_2}+\frac{1}{sC_1}} \tag{1}$$

$$H_2(s)=\frac{\frac{1}{sC_p}}{\frac{1}{sC_c}\,||\,(R_c+sL_c)+\frac{1}{sC_p}} \tag{2}$$

$$H_3(s)=\frac{R_m||sL_m}{R_m||sL_m+R_{T,1}+sL_{T,1}} \tag{3}$$

$$H_4(s)=\frac{R_D+sL_D}{R_D+sL_D+R_{T,2}+sL_{T,2}} \tag{4}$$

Then the transfer function of harmonic equivalent circuit of CVT is:

$$H(s)=\frac{NU_2(s)}{U_1(s)}=H_1(s)\cdot H_2(s)\cdot H_3(s)\cdot H_4(s) \tag{5}$$

3. Influence of key parameters on harmonic transmission characteristics of CVT

3.1. Influence of internal manufacturing parameters

Using the transfer function of CVT and changing the parameters by simulation, the harmonic transfer characteristic curves of CVT under different distribution parameters can be obtained, as shown in figure 2 and figure 3:

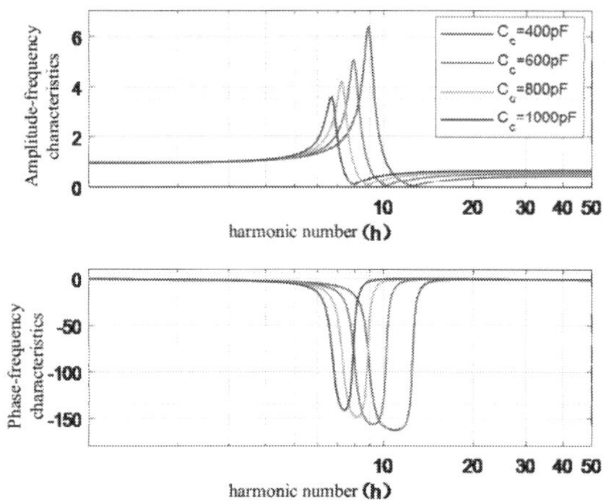

Figure 2. CVT harmonic transfer characteristic curves under different distribution parameters C_c of compensation reactor.

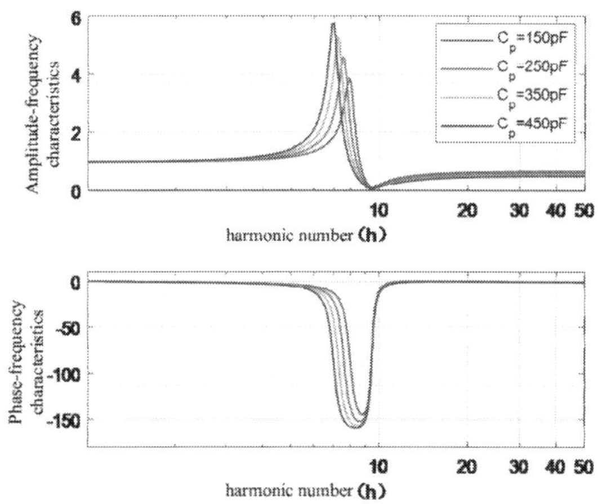

Figure 3. CVT harmonic transfer characteristic curves under different distribution parameters C_p of intermediate transformer.

It can be found from figure 2 and 3 that the change of distribution parameters has a great impact on the harmonic transmission characteristics of CVT. Thus, there will be great differences in the harmonic transmission characteristics of different types of CVT. Therefore, it is difficult to correct the harmonic transmission characteristics of different types of CVT through a certain frequency characteristic. It is reasonable to obtain different transmission characteristics to correct the harmonic measurement results for different types of CVT.

3.2. Influence of external environmental factors

For the same type of CVT, there is little difference in its internal manufacturing parameters, which provides a theoretical possibility to correct CVT harmonic transmission error by using transmission characteristics. However, the influence of field operating environment on the harmonic transmission characteristics of CVT still needs to be further analysed.

3.2.1. Influence of the temperature.

Since CVT works outdoors all year round, the influence of ambient temperature change must be considered when analyzing the harmonic transmission characteristics of CVT.

The reason for the change of voltage divider capacitor caused by temperature is that the film capacitor or all film capacitor used in CVT will be affected by temperature. When the ambient temperature changes, the capacitor value will also change correspondingly. The degree of change mainly depends on the temperature coefficient of the capacitor. Through investigation, it is found that the temperature coefficient of most types of voltage dividing capacitors is negative, that is, the high and medium voltage equivalent capacitor of CVT will decrease with the increase of ambient temperature. However, because high and medium voltage capacitors are generally made of the same material, their temperature coefficients are basically the same. Therefore, the influence law of ambient temperature on voltage dividing capacitor will be the same and the capacitor voltage dividing ratio will not change. Although the voltage divider ratio of the capacitor does not change, the overall equivalent capacitance of the capacitor voltage divider will change. In order to simplify the analysis, the equivalent capacitance of the capacitor voltage divider is changed according to the condition that the capacitance value of the capacitor voltage divider will change by about 0.25% ~ 0.5% every 25°C change in temperature. The capacitance temperature coefficient α_c is taken as -200 and the definition of α_c is as follows:

$$\alpha_c = \frac{\Delta C}{C_0 \Delta T} \times 10^6 \tag{6}$$

where ΔC is the change of capacitance; ΔT is the temperature difference between the current temperature and room temperature; C_0 is the capacitance measured at room temperature. The capacitance values at different temperatures can be obtained:

$$C' = C_0 (1 + \frac{\alpha_c \Delta T}{10^6}) \tag{7}$$

The reason for the change of stray capacitance of internal coil caused by temperature is different from that of voltage divider capacitor. The calculation formulas of stray capacitance of winding coil deduced in [16] are as follows:

$$C_z = \frac{(n-1)\varepsilon_r \varepsilon_0 lND}{n^2 d} \tag{8}$$

$$C_u = \frac{4(n-1)\varepsilon_r \varepsilon_0 lDN}{3n^2 d} \tag{9}$$

where N is the number of turns of the winding; n is the number of layers of winding coil; l is the average turn length of winding; d is the distance between winding layers; ε_r is the relative permittivity; ε_0 is the vacuum dielectric constant. It can be seen that the influence of temperature on the stray capacitance of the internal coil is mainly due to the change of the ε_r of the winding coil with the change of temperature. Generally, within 100°C, the changes by about 1% for every 20°C, i.e.:

$$C_z' = C_{z\text{-}20^\circ C}(1 \pm 0.0005 \times \Delta T) \tag{10}$$

$$C_u' = C_{u\text{-}20^\circ C}(1 \pm 0.0005 \times \Delta T) \tag{11}$$

The reason for the change of the inductance parameters of the coil caused by the temperature is that the magnetic permeability of the cold-rolled silicon steel, which is the common material of the core of the CVT compensation reactor and the intermediate electromagnetic unit, changes with the temperature. In general, this change process is nonlinear, but if only qualitative analysis of the influence of temperature is required, the analysis can be appropriately simplified. It is approximately considered that the change of permeability is about 0.3% for every 25°C change in core temperature[15], that is:

$$\mu_r' = \mu_r (1 + 0.00012 \times \Delta T) \tag{12}$$

Combined with the calculation formula of inductance of inductance coil, the inductance at different temperatures can be obtained:

$$L = \frac{N^2 D\mu_0}{l}\mu_r(1+0.00012\Delta T) \tag{13}$$

The variation law of resistance parameters of inductance coil is similar to that of inductance parameters. Generally, the resistance value of copper resistance changes by about 10% every 25°C, that is:

$$R' = R_{20℃}(1+0.004\times\Delta T) \tag{14}$$

According to the above analysis, the frequency response curve of CVT with temperature can be obtained, as shown in figure 4.

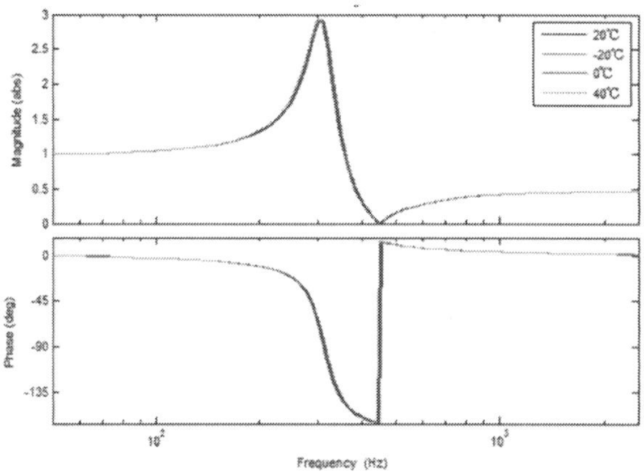

Figure 4. Harmonic transfer characteristic curves of the same type of CVT under different operating temperatures

It can be seen from figure 4 that the change of internal parameters of CVT caused by temperature has a certain impact on the harmonic transmission characteristics of CVT, but the impact is not obvious on the whole.

3.2.2. Influence of external electric field.

The capacitor voltage divider of CVT has no capacitance shielding. Therefore, when CVT works in a complex electromagnetic environment, the capacitor voltage divider and the live parts of the substation close to each other form stray capacitance through the space electric field. These coupling capacitors at different locations will flow capacitive currents. Capacitive current will flow into the capacitor voltage divider, resulting in a change in the voltage divider ratio. However, when the main capacitor of CVT is large enough, the stray capacitance has less influence on it. Moreover, only when the main capacitor changes, it has a great impact on the measurement results of CVT at power frequency, but has little impact on the harmonic transmission ratio, so this paper will not discuss it in depth.

3.2.3. Influence of the operating conditions.

The CVT harmonic test platform is used to test various operating conditions that may be encountered by the same type of CVT, as shown in figure 5.

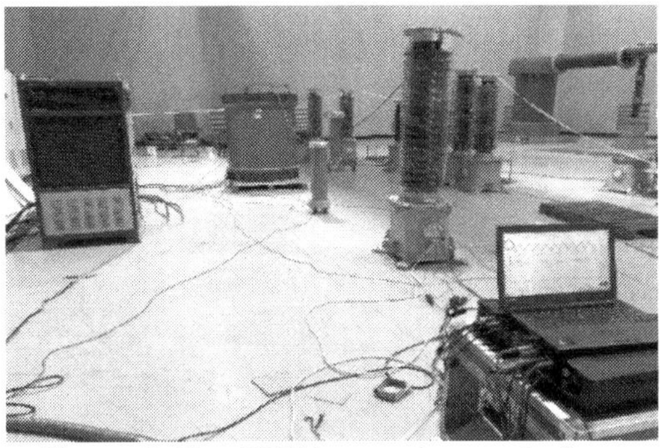

Figure 5. The CVT harmonic test platform

Figure 6 and figure 7 show the test results of changing the input voltage amplitude, input harmonic content and operating ambient temperature for multiple CVTs of the same batch and type.

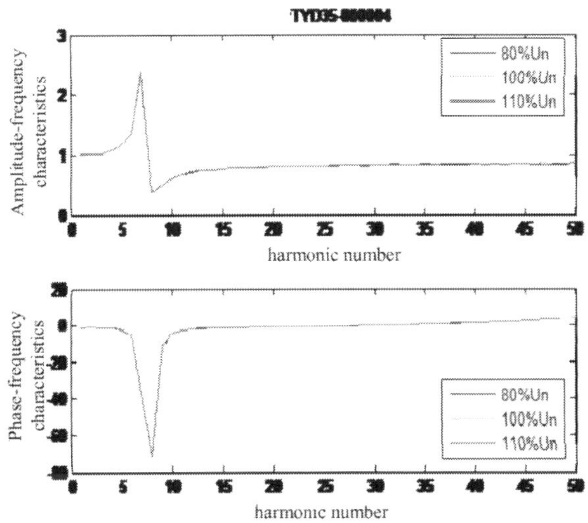

Figure 6. Harmonic transfer characteristic curves of the same type of CVT under different input voltage amplitudes

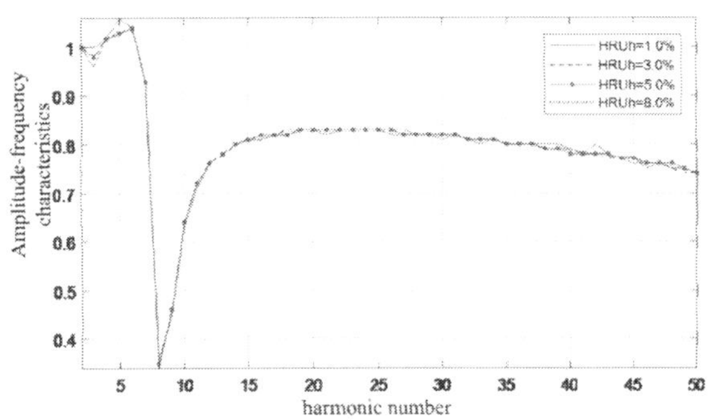

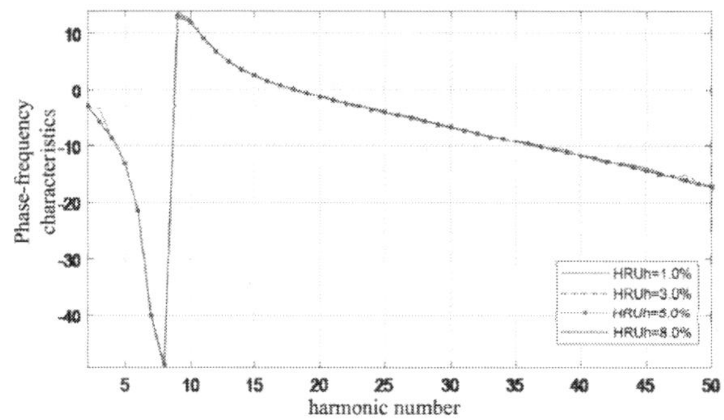

Figure 7. Harmonic transfer characteristic curves of the same type of CVT under different harmonic content

It can be seen from figure 6 and 7 that under certain operating conditions, CVTs produced in the same batch have the same and stable frequency characteristics. Therefore, the harmonic measurement results of the actual operating CVT can be corrected by measuring the harmonic transmission characteristics in the laboratory, so as to realize the accurate transmission of harmonics by the CVT.

4. Correction method of harmonic measurement error characteristics

4.1. Error characteristic analysis method of CVT harmonic measurement based on MLE

According to the characteristics that CVT with the same model and parameters has stable harmonic error characteristics under certain operating conditions, a new analysis method of CVT harmonic measurement error characteristics is proposed in this paper. Firstly, obtain the CVT voltage test signal with comparison and analysis conditions and the voltage test signal of standard source; Secondly, each harmonic content analysis sample with time scale is obtained after harmonic analysis. Based on the number of remaining samples and the interval span of sample probability distribution, the analysis samples are screened step by step and the effective data of harmonic error analysis are obtained; Then, the effective data are analysed based on the maximum likelihood estimation (MLE) method to obtain the harmonic error characteristics of the CVT; Finally, the measured results are corrected by using the obtained CVT harmonic error characteristics. The process is shown in figure 8.

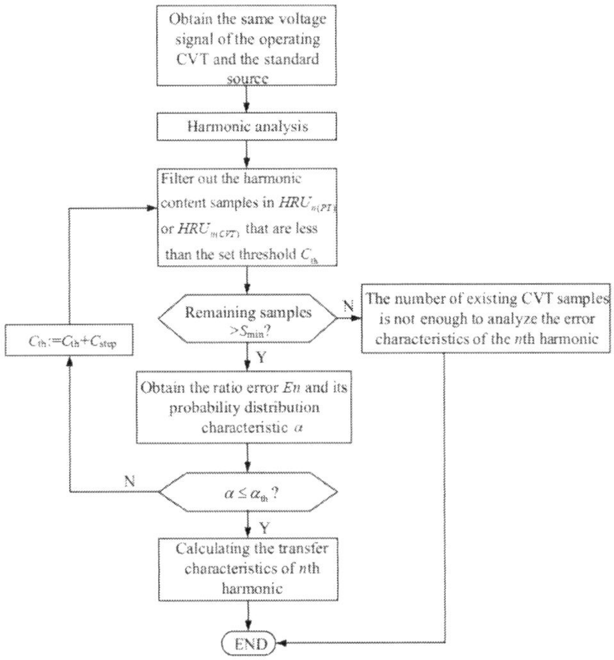

Figure 8. Flow chart of CVT harmonic measurement result processing based on MLE algorithm

4.2. Field experimental verification

The 1000kV CVT of a UHV substation in Shandong Province is used for analysis and verification. A 1000kV bus tank electromagnetic voltage transformer (hereinafter referred to as bus PT) is installed in phase A, B and C of bus I and bus II in GIS, with an accuracy level of 0.5. The research shows that PT has good frequency characteristics for harmonics below 1kHz. In the verification work, the PT of A-phase bus II of GIS is used as the error characteristic analysis standard source of harmonics within 20 times measured by CVT, and the harmonic error characteristic of A-phase line CVT is obtained by using the method proposed in this paper. As mentioned above, the harmonic error characteristics of CVT with the same parameters from the same manufacturer are basically the same. Using the obtained harmonic error characteristics, the harmonic error of B-phase line CVT is corrected. Then the corrected harmonic is compared with its corresponding standard source (PT harmonic measurement value of GIS bus II) to verify the effectiveness of the proposed method.

Taking the quintuple harmonic as an example, two channels of a power quality analysis device are used to test the quintuple harmonic of 1000kV A-phase line CVT and that of A-phase GIS bus at the same time, so as to ensure that the data time scale is strictly corresponding. The analysis interval is taken as 3s. The quintuple harmonic transmission characteristic of CVT is represented by E_5, calculate the ratio error of $HRU_{5(CVT)}$ and $HRU_{5(PT)}$ at each same time interval to obtain E_5 samples at all analysis time intervals and its probability distribution is shown in figure 9.

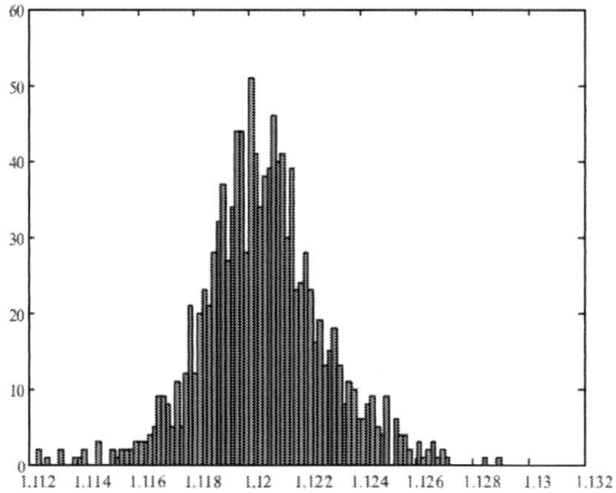

Figure 9. Sample probability distribution of all analysis time interval of quintuple harmonic

As can be seen from figure 9, the harmonic error characteristics of CVT in field operation show normal distribution characteristics. The minimum ratio error of 95% probability interval is 1.116 and the maximum ratio error is 1.125. The set threshold is α_{th} =0.05 and the span value of E_5 analysis sample is α =0.009, which is far less than the set threshold. It shows that the distribution of E_5 sample data is relatively concentrated and the probability distribution of sample data can reflect the actual harmonic test error level. The expected value of E_5 sample data distribution is μ=1.1205, that is, the quintuple harmonic error characteristic of the 1000kV A-phase CVT is 1.1205 times amplification.

The above conclusions are applied to the B-phase CVT harmonic measurement results for correction. Compare the harmonic data corrected by this method with the data measured by the standard source (i.e. measured value of GIS bus B-phase PT) and obtain the comparison results of all analysis interval data, as shown in figure 10.

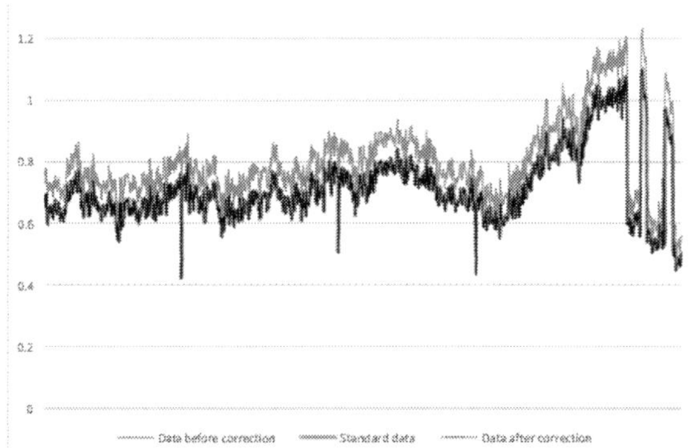

Figure 10. Comparison curves of quintuple harmonic before and after correction

As shown in the figure 10, the method proposed in this paper can be applied to the harmonic measurement error analysis of CVT with the same model and parameters. After analysing and correcting the error characteristics of B-phase CVT harmonic measurement data, it can be found that the results are very close to the standard source measurement data. The average relative error after correction is only 0.3% while the average relative error before correction is 10.75%.

5. Conclusions

The inevitable harmonic measurement error of CVT Seriously affects the observability of power quality of power grid. Aiming at this problem, this paper takes the harmonic measurement error characteristics of UHV CVT as the research object and puts forward the analysis and test scheme of harmonic error characteristics of CVT in operation. At the same time, the screening method and analysis method of effective data of harmonic error characteristics are proposed. The main conclusions are as follows:

Different brands and batches of CVTs have different harmonic transmission characteristics while CVTs of the same batch have similar harmonic transmission characteristics;

The analysis method of CVT harmonic error characteristics proposed in this paper does not need power failure, CVT equipment transformation or additional partial voltage measurement device. The security risk is relatively small compared to traditional analysis methods.

Compared with laboratory analysis or delivery test methods, the CVT harmonic error characteristic analysis method proposed in this paper considers the impact of field installation conditions, electromagnetic environment and other actual operating environment, which can reflect the actual harmonic error characteristics of CVT in field operation.

In this paper, the accuracy and effectiveness of the proposed method are verified by the experimental results of 1000kV CVT. In the next step of work, the harmonic measurement error correction method will be used in online harmonic monitoring data correction.

Acknowledgments
This work is supported by State Grid Science and Technology project (No.5200-201924064A-0-0-00).

References

[1] Li YS. (2016) Research on Influencing Factors of Error Characteristics of Capacitor Voltage Transformer. Chongqing: Chongqing University.
[2] Power quality public grid harmonics： GB/T 14549—1993 [S]. 1993.Quality of electric energy supply Harmonics in public supply network： GB/T 14549-1993[S]. 1993.
[3] Duan XB. (2005) New measurement system and test method for solving harmonic measurement signal distortion in HV power network. Electrical Equipment,6(6):37-39.
[4] Duan XB. (2004) Harmonic test method for 500 kV substation. Electrical Equipment, 5(12):43-44.
[5] Wang X, Shen X, CAO M, et al. (2016) Research on capacitive voltage transformer harmonic measurement features. Electrical Measurement & Instrumentation, 53(S1):1-9.
[6] Fu CZ, Liu ZL, Yang LH, et al. (2015) Transient over-voltage online monitoring technology based on 500 kV capacitive voltage transformer with interior low-voltage capacitor C3. High Voltage Apparatus, 51(9):77-84.
[7] Ghassemi F, Gale P, Cumming T, et al. (2005) Harmonic voltage measurements using CVTs. IEEE Transactions on Power Delivery, 20(1):443-449.
[8] Feng Y, Wang XQ, Chen XM, et al. (2014) Influences of circuit parameters of capacitor voltage transformer ongrid harmonic voltage measurements. Proceedings of the CSEE, 34(28): 4968-4975.
[9] Duan XB, Zhu MX, Hu WP, et al. (2014) Key parameters affecting harmonic transfer characteristics of capacitor voltage transformer. Power System Technology,38(11):3158-3164.
[10] Ghassemi F, Gale P F, Clegg B, et al. (2002) Method to measure CVT transfer function. IEEE Transactions on Power Delivery, 17(4):915-920.
[11] Gao HL, Li QL, Yu XP, et al. (2013) Harmonic transfer characteristic of capacitor voltage transformer. Power System Technology, 37(11):3125-3130.
[12] Wang R. (2008) Research on distortion of CVT harmonic measurement. Baoding: North China Electric Power University.
[13] Feng Y, Wang XQ, Chen XM, et al. (2014) The influence of capacitance voltage transformer circuit parameters on the measurement of harmonic voltage in power grid. Proceedings of the Chinese Society of Electrical Engineering, 28:4968-4975.
[14] Zhao ZY， Gong CY， Qin HH. (2008) Effect factors on stray capacitances in high frequency transformers. Proceedings of the CSEE, 28(9):55-60.

[15] Wang LR. (2010) Verification and application of power transformers. China Metrology Publishing House, Beijing.
[16] Chen L. (2014) Study on distributed capacitance of inductance coil and its inductance in alternating magnetic field. Nanjing Normal University.

Research on load loss prediction of distribution network outage based on hybrid neural network

Jiamin Liao*

CHINA SOUTHERN POWER GRID CO., LTD.

*Corresponding author's e-mail: s2120334@siswa.um.edu.my

Abstract—In order to fully mine the relationship between temporal characteristics in load data and improve the accuracy of load forecasting, a load forecasting method based on convolutional neural networks (CNN) and gated recurrent unit (Gru) hybrid neural network is proposed. Taking date factors, climate factors and similar daily load factors as input, the sample data sets in the region are grouped by K-means clustering method; Then cnn network is used to extract the relationship between features and load in high-dimensional space, construct the high-dimensional feature vector of time series, and input the results into Gru network; Finally, the parameters of Gru network model in each group are trained and the load forecasting value is output. The results show that the proposed load forecasting method has significant advantages in forecasting accuracy and efficiency compared with long short term memory (LSTM) network model, Gru network model, cnn-lstm network model, support vector machine regression model and decision tree model.

1. INTRODUCTION

The stable operation of power system requires real-time dynamic balance between power generation and load change. However, the randomness of load fluctuation is strong, the nonlinear characteristics are obvious, and the influencing factors are diversified [1]. Therefore, it is necessary to deeply explore the law of load change and improve the accuracy of load forecasting, so as to reasonably formulate power production plan and reduce power production cost, Effect of maintaining the balance between supply and demand of power resources [2-3].

At present, load forecasting methods at home and abroad are mainly divided into three categories: statistical method, time series forecasting method and artificial intelligence method. The statistical method is represented by Kalman filter method [4]; The time series method is divided into regression analysis method [5], multiple linear regression method [6], Fourier expansion method [7], etc..

2. CONSTRUCTION OF LOAD FORECASTING FEATURE SET

Reasonable construction of prediction feature set is very important to the convergence and prediction accuracy of the model.

1) Considering the date factor, the industrial load accounts for a large proportion in the urban power load, while the date factors such as weekends and holidays have a great impact on the industrial load. Based on this, a 6-D date feature including season, month, day, week date, whether it is a working day and whether it is a holiday is constructed. For \ ({{l}_{\ Text {r} (T) \) for modeling.

2) Considering the climate factors, the changes of temperature, weather type and other climate factors will affect the residential power load, this paper constructs a six-dimensional climate feature including weather type, wind force, maximum temperature, minimum temperature, maximum humidity and minimum humidity_{\ Text {w} (T) \) for modeling.

3) Considering the similar daily load factors, because the load has time series characteristics, the model can learn the recent change law of load according to the daily load data close to the date to be measured, so as to enrich the prior information of the prediction model. Based on this, this paper selects the load data of four similar characteristic days before the forecast day to form a four-dimensional vector pair \ ({l})_{\ Text {n} (T) \) for modeling.

To sum up, a 16 dimensional feature vector including date factors, climate factors and similar daily load factors is constructed as the input of the prediction model, as shown in Table 1.

3. CNN-GRU HYBRID NEURAL NETWORK MODEL

In order to verify the scientificity and accuracy of the proposed load forecasting method, 731 days of power load data in a region of Zhejiang Province from January 1, 2016 to December 31, 2017 are selected as the data set, which is divided into training set and test set according to the ratio of 8:2, including 585 days of training set and 146 days of test set. The data is collected once a day, and the input feature dimension is 16 dimensions, as described in Table 1.

The cnn-gru hybrid neural network is used to train on the data set. The prediction results are compared with LSTM network, Gru network, cnn-lstm network, SVR regression model and decision tree model under the same conditions in terms of prediction efficiency and prediction accuracy. It can be known that the model has good prediction accuracy while maintaining high-efficiency training.

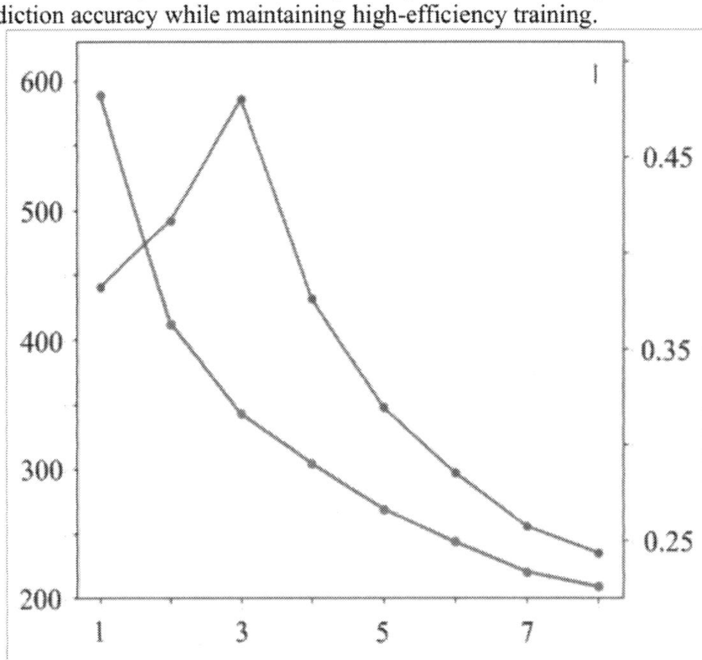

Fig. 1 Variation trend of sum of squared errors and contour coefficient

3.1 experimental environment configuration

The experimental environment of this paper adopts Intel i7-8750h processor and nvidia-gtx-1060 graphics card. The algorithm model adopts Python 3.6 as the programming language, the software architecture is based on tensorflow framework, python framework and scikit learn algorithm framework, and the drawing tool adopts Matplotlib drawing module.

3.2 load cluster analysis

Carry out K-means clustering on the load data, and draw according to formula (7) (8). When the clustering category \ (K \) increases, the variation trend of the sum of squares of errors and contour coefficient is shown in Figure 4. As can be seen from Fig. 4, according to the cluster evaluation elbow method [25], when the cluster category \ (K \) is set to 2 or 3, the absolute value of the second derivative of the sum of squares of errors is large, which is suitable for selecting the cluster category; For the contour coefficient, when \ (K \) is 3, the contour coefficient \ (s (I) \) gets the maximum value of 0.48. Based on this, the clustering category \ (K \) is set to 3, and the 585 day training set is divided into three categories: A, B and C. the capacity of the three categories of samples is 196, 135 and 254 respectively. Of which,The load clustering results from January 1 to December 31, 2016 are shown in Table 1 in the appendix. After training the three kinds of samples respectively, the final prediction model is integrated. The coordinate system is established based on the normalized maximum temperature, maximum humidity and weekly date, and the load clustering distribution is obtained, as shown in Figure 5. It can be seen from Figure 5 that the sample has a clear clustering effect in high-dimensional space.

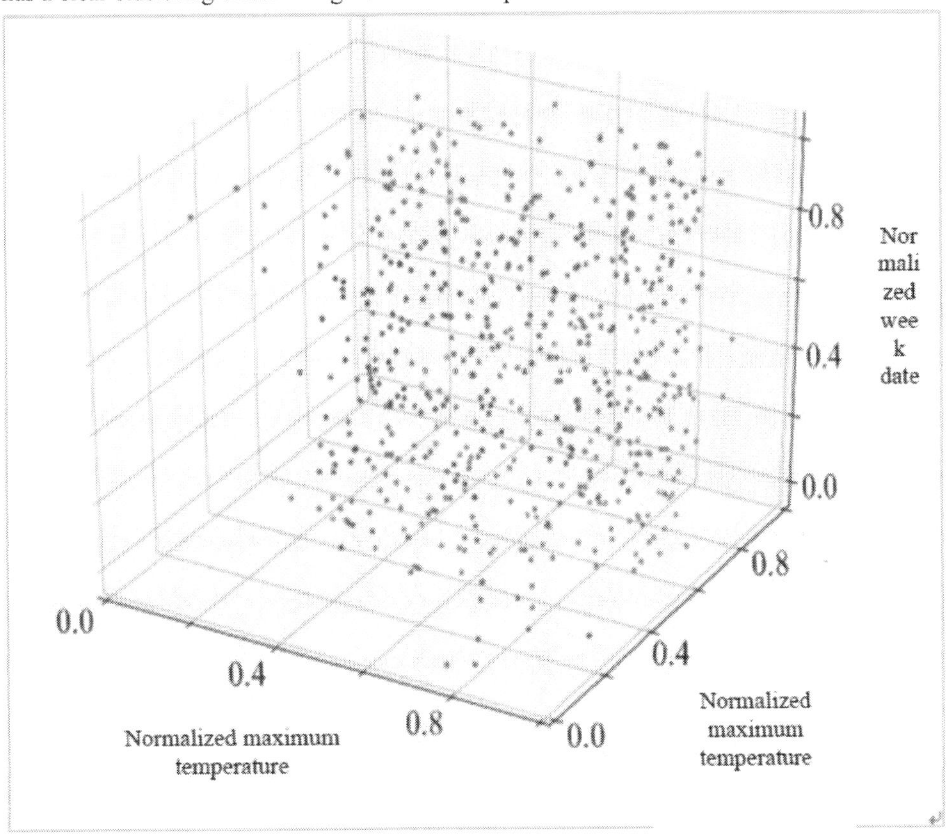

Fig. 2 Distribution of load clustering features

3.3 determination of network architecture super parameters

This paper considers the super parameters of cnn-gru hybrid neural network training, including the number of layers of Gru network D and L2 regularization weight coefficient μ in CNN training. Using the control variable method, based on 3000 iterations, the number of layers of Gru network is deepened step by step without setting the regularization term, so as to find the optimal network layer ${D}^{*}$ to minimize the empirical risk of the model. On this basis, the L2 regularization weight coefficient μ is adjusted to determine the L2 regularization weight coefficient ${\mu}^{*}$ to minimize the structural risk of the model.

The sum of the electric power taken from all operating power equipment of the power customer to the power system at the same time is called the user's electric load at that time. Then the sum of the power loads supplied by a power grid to all users of the power grid at the same time is called the load of the power grid. The maximum load borne by the power grid in the same natural year is called the maximum load. The common division of load types is the power consumption of the whole society, which is divided into four categories: the first, the second, the third industry and residential power consumption. It can also be subdivided into four categories according to agriculture, forestry, animal husbandry and fishery, industry, geological survey and exploration, construction, transportation, posts and telecommunications, commercial food supply, marketing and storage, residential power consumption and others

types of power load forecasting

Load forecasting is divided into system load forecasting and bus load forecasting. System load forecasting is mainly used for load control, safety monitoring, preparation of power grid operation plan and arrangement of power grid maintenance plan. Bus load forecasting is to convert the load forecasting value of power system into the calculated value of active load and reactive load of each bus in the system at a specified time.(1) Divided by forecast time. According to the length of load forecasting time, power load forecasting can be divided into long-term load forecasting, medium-term load forecasting and short-term load forecasting.(2) According to the characteristics of predicted load, it is divided into maximum load forecasting, minimum load forecasting, average load forecasting, peak valley difference forecasting, average peak load forecasting, average underestimated load forecasting, bus load forecasting, load rate forecasting, etc.(3) It can be divided into urban civil load forecasting or commercial load forecasting, rural load forecasting, industrial load forecasting, etc.

characteristics of power load forecasting

Because the power system load forecasting has a series of characteristics such as inaccuracy, conditionality, timeliness and multi scheme, in order to accurately obtain the change of future load, on the one hand, we should fully grasp and make use of its characteristics, at the same time, we should also consider the influence of various factors and select the corresponding forecasting method according to the specific factors, Finally, the load forecasting model is established combined with the forecasting method. The characteristics of load forecasting are as follows:(1) InaccuracyIn the process of power system load forecasting, it is easy to be affected by many factors (such as temperature and season), and the related influencing factors are also developing and changing. In addition, it may be affected by emergencies, so the prediction results are inaccurate.(2) Conditionality.Conditionality means that the influence of various factors must be comprehensively considered when forecasting the future power load data. The said conditions can be divided into the inevitable conditions that can reliably and directly affect the load forecasting results and some assumptions that we need to add in order to obtain more accurate forecasting because it is difficult to control the development and change regularity of future load.(3) Temporality.Timeliness means that the power system load forecasting is completed within a certain time range, that is, the power system load forecasting is required to be real-time, and the prediction time corresponding to the predicted value and the time range of this prediction shall be specifically indicated when making the power system load forecasting.(4) Multi scheme.In power system load forecasting, different forecasting methods can be selected according to different forecasting conditions, and then an appropriate load forecasting model can be established. Due to the inaccuracy and conditionality of load forecasting, different forecasting methods should be adopted for loads with different characteristics under the condition of comprehensive consideration of various influencing factors, Different mathematical models are established to make load forecasting more accurate

With the deepening of Gru network layers, the change of square difference loss is shown in Table 2. It can be seen from table 2 that when the number of layers \ (D \) of Gru network is 3, the square difference loss gets the minimum value. When the number of layers is greater than 3, the square difference loss does not decrease but increases due to over learning of the model. Therefore, the optimal number of layers \ ({D} ^ {*} \ text {=} 3 \); Adjust the L2 regularization weight coefficient \ (\ mu \) of CNN network training to obtain the square difference loss, as shown in Table 3. It can be seen from table 3 that when the L2 regularization weight coefficient \ ({\ Mu} ^ {*}} \) is set to 0.001, the square difference loss after iteration is the smallest. Based on this, in this experiment, the L2 regularization weight coefficient \ (\ mu \) trained by cnn-gru network model is 0.001, and the number of Gru network layers \ (D \) is set to 3.

3.4 analysis of prediction results

For the three types of training set data, the CNN Gru hybrid neural network model proposed in this paper, LSTM model, Gru model, cnn-lstm model, SVR regression model and decision tree model are used to fit the training data. Classify the 146 day test set according to the clustering results obtained in Section 3.2, and then select the corresponding model for prediction. In the above network model, the LSTM network model is composed of three-layer LSTM units; The Gru network model consists of three layers of Gru units; The CNN architecture setting in cnn-lstm model is the same as the model proposed in this paper; The kernel function of SVR model adopts Gaussian kernel function, and the penalty parameter is set to 5.0; The penalty parameter of pre cut branches and leaves node of decision tree model is set to 0.01.

For the above prediction model, calculate the maximum percentage of prediction error on three types of test samples \ ({y}_{\ Text {E-MAX} \), prediction accuracy \ ({y}_{\ Percentage error between average value and average absolute value of text {FA} \ \ ({y})_{\ The average value of text {MAPE} \), and the comparison results are shown in Table 4. The comparison of prediction efficiency of depth model is shown in Table 5. According to the information in tables 4 and 5, compared with the traditional Gru network, the maximum percentage of prediction error of the cnn-gru neural network model proposed in this paper is reduced by 4.79%, and the prediction accuracy is improved by 1.11%; Compared with LSTM neural network, SVR regression and decision tree model, the maximum error percentage decreased by 2.9%, 12.05% and 15.81% respectively, and the prediction accuracy increased by 1.03%, 3.12% and 4.44% respectively; Compared with cnn-lstm model, on the basis of almost the same prediction accuracy, the proposed model can shorten the model training time by 50.43%, and greatly improve the prediction efficiency of the model.

4. CONCLUSION

A regional medium-term load forecasting method based on cnn-gru hybrid neural network model is proposed. Firstly, K-means clustering method is used to divide the sample data set, then CNN network is used to extract the relationship between features and load data in high-dimensional space, construct the feature vector of time series and input it into Gru network. Finally, Gru network is trained to output load forecasting value, which has the following advantages:

1) A feature set including date factors, climate factors and similar daily load factors is constructed, which can give full play to the advantages of CNN network in the field of data mining and extract the potential relationship of discontinuous data in high-dimensional space.

2) The Gru cyclic network model can fully consider the time series characteristics of load characteristics, has good time series data fitting and regression ability, and has high prediction efficiency.

3) Cnn-gru hybrid neural network combines the advantages of CNN network and Gru network. The experimental results show that compared with the existing artificial intelligence prediction methods, the proposed method not only maintains a faster model training speed, but also has higher prediction accuracy.

The load forecasting feature set established in this paper does not take into account the diversified load types, and the price factor is not included in the feature set. Therefore, in the next work, this paper will further study the impact of load characteristic classification on load forecasting, build a richer feature set including electricity price factors, explore the internal relationship of input characteristics, and further improve the accuracy of load forecasting.

References

[1] North American Blackout Time Series Statistics and Implications for Blackout Risk[J] . Benjamin A. Carreras,David E. Newman,Ian Dobson. IEEE Transactions on Power Systems . 2016 (6)

[2] On Hurst exponent estimation under heavy-tailed distributions[J] . Jozef Barunik,Ladislav Kristoufek. Physica A: Statistical Mechanics and its Applications . 2010 (18)

[3] Power-Law Distributions in Empirical Data[J] . Aaron Clauset,Cosma Rohilla Shalizi,M. E. J. Newman. SIAM Review . 2009 (4)

[4] Large blackouts in North America: Historical trends and policy implications[J] . Paul Hines,Jay Apt,Sarosh Talukdar. Energy Policy . 2009 (12)

[5] Risk-management and risk-analysis-based decision tools for attacks on electric power.[J] . Simonoff Jeffrey S,Restrepo Carlos E,Zimmerman Rae. Risk analysis : an official publication of the Society for Risk Analysis . 2007 (3)

[6] Evidence for self-organized criticality in a time series of electric power system blackouts[J] . B.A. Carreras,D.E. Newman,I. Dobson,A.B. Poole. IEEE Transactions on Circuits and Systems I: Regular Papers . 2004 (9)

[7] Rescaled variance and related tests for long memory in volatility and levels[J] . Liudas Giraitis,Piotr Kokoszka,Remigijus Leipus,Gilles Teyssière. Journal of Econometrics . 2003 (2)

[8] Long-Term Memory in Stock Market Prices[J] . Econometrica . 1991 (5)

[9] Robustness of the rescaled range R/S in the measurement of noncyclic long run statistical dependence[J] . Benoit B. Mandelbrot,James R. Wallis. Water Resources Research . 1969 (5)

[10] Fractional Brownian Motions, Fractional Noises and Applications[J] . Benoit B. Mandelbrot,John W. Van Ness. SIAM Review . 1968 (4)

Optimization of integrated energy system for combined cooling, heating and power supply of new energy based on energy storage

Tieyan Zhang[1*], Junbao Yang[1], Yaru Wang[2,3], Min Li[4], Dawei Zhang[5] and Yuri Wang[6]

[1]Shenyang Ligong University

[2]Shenyang Institute of Engineering

[3]Key Laboratory of Regional Multi-energy System Integration and Control of Liaoning Province

[4]GD Hefeng Wind Power Development Co., Ltd.

[5]Liaoning Shengfang Electric Power Technology Co., Ltd.

[6]Benxi High-tech Industrial Development Zone Power Supply Branch, State Grid Liaoning Electric Power Co., Ltd.

*Corresponding author's e-mail: zhangty@sylu.edu.cn

Abstract. With the development of science and technology, people also pay more and more attention to the development of new energy. Although there are also many studies on integrated energy systems now, integrated energy systems containing energy storage should also be further studied. This paper proposes an optimization of integrated energy system for combined cooling, heating and power supply of new energy based on energy storage, which analyzes the gas turbine, absorption refrigerating machine, electric refrigerator, photovoltaic power generation units, wind turbine and the work characteristics of the energy storage device. In this paper, an integrated energy system optimization model of new energy cogeneration with energy storage equipment is established. An example shows that the integrated energy system with energy storage can effectively solve the independent decoupling operation relationship among cool, heat and electricity. At the same time, the proposed model can also solve the energy interaction among cool, heat and electricity. In this way, the optimal operation of the integrated energy system can be realized.

1. Introduction

Nowadays, the social demand for energy is increasing, but the existing traditional energy is increasingly exhausted, so the development and utilization of new energy are also urgent. In the development and utilization, we should also pay attention to environmental problems[1]. At present, the research on integrated energy system mainly includes three aspects: system modeling, planning and operation optimization[2]. In paper[3], Integrated energy systems are connected with the heat network and then optimized, but the randomness of renewable energy is not taken into account. Therefore, the description of energy storage components in this paper is not detailed, which is far from the actual operation results. In paper[4], this article optimizes multiple regional integrated energy systems through the thermal network. This article does not fully consider the energy storage equipment, which has a certain distance from the actual operation results.

This paper proposes an optimization of integrated energy system for combined cooling, heating and power supply of new energy based on energy storage, which analyzes the gas turbine, absorption refrigerating machine, electric refrigerator, photovoltaic power generation units, wind turbine and the work characteristics of the energy storage device[5]. In this paper, an integrated energy system optimization model of new energy cogeneration with energy storage equipment is established. An example shows that the integrated energy system with energy storage can effectively solve the independent decoupling operation relationship among cool, heat and electricity. At the same time, the proposed model can also solve the energy interaction among cool, heat and electricity. In this way, the optimal operation of the integrated energy system can be realized[6].

2. Structure of Integrated energy system

Integrated energy system refers to the energy system that is highly coordinated in the development, transformation, reserve, transportation and use of different forms of energy, and jointly meet the terminal energy needs[7]. In this paper, an optimization of integrated energy system for combined cooling, heating and power supply of new energy based on energy storage is proposed, and its structure is shown in figure 1.

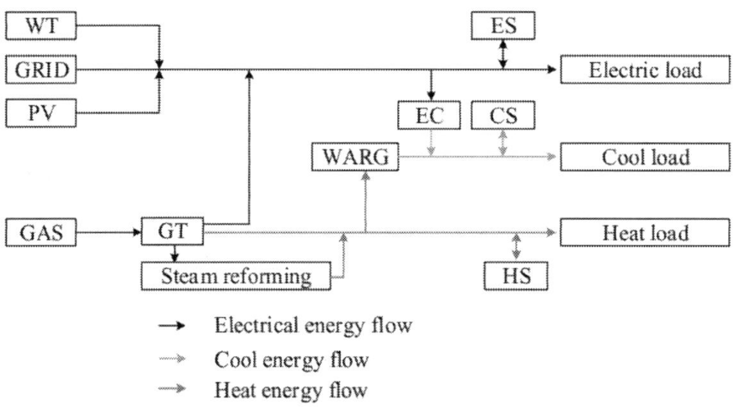

Figure 1. Integrated energy system structure diagram for combined cooling, heating and power supply of new energy based on energy storage.

3. Objective function and units model of integrated energy system

3.1. Objective Function

Under the conditions of load balance of cool, heat and power and related constraints, and on the premise of absorbing new energy, the energy scheduling and transformation among cool, heat and power should be coordinated, and the operating cost of the integrated energy system of cool, heat and power should be the lowest as the objective function[8]. The objective function of the optimization model can be expressed as:

$$\min M = \sum_{t=1}^{T}[M_F(t) + M_G(t) + M_m(t)] \tag{1}$$

$$M_{GT}(t) = M_{GAS}(t)V_{GT}(t) \tag{2}$$

$$\begin{aligned} M_m(t) = {} & L_{GT,m}P_{GT}(t) + L_{AC,m}P_{AC}(t) + L_{EC,m}P_{EC}(t) + L_{WHB,m}P_{WHB}(t) \\ & + L_{ES,m}P_{ES}(t) + L_{CS,m}P_{CS}(t) + L_{HS,m}P_{HS}(t) \end{aligned} \tag{3}$$

$$M_G(t) = M_B(t)P_G(t) \tag{4}$$

where $M_m(t)$ is the maintenance cost in time period; $M_F(t)$ is the system raw material cost in time period; $M_G(t)$ is the cost of purchasing electricity from the grid in time period; $L_{GT,m}$, $L_{AC,m}$, $L_{EC,m}$, $L_{WHB,m}$, $L_{ES,m}$, $L_{CS,m}$ and $L_{HS,m}$ represent the maintenance cost coefficient of gas turbine, absorption chiller, electric refrigerator, waste heat boiler, electric storage equipment, cool storage equipment and heat storage equipment respectively; $P_{GT}(t)$, $P_{AC}(t)$, $P_{EC}(t)$, $P_{ES}(t)$, $P_{ES}(t)$, $P_{CS}(t)$ and $P_{HS}(t)$ represent its output power at time period; $M_{GT}(t)$ is the unit price of natural gas; $M_{GT}(t)$ represents the consumption of natural gas in time period; $M_B(t)$ is the electricity purchase price in time period.

3.2. *Wind turbine model*

The relationship between fan output power and wind speed can be expressed by equation (5):

$$P_{out} = \begin{cases} 0 & s < s_1 \ or \ s > s_2 \\ P_r \cdot \dfrac{s - s_1}{s_r - s_1} & s_1 \le s \le s_2 \\ P_r & s_r \le s \le s_2 \end{cases} \tag{5}$$

where P_{out} represents the power output of the wind power; P_r represents the rated power that the fan can reach; s represents the actual speed of the fan, its unit is m/s; s_1 is the speed of entering the fan; s_2 is cut out wind speed; s_r is rated wind speed.

3.3. *Photovoltaic generator set model*

The output power of the photovoltaic system is affected by solar radiation intensity, environmental temperature, wind speed and other factors, and can be generally adjusted within the range of output power under standard test conditions[9]. The output power can be expressed by equation (6) and equation (7):

$$P_{PV} = \partial_{PV} P_S \frac{R}{R_S}[1+m(T_b - T_S)] \tag{6}$$

$$T_b = T_Z + 0.0138(1+0.0138T_Z)(1-0.042W_v)R \tag{7}$$

where P_{PV} represents the power output of the photoelectric unit; P_S represents the maximum power that the photoelectric unit can output under standard test conditions; ∂_{PV} represents the generation factor of photounits. Generally, we take this factor to 0.9; R indicates the actual irradiation intensity of the sun, and its unit is W/m^2; R_S, T_S and T_Z indicate the irradiation intensity of the sun, the temperature of the environment, and the actual temperature around the surrounding environment under the standard test conditions; T_b represents the temperature of the photoelectric unit surface; m represents the temperature factor on the surface of the photoelectric, its unit is $\%/°C$. Generally, it can be taken as $-0.47\%/°C$; W_v indicates the wind speed on the surface of the photoelectric, and its unit is m/s.

3.4. *Gas turbine model*

The heat and electricity production of a gas turbine is related to its equipment efficiency and gas intake, and their relationship is shown in equation (8) and equation (9):

$$P_{GT,h}(t) = \beta_1 F_{GT}(t) \cdot G_C \tag{8}$$

$$P_{GT,e}(t) = \beta_2 F_{GT}(t) \cdot G_C \tag{9}$$

where $P_{GT,h}(t)$ and $P_{GT,e}(t)$ are the thermal power and electrical power of the gas turbine; β_1 and β_2 are the thermal conversion coefficient and electric conversion coefficient of gas turbine; G_C indicates the combustion heat value of natural gas; $F_{GT}(t)$ represents the volume of gas consumed by the gas turbine.

3.5. Electric storage element model

Assuming that the discharge efficiency of the storage device is equal to the charging efficiency, then:

$$S_e(t+1) = S_e(t)(1-\phi_{ES}) - \frac{\phi_{ES} P_{ES} \cdot 1}{V_{ES}} \quad (10)$$

$$S_{e,\min} \le S(t+1) \le S_{e,\max} \quad (11)$$

where $S_e(t+1)$ and $S_e(t)$ are respectively the state of charge of the storage device at time points $t+1$ and t; $S_{e,\min}$ and $S_{e,\max}$ represent represents the minimum and maximum of the storage capacity; ϕ_{ES} is self-discharge rate and φ_{ES} charge-discharge efficiency.

3.6. Solution method

In this paper, the integrated energy system model for combined cooling, heating and power supply of new energy based on energy storage is a mixed integer nonlinear programming problem, which is transformed into a mixed integer linear programming model by genetic algorithm[10].

4. Case Analysis

This article takes a certain building as an example. Below figure are the output of the integrated energy system without the energy storage elements and the integrated energy system containing the energy storage elements.

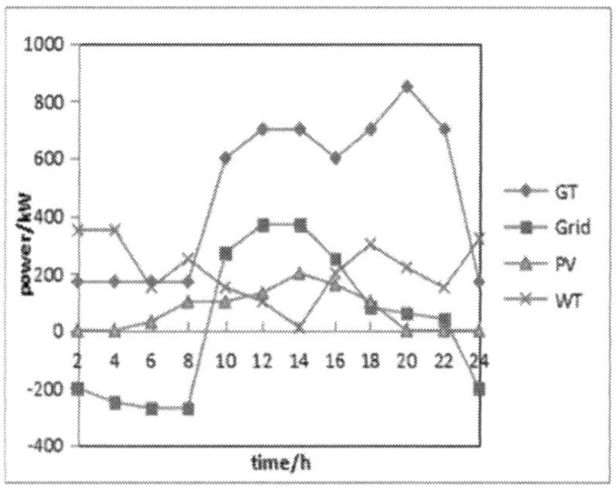

Figure 2. Integrated energy system output without energy storage elements.

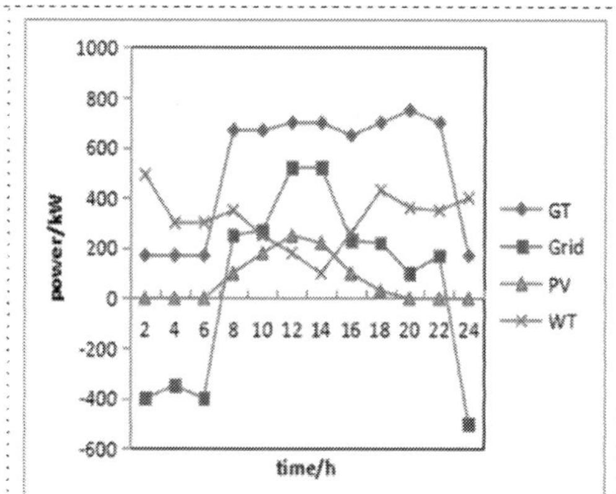

Figure 3. Integrated energy system output with energy storage elements.

According to figure 2 and figure 3, when the cool and heat demand are large, the heat discharge of the heat storage equipment greatly reduces the peak heat of the waste heat boiler. Due to the existence of power storage equipment, the trough power can be fully utilized, increasing the trough power purchased from the power grid. Discharging at peak power consumption increases the selling power to the power grid at peak, and effectively moves the peak to fill the valley of the power grid.

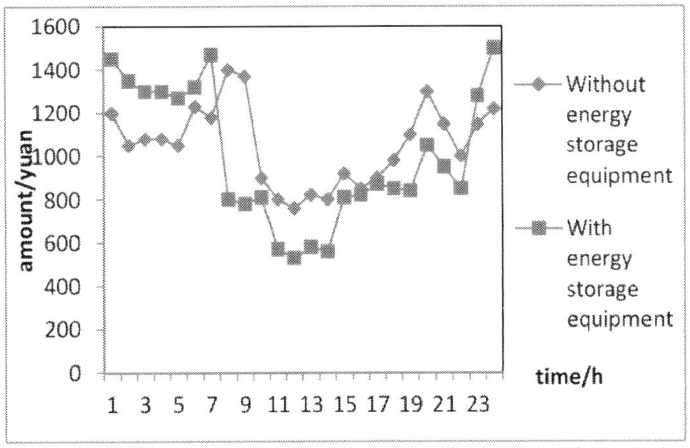

Figure 4. Comparison of operating costs of integrated energy without and with storage.

From figure 4, we can clearly see that the comprehensive energy system containing energy storage elements has good economic benefits and can effectively reduce the systematic operating cost.

5. Conclusion

This paper analyzes the working characteristics of gas turbine, absorption chiller, electric refrigeration, photovoltaic generator set, wind turbine generator unit and energy storage equipment. The optimization model for integrated energy systems proposed here includes energy storage elements and outputs of cold, thermal and electrical. Furthermore, this paper proposes to use genetic algorithms to transform mixed nonlinear mathematical models into mixed linear models, then illustrated by an analytical example. In this paper, an integrated energy system optimization model of new energy cogeneration with energy storage equipment is established. An example shows that the integrated energy system with energy storage can effectively solve the independent decoupling operation relationship among cool, heat and electricity. At the same time, the proposed model can also solve the energy interaction among cool, heat and electricity. In this way, the optimal operation of the integrated energy system can be realized.

Acknowledgments

This work was supported by Key R&D Program of Liaoning Province (2020JH2/10300101, 2018220017), Technology Innovation Talent Fund of Shenyang (RC200252, RC190360) and Key R&D Program of Shenyang (GG200252).

References

[1] Kang C Q, Yao L Z. (2017) Key scientific issues and theoretical research framework for power systems with high pro-portion of renewable energy. J. Automation of Electric Power Systems, 9: 2–11.

[2] Li Z, Wang C F, Liang J, et al. (2018) Expansion Planning Method of Integrated Energy System Considering Uncertainty of Wind Power. J. Power System Technology, 211: 3477–3487.

[3] Liu D C, Ma H R, Wang B, et al. (2018) Operation optimization of regional integrated energy system including cool-heat-electricity co-generation and energy storage. J. Automation of Electric Power Systems, 4: 113–120, 141.

[4] Feng W, Lu H Y, CHRIS M, et al. (2019) Development of distributed energy microgrid system in US. J. Electric Power, 6: 45–51.

[5] Xu H H, Wang X D, Zhu X Y, et al. (2020) Research on the optimal configuration of energy storage in user side integrated energy system. J. Power Demand Side Management, 2: 13–20.
[6] Xue S H, Li N, Zhou X M, et al. (2020) Optimal operation of integrated energy system considering integrated demand response. J. Power Demand Side Management, 5: 7–12.
[7] Xiong W, Liu Y Q, Su W H, et al. (2019) Optimal configuration of multi-energy storage in regional integrated energy system considering multi-energy complementation. J. Electric Power Automation Equipment, 1: 124–132.
[8] Jiang C F, Ai X. (2019) Integrated energy system operation optimization model considering uncertainty of multi energy coupling units. J. Power System Technology, 8: 2843–2852.
[9] Zeng M, Liu Y X, Zhou P C, et al. (2018) Overview and prospect of integrated energy system modeling and benefit evaluation system. J. Power System Technology, 6: 1697–1708.
[10] Huo X X, Wu M, Xu K, et al. (2018) Capacity allocation of micro-grid energy storage system considering uncertainty. J. Renewable Energy, 4: 544–549.

Design of meteorological intelligent sensor based on POE Technology

Qiangjun Liu[1*], Yun Liang[1], Junlin Zhang[2], Liangbo,Qi[3]

[1]Jincheng Meteorological Bureau, Jincheng, Shanxi Province, China

[2]China Environmental Research Institute, Beijing, China

[3]Shanghai Meteorological Bureau, Shanghai, China

[*]Corresponding author email:2016150024@jou.edu.cn

Abstract: An application scheme of Poe based Ethernet technology in meteorological intelligent sensor system is designed. The working principle and implementation method of Ethernet power supply system (POE) based on IEEE802.3af Ethernet power supply industry standard are experimentally analyzed. The power supply part of the meteorological intelligent sensor makes full use of Poe technology to provide current on the network cable transmitting data, which greatly reduces the complexity of the power supply system and improves the reliability of the system power supply design. Through the test of the actual system, the function and performance of the meteorological intelligent transmission system have achieved the expected results.

1. Introduction

At present, POE(Power Over Ethernet) power supply technology has been widely used [1]. Yin[2] studied the application of POE technology in security network monitoring. Riley[3] and Ding[4] studied the application of POE in the Internet of things. Lin et al. [5] studied the safety performance of POE in lightning surge protection. Wang[6], Zeng[7] and Xu[8] studied the application of POE technology in LED lighting system. Li[9] and Xing[10] studied the application of POE technology in weak current engineering and intelligent substation. Xuet al.[11] Application of POE technology in clean environment system. Yan et al. [12] studied the application of POE technology in 5G and wireless communication. In recent years, the increase of global extreme climate has posed a great threat to industrial and agricultural production and the safety of people's lives and property, which puts forward higher requirements for meteorological forecasting. The most important thing is to improve the intelligent level of relevant technologies and equipment of meteorological forecast, and the meteorological intelligent sensor is the basis to improve the ability of meteorological forecast, disaster prevention and reduction, and improve the accuracy, accuracy and density of observation data. The "China Meteorological Observatory ground integrated observation business platform" being implemented by the meteorological department puts forward higher requirements for meteorological intelligent sensors. Figure 1 is the schematic diagram of the ground comprehensive observation business platform of China's National Meteorological Observatory. It can be seen from the figure that the intelligent sensor equipment, as the infrastructure of the meteorological observatory, is composed of intelligent sensors that collect meteorological elements such as rainfall, temperature, humidity and air pressure. The intelligent observation equipment is located in the observation field, The collected data information is transmitted to the database service system through the network switch. The database service system performs statistical processing, fusion and other processing on the collected meteorological data, and the processed data is uploaded to the "Tian Qing" database. Due to the dispersion and uncertainty of the installation location of the meteorological intelligent sensor, the intelligent temperature sensor may be installed in the louver box or underground soil, and the wind speed sensor must be installed on the wind pole to collect data. Therefore, the power supply of the intelligent sensor has become a thorny problem in the system design. If the mains rectifier power supply is adopted, the cost is high and the stability cannot be guaranteed, The wiring of each intelligent sensor is also complex. POE technology is used to solve the power supply problem of intelligent sensor, which can be popularized and applied in meteorological intelligent sensor.

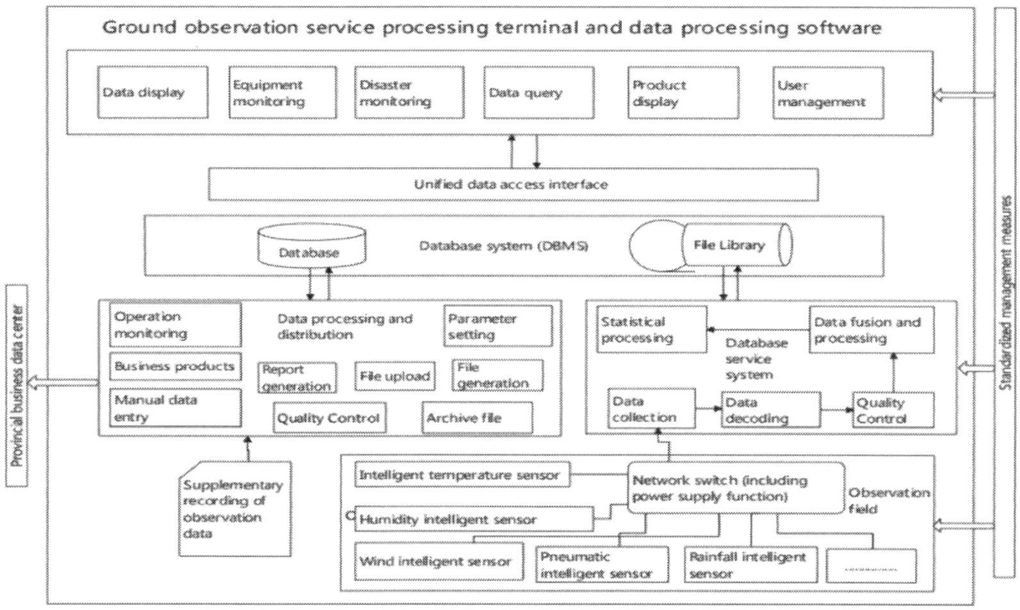

Figure 1. Flow chart of ground comprehensive observation business platform of China's national meteorological stations

2. Overall structure design

2.1. IEEE802.3af standard

The International Institute of Electronic Engineers (IEEE) approved the 802.3af standard for providing power over standard Ethernet cables on June 23, 2003. The standard defines a method that allows 48VDC power supply to be transmitted while transmitting data through Ethernet. At present, the Ethernet cable used is class 6 or super class 5. It uses one pair to send data, one pair to receive data, and the other two pairs are standby pairs. In IEEE802.3af standard, two ways can be used to supply power to PD (Power Device) equipment, One is to use two spare pairs; The other is to use the data line to transmit DC power, that is, the center tap of two groups of isolation transformers at both ends of the cable is used to provide nominal 48VDC. It must be pointed out that PSE (Power Supply Equipment) equipment can choose one of the power supply methods to supply power to PD equipment, and PD equipment must be compatible with two modes of power supply, This power supply method can safely and reliably introduce the Ethernet Power Supply (POE) technology into the existing network infrastructure and is compatible with the original network equipment. At present, it can provide about 12.95w power at the maximum. Therefore, some small network equipment such as IP phone, network camera Wireless access points and data terminals can be connected to supply power through Ethernet without using AC power socket, which greatly simplifies wiring, reduces the construction cost of network infrastructure, and will further expand the application of Ethernet technology.

2.2. PSE equipment

In POE system, PSE mainly completes the power management function of the whole POE system. Power management is to carry out equipment detection, power classification, power supply monitoring and module test on PE equipment, and obtain various technical indexes of PE equipment. The system constructed with these technical indexes can realize Ethernet power supply while protecting the reliability of network data transmission, And prevent various possible hazards of PSE equipment and power receiving equipment due to power supply to equipment not suitable for network power supply.

According to the network topology and implementation mode of PSE providing power to PD, PSE equipment is divided into END-POINT PSE and MID-POINT PSE. Terminal PSE uses Ethernet devices supporting POE function, such as switches and routers, to provide power, as shown in Figure 2; Mid span PSE is a way to add a mid span hub that provides power to Ethernet switches and PD devices, as shown in Figure 3. The system adopts the terminal Poe system, directly adopts the Ethernet switch supporting POE technology.

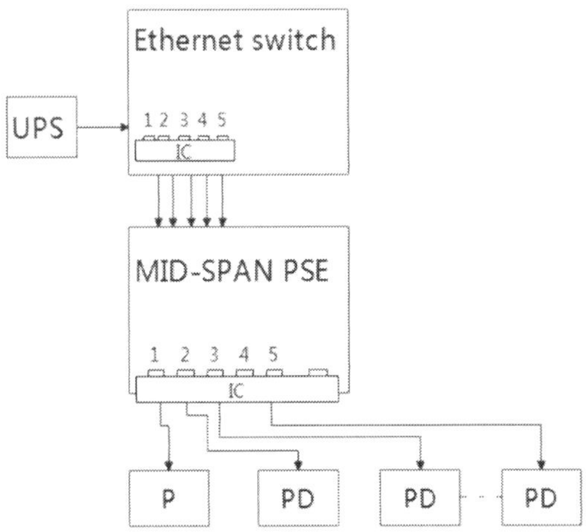

Figure 2. Terminal POE system

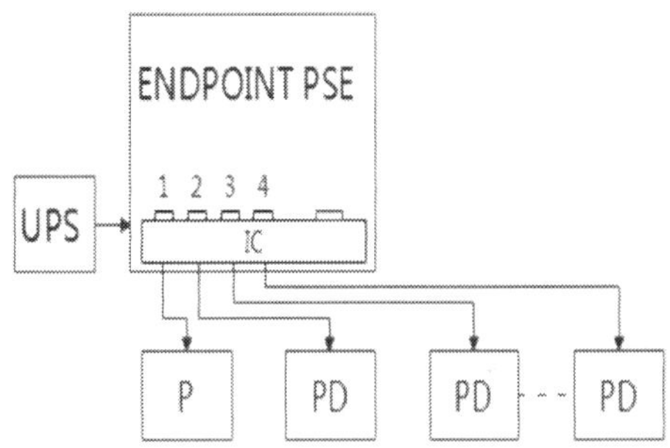

Figure 3. Mid-SPAN POE system

2.3. Design and implementation of PD equipment

In POE system, PD equipment is power receiving equipment. The design of power receiving equipment requires an interface chip conforming to Poe protocol to provide PSE equipment with PD equipment detection characteristic signal, hierarchical characteristic signal and program-controlled surge current signal. In this system, compatible standard high-power PD controller TPS2376-H introduced by TI company is adopted, as shown in Figure 4, as specified in IEEE802.3af standard, PSE equipment only needs 1,2 line pairs, 3,6 line pairs or 4,5,7,8 line pairs to supply power, but it can not supply power at the same time. PD equipment needs to deal with these two pairs of line pairs. Therefore, two rectifier bridges are selected to prevent reverse polarity of DC voltage between line pairs. Diode D1 is designed to eliminate over voltage caused by lightning strike on the network line, The resistance between IIIM pin and VSS pin is used to the limit value of surge current. The setting formula is: $I_{LIM} = 40000 / R_{ILIM}$. It is recommended to select 287 kΩ when connecting with general PSE equipment. The setting of surge current limit value can prevent the output voltage from being lower than the lowest UVLO value by the subsequent step-down capacitor. When adjusting the surge current value, the capacitance of the step-down capacitor can be larger, which is beneficial to the design of the later step-down circuit. Pin ILIM is not allowed to be suspended or short circuited with VSS.

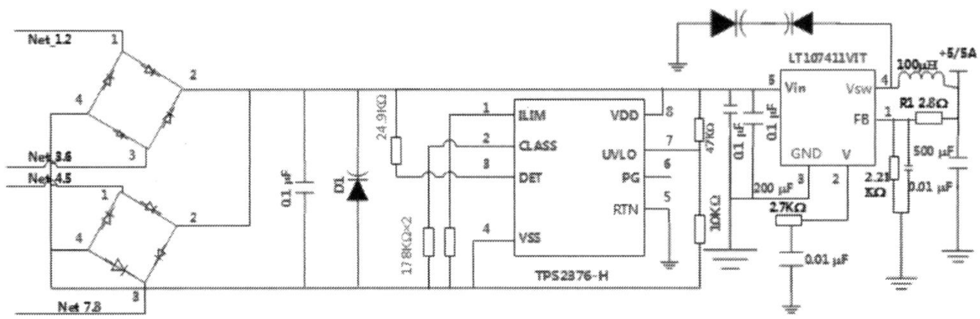

Figure 4. PD interface circuit designed by TPS2376-H

DET pin is the detection resistance of PSE equipment to detect PD equipment, and the resistance value is 24.9 (1 + 1) K12. When PSE equipment detects, measure two V_ I point or the slope between them to calculate the resistance value and judge the existence of the common mode terminal of the port. For Ethernet power supply, detection is critical because it can ensure that 48V DC voltage is only applied to the effective PD and will never damage the equipment that does not need to receive 802.3af power supply. PD detection must be completed within 500ms, otherwise if the PSE waiting time is too long, the Ethernet equipment plugged in the original PD location that does not need power supply will be damaged. After successfully completing the detection, PSE provides 15.5 ~ 20.5v voltage to the port and puts PD in hierarchical mode. In particular, if the PD terminal does not need to be powered due to special reasons, the PD equipment must try to reduce its impedance below 12kfz or above 45kq, so as to ensure that it will not be detected by PSE equipment. Due to the limited power supply resources of PSE equipment, it is impossible to provide electric energy of any size to PE equipment. Therefore, after PSE successfully detects PE equipment, it is necessary to carry out hierarchical operation. The class pin is externally connected with the hierarchical resistance of PD equipment. As shown in Table 1, Pd equipment is divided into five levels, and each level corresponds to a power range of PD equipment, ranging from 0 to 12.95w, The resistance value of this resistor can be selected according to the power value required by the later stage equipment. In the grading stage, PSE will apply 15 ~ 20V voltage to PD, and determine the specific level of PD by measuring the current. At this stage, the power supply part of the PD will be maintained in the passive state by the undervoltage locking (UVIO) circuit in order to isolate the switch stage until the characteristic and classification stage is completed.

Table 1. Energy level of power receiving equipment

Gradation	PD power（W）	Graded resistance （Ω）	Grade current （mA）	Remarks
0	0.45～12.96	4422±1%	0～4.5	default
1	0.45～3.85	952±1%	9.0～12.0	optional
2	3.85～6.48	548±1%	17.0～20.0	optional
3	6.48～12.96	356±1%	26.0～30.0	optional
4	retain	254±1%	36.0～44.0	retain

If the detection and grading operations are successfully completed, the PSE equipment will continuously supply power to the PD equipment according to the grading power. While continuously supplying power, the PSE equipment will also prepare to supply power to the next PD equipment, and will continuously detect the current PD equipment to ensure that the PD equipment is still connected. 802.3af standard specifies that PSE uses DC disconnect and AC disconnect to judge whether it remains connected with a PD. If the current PD equipment is detected to be disconnected, Poe power supply will be stopped immediately.

In this system, the post PD equipment adopts the arm core. The start-up of the arm requires an instantaneous large current. If the instantaneous current is not enough, the arm cannot be started and the POE power supply will not succeed. At this time, the resistance of Ilim pin can be adjusted to increase the surge current, or the capacitance of DC / DC pre stage capacitor can be reduced. Similarly, the instantaneous current can be increased to make the arm start normally, After actual test, POE system works normally.

3. Conclusion

This paper presents a meteorological intelligent sensor system based on POE Ethernet technology. The working principle and implementation method of Ethernet power supply system based on IEEE802.3af Ethernet power supply industry standard are experimentally analyzed. After POE power supply is adopted, the power supply part of the meteorological intelligent sensor system is very simple, and the node part does not need to be close to the power supply part, so the distribution of terminal nodes is flexible and diverse. The communication protocol adopts Ethernet protocol, which greatly reduces the difficulty and complexity of hardware and software design, and enhances the reliability and scalability of the system. Through the actual system test, all the initial assumptions are realized.

Acknowledgments

This work is financially supported by the Research project of Shanxi Province (20188243003).

References

[1] Yuan P.(2021)Discussion on Ethernet power supply technology and application.Electrical application, 40:4-8.
[2] Yin S.(2021)Analysis on the application of POE technology in security network monitoring. Network security technology and application, 12:18-19.
[3] Riley B.(2020)Ethernet power supply and receiving equipment designed for the Internet of things. Application of single chip microcomputer and embedded system,20:90-91.
[4] Ding Q.W.(2017) Design of power interface module of Internet of things terminal based on POE technology.Computer knowledge and technology, 13:29-30.
[5] Lin T., Li S., Tao D., Cai J.h. (2020)Research on improving lightning surge protection reliability of POE power supply system.Jiangsu communication, 36:67-71.
[6] Zeng F.H.(2019) Application of POE power supply in communication system.Low carbon world, 9 : 67-68.
[7] Wang N., Wang Y.Y., Liu Z.F., Yu H.Y. (2018) Power over Ethernet (POE) technology and its application in LED lighting system.Intelligent building, 10:20-25.
[8] Xu H.L. (2019) Research on solar LED intelligent lighting system based on POE technology. China Equipment Engineering, 17:145-146.
[9] Li D. (2019)Application of POE power supply in weak current engineering.Communication world, 26:242-243.
[10] Xing Z.L. (2019)Design and implementation of intelligent substation monitoring system.Dalian University of technology Publishing , Dalian.
[11] Xu X.B., Cheng Z.F., Yu T.,Xie J.F. (2019)Development of steam generator cleaning environment monitoring system based on POE technology. Cleaning world, 35:41-43.
[12] Yan C.W., Sun F.J. (2019)Research on POE cabling technology and 5G application.Intelligent building and smart city, 2:89-92.

Analysis of power frequency electric field distortion caused by substation inspection robot

Quanquan Gong[1], Yunqi Li[2], Zhongju Yang[3], Dandan Dou[1], Fan Yang[3*], Hao Yang[3]

[1] State Grid Shandong Electric Power Company Electric Power Research Institute, Jinan, Shandong 250003, China

[2] UC United College, Chongqing University, Chongqing 400044, China

[3] College of Electrical Engineering, Chongqing University, Chongqing, 400044, China

*Corresponding author's e-mail: yzj_828@163.com

Abstract. In the era of intellectualization, many substation inspection robots are being put into operation, but they will cause electric field distortion in the surrounding space during operation. This paper establishes a three-dimensional simplified simulation calculation model for 500kV substations and inspection robots. Based on the boundary element method, we simulate the power frequency electric field of the substation switchyard and analysis the distortion electric field when the inspection robot is in the station. The results show that when the electric field intensity is strong, the distortion coefficient caused by the robot will be larger, and the maximum can reach 2.25 times, and when the electric field intensity is small, the distortion coefficient will be smaller, and the distortion coefficient is 1.42 times on average.

1. Introduction

Substation inspection robot can be used to inspect the large power equipment in the station, which can find out whether the power equipment in the substation is abnormal operation in time, so that the staff can deal with the hidden trouble in operation in time, which can greatly improve the work efficiency and substation inspection quality[1-3]. However, when the inspection robot conducts inspection work in the station, the complex electromagnetic environment in the station will disturb the information transmission and other functions of the robot, and even damage the robot equipment in serious cases, resulting in permanent damage to the inspection robot. Therefore, it is of great significance to the construction and development of smart power grid in China to calculate and study the electromagnetic environment distribution in substation and the field intensity distribution around inspection robot.

A lot of research has been done on the distribution and measurement of power frequency electric field in substations at home and abroad. Literature [4] measured the power frequency electromagnetic field and radio interference level inside and outside multiple 500kV high-voltage substations, and conducted research and analysis on the electromagnetic environment distribution inside and outside the substation. According to the GB/T12720-91 "Power Frequency Electric Field Measurement" standard, Literature [5] used the HI3604 type low frequency electric field and magnetic field measuring instrument to measure the power frequency electromagnetic field in a substation in Northeast China, and measured the electromagnetic field distribution of the 500kV and 220kV switching field in the station respectively. Although the above-mentioned literature studied and analyzed the distribution of power frequency electric fields in substations, there are few studies on the distribution of the field intensity in the surrounding space and the changes caused by the inspection robot during operation.

Based on the above problems, this paper establishes a three-dimensional simplified simulation calculation model for the 500kV substation switchyard and inspection robot. Based on the boundary element method, we simulate the power frequency electric field in the substation 1.5m above the ground and calculate the space power frequency electric field around the inspection robot. Through the simulation calculation results, we study and analyze the distribution of the power frequency electric field in the substation, and the power frequency electric field distortion caused by the inspection robot under different electric field intensity, which can provide guidance for the operation and use of the inspection robot in the substation and is of great significance to the construction of smart grids.

2. Establishment of simulation model

2.1. 500kV substation model

This paper mainly takes 500kV substation switchyard as an example to establish a simulation calculation model, as shown in Figure 1 below. In order to obtain the overall power frequency electric field distribution of the switching yard in the substation, we use the following simplified process to deal with model[6-7]:

- Ignore the influence of external composite insulator umbrella skirts such as insulation pillars and current transformers, that is, the surface of the equipment is regarded as a smooth surface.
- The separate conductor adopts equivalent radius.
- Ignore the influence of transformer and structure.
- Regarding the earth as an infinite plane, set its potential to be zero.

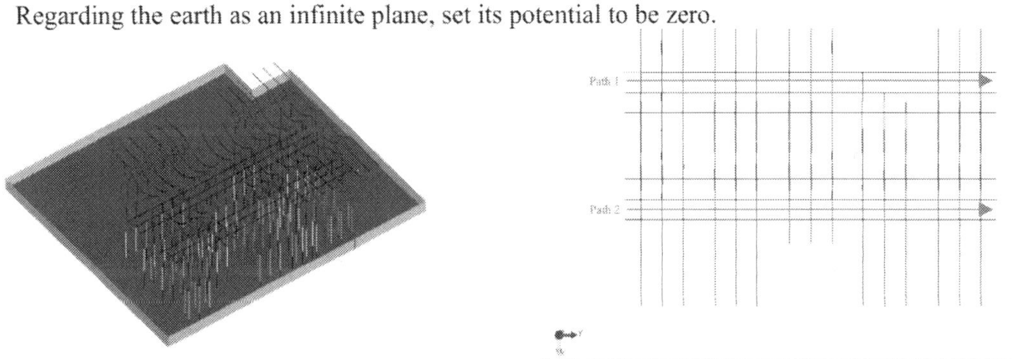

(a) Three-dimensional simulation diagram of substation (b) Substation layout diagram

Figure 1. Substation model

2.2. Substation model parameters

The conductor type is 2×LGJQT-1400. There are 5 overhead conductors and 2 bus conductors in the 500kV switchyard of the substation. The average height of the overhead conductor s is 26m, and the phase distance is 8m; the height of the bus conductor is 15.1m, and the phase distance is 6.5m. During normal operation, the phase voltage is 303kV, and the phases of A, B, and C three-phase voltage differ by 120 degrees in sequence.

2.3. Inspection robot model

When we simplify the modeling of the inspection robot, the bottom drive system has little effect on the power frequency electric field value at a height of 1.5m. Therefore, we mainly carry out simplified modeling of the robot measurement system and ignore the influence of the bottom wheel, the model is shown in Figure 2 below. we can see it from Figure 2 that the simplified model of the robot measurement system is mainly composed of a measuring probe, a support arm and a connecting base. Its constituent materials are mainly PVC and ABS, the Relative dielectric constant of PVC is taken as 4, and the Relative dielectric constant of ABS is taken as 3.5.

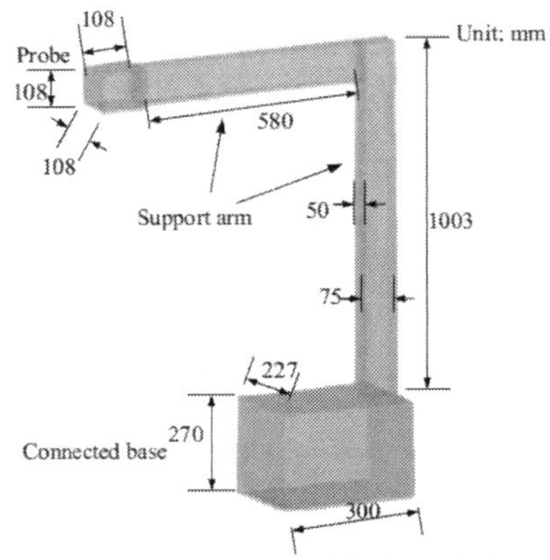

Figure 2. Robot simplified model

3. Simulation calculation and result analysis

3.1. Simulation and analysis of power frequency electric field in substation

Based on the above-mentioned model of the substation, the power frequency electric field distribution of the substation is simulated and calculated. The power frequency electric field distribution in a plane 1.5m above the ground is shown in Figure 3.

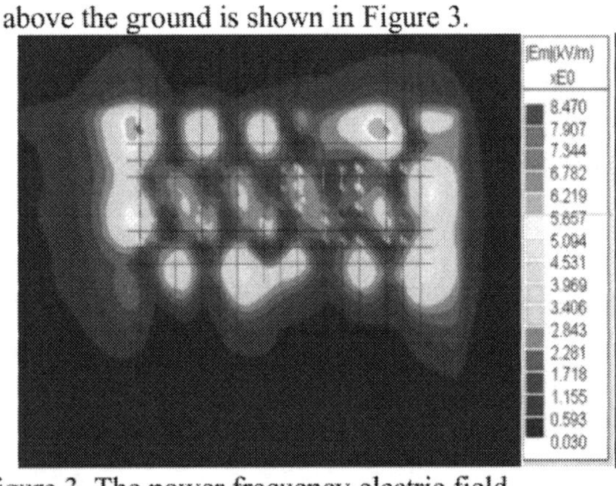

Figure 3. The power frequency electric field intensity distribution of the substation

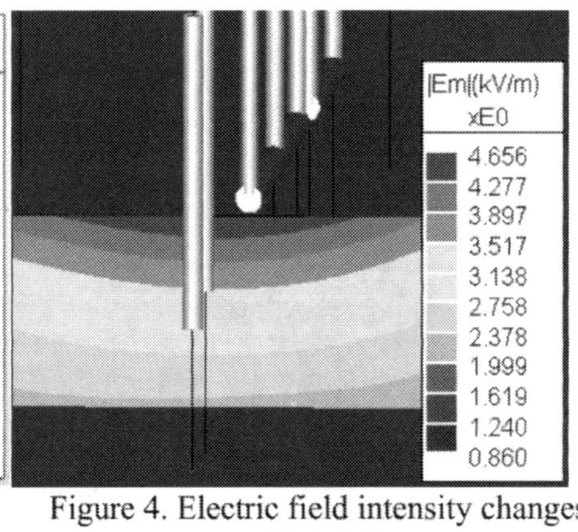

Figure 4. Electric field intensity changes with height

It can be seen from Figure 3 that the power frequency electric field intensity in the 500kV switchyard area of the substation is mostly concentrated between 2kV/m and 6kV/m. The maximum electric field intensity appears at the upper left of the intersection of the bus conductor and the overhead conductor, and the value is 8.47kV/m, the reason mainly is that the bus conductor and overhead conductor are close to the ground here, and the lines are interlaced here. Larger values of electric field intensity are mostly concentrated near high-voltage equipment, such as circuit breakers, isolating switches, which are mainly caused by the combined influence of overhead conductors, low ground of equipment, and the calculation point's height. The electric field intensity under the bus conductor of the middle phase is relatively small, and the electric field intensity outside the edge phase has a locally large value.

As shown in Figure 4, taking a rectangular area perpendicular to the ground, we can see that as the distance from the ground increases, the electric field intensity increases; the closer the distance to the switch field device, the higher the electric field intensity.

3.2. Verification of simulation results

To verify the correctness of the model, a point is selected every 5m in path 1 for power-frequency electric field measurement and calculation, as shown in Figure 1 (b), and the comparison results are shown in Figure 5 below. It can be seen from Figure 5 that the change trend of the calculated value and the measured value is basically the same, and the results are in good agreement, which verifies the correctness of the model built.

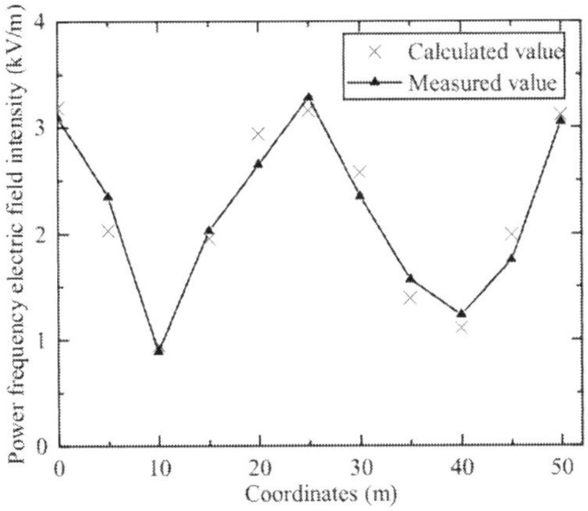

Figure 5. Comparison of calculated and measured electric field intensity

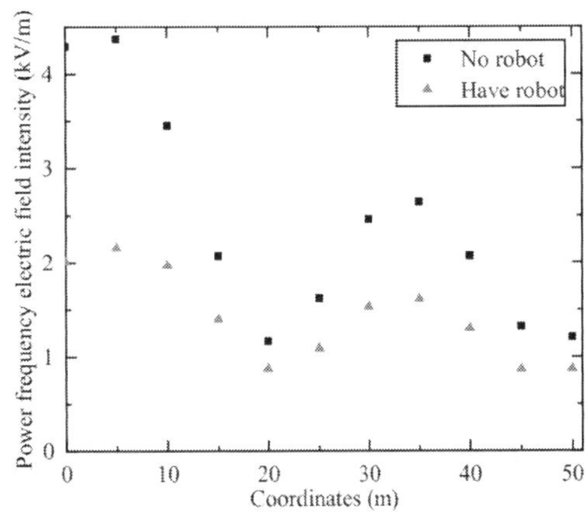

Figure 6. Power frequency electric field intensity

3.3. The impact of the inspection robot on the power frequency electric field in the surrounding space

In order to study the electric field distortion of the surrounding space caused by the inspection robot working in the station, a number of points are randomly selected in the switchyard corridor of the built model to calculate the power frequency electric field. According to whether there is an inspection robot working at the point, the difference of calculation results is compared, and the electric field distortion intensity caused by inspection robot operating in the surrounding space is obtained. According to the measurement method stipulated in the relevant national industry standards[8-9], a point is selected every 5m on the center line of the switching field corridor for calculation, as shown in Path 2 in Figure 1 (b), and the power-frequency electric field with and without robot 1.5m away from the ground is calculated. The results are shown in Figure 6.

4. Analysis of distortion results caused by inspection robots

From Figure 6 above, we can see that because the dielectric constant of the materials used by the inspection robot (PVC, ABS) is very different from air, the location of the inspection robot will cause distortion of the power frequency electric field in the surrounding space. The electric field intensity distortion is shown in Table 1 below.

Table 1. Distortion coefficients at different field intensities

Electric field intensity (kV/m)	Maximum distortion coefficient	Average distortion coefficient
3~5kV/m	2.25	2.01
2~3kV/m	2.06	1.7
1~2kV/m	1.58	1.42

From Table 1, we know that the electric field distortion intensity in the surrounding space caused by the inspection robot varies with the magnitude of the field intensity value. When there is no inspection robot in the station for work, the electric field intensity is greater than 3kV/m at some measurement point, the maximum distortion coefficient caused by inspection robot can reach 2.25, and the average distortion coefficient is 2.01; At some measuring points where the electric field intensity is 2~3kV/m, the maximum distortion coefficient can reach 2.06, and the average distortion coefficient is 1.7; At some measuring point where the electric field intensity is 1~2kV/m, the maximum distortion coefficient can reach 1.58, and the average distortion coefficient is 1.42.

From the above analysis, we know that when the inspection robot work in the substation, it can greatly improve the work efficiency, but at the same time it will also cause the electric field distortion in the surrounding space. Therefore, when we design the robot shell material, the dielectric constant should be as close as possible to the dielectric constant of air, so that to reduce the influence of electric field distortion caused by the robot.

5. Conclusion

- This paper establishes a 500kV substation simulation calculation model, simulates and calculates the power frequency electric field intensity at a height of 1.5m above the ground, and analyzes the distribution of the power frequency electric field in the station. The results show that the areas with large electric field intensity are mainly concentrated in high voltage live equipment, the maximum field intensity value can reach 8.47kV/m, while the electric field intensity under the bus conductor of the middle phase is relatively small, and the average field intensity is about 2kV/m.
- The three-dimensional simulation model of the robot is established, and the distortion of the electric field around the robot is analyzed. The results show that the distortion intensity varies with the electric field intensity. When the electric field intensity is 3-5kV/m, the distortion coefficient is larger, up to 2.25; while when the electric field intensity is 1-2 kV/m, the distortion coefficient is smaller, averaging 1.42.

References

[1] Yu Y, Dong MZ. (2021) Design and implementation of intelligent inspection robot in photovoltaic substation. Electrical Measurement & Instrumentation. 2: 1-8.
[2] Dong SH, Niu CW, Dai K. (2021) Study on automatic control method of Substation Inspection Robot Based on Deep Reinforcement Learning. High Voltage Apparatus, 57(02): 172-177.
[3] Li YM, He SY, Xun LW, Wang XX, Zou AX. (2020) The influence of inspection robot on the electric field distribution of human body in 500 kV substation. Advanced Technology of Electrical Engineering and Energy, 39(09): 74-80.
[4] Sun T, Wan BQ. (2006) Measurement of the electric magnetic environmental for 500kV substation. High Voltage Engineering, 6: 51-55.
[5] Rao ZQ, Guo QG, Zhao DQ. (2004) Measurement of power frequency electric magnetic field in 500kV substation. High Voltage Engineering, 9: 41-43.
[6] Yun YX, Zhao XX, Luo RT. (2009) Calculation of power-frequency electric fields at HV Substation by BEM. Shandong Electric Power, 5: 17-20.
[7] Tang KD. (2018) The study on electromagnetic field distribution characteristic of typical universal design standard substation. Chongqing University, Chongqing.
[8] National Development and Reform Commission, PRC. (2005) Methods of measurement of power frequency electric field and magnetic field from high voltage overhead power transmission line and substation. http://www.360doc.com/document/16/1121/05/10819955_608139083.shtml.
[9] Ministry of Ecology and Environment, PRC. (2020) Technical guidelines for environmental impact assessment of electric power transmission and distribution project. https://www.mee.gov.cn/ywgz/fgbz/bz/bzwb/hxxhj/xgjcffbz/202012/t20201218_813935.shtml.

Sorting path optimization of parallel robot based on improved Genetic Algorithm

Feize Xia[1], Yuan Sun[2*] and Meng Wang[3]

[1,2,3] School of mechanical, Shanghai Dian Ji University, Shanghai, Shanghai, 201306, China

*Corresponding author's e-mail: suny@sdju.edu.cn

Abstract. In industrial production, parallel robot is often used for sorting, battery clamp is divided into two structures, the classification and sorting, , for multi-target sorting problems. The motion path of end-effector is planned, the overall sorting path is planned, and the practical problem is transformed into a similar traveling salesman problem. The improved genetic algorithm is proposed to optimize the sorting time sequence, and the optimized path shortens 10.23% total distance on average compared with random sorting and 5.37% total distance compared with fixed longitudinal sorting. The shortening of the total distance can lead to higher sorting efficiency in sorting and increase productivity.

1. Introduction

At the present stage, it has become a trend to introduce technologies such as robots and machine vision into industrial production to improve production efficiency[1]. Delta parallel robot has the advantages of small footprint, strong bearing capacity, high accuracy and fast speed. The sorting system of Delta parallel robot combined with machine vision can adapt to diversified operating environments, significantly improve production efficiency and reduce labor intensity[2]. However, in high-speed motion, problems such as reduced grasping accuracy will occur, resulting in loss of the robot.

In this paper, the Delta parallel robot is used to sort the workpiece. The trajectory design of the end-effector of Delta parallel robot is carried out to improve the grasping speed, reduce the jitter of the robot and reduce the loss and improve the overall sorting efficiency.

2. Single trajectory optimization

2.1. Trajectory optimization

As shown in Fig.1, the Delta parallel robot sorting system consists of a conveyor belt, a parallel robot, a vision module and a control system. The control system adopts upper and lower computers[3]. Fig.2 shows the workpiece to be sorted.

Fig.1 Sorting structure diagram

Fig.2 Sorting workpiece

In order to ensure smooth operation, must optimize the operation of the parallel robot trajectory, the Delta actuator at the end of the parallel robot trajectory approximate portal, arc line method[4], as shown in Fig.3, to be broken down into five parts, including P_1-P_2, P_3-P_4, P_5-P_6 straight-line interpolation, and P_2-P_3, P_4-P_5 respectively using circular arc interpolation.

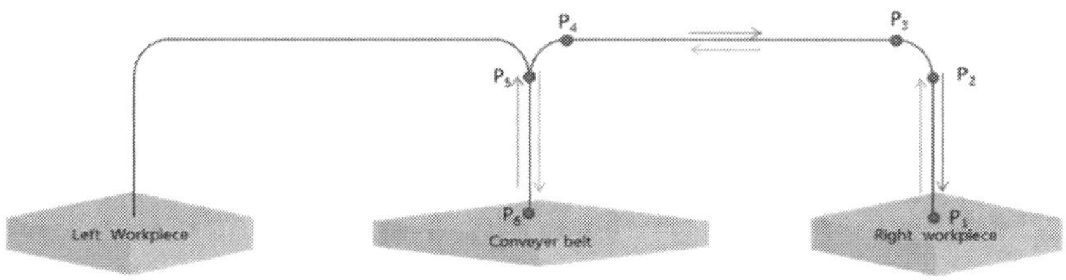

Fig.3 Sorting trajectory planning

Because the interpolation method will lead to excessive acceleration at the end of the trajectory, a positive modified trapezoidal acceleration change method is proposed for this problem[5]. This method has relatively fast acceleration characteristics[6], but it's easy to cause the inertia impact of the end-effector. In order to optimize its dynamic characteristics, High order derivative smooth continuous polynomial is used as the transition segment to correct trapezoidal acceleration. The plan is as follows:

$$S=\begin{cases}\left(\frac{t^5}{2\tau^3}-\frac{t^6}{2\tau^4}+\frac{t^7}{7\tau^5}\right)a_{\max} & \left[0,\ \frac{T}{8}\right) \\ \left(\frac{t^2}{2}-\frac{\tau}{2}t+\frac{\tau^2}{7}\right)a_{\max} & \left[\frac{T}{8},\ \frac{3T}{8}\right) \\ -\left[\frac{(t-4\tau)^5}{2\tau^3}+\frac{(t-4\tau)^6}{2\tau^4}+\frac{(t-4\tau)^7}{7\tau^5}-3\tau\cdot t+6\tau^2\right]a_{\max} & \left[\frac{3T}{8},\ \frac{T}{2}\right) \\ -\left[\frac{(t-4\tau)^5}{2\tau^3}-\frac{(t-4\tau)^6}{2\tau^4}+\frac{(t-4\tau)^7}{7\tau^5}-3\tau\cdot t+6\tau^2\right]a_{\max} & \left[\frac{T}{2},\ \frac{5T}{8}\right) \\ -\left(\frac{t^2}{2}-\frac{15\tau}{2}t+\frac{113\tau^2}{7}\right)a_{\max} & \left[\frac{5T}{8},\ \frac{7T}{8}\right) \\ \left[\frac{(t-8\tau)^5}{2\tau^3}+\frac{(t-8\tau)^6}{2\tau^4}+\frac{(t-8\tau)^7}{7\tau^5}+12\tau^2\right]a_{\max} & \left[\frac{7T}{8},\ T\right]\end{cases} \tag{1}$$

When $t=8\tau=T$,
$$T=\sqrt{\frac{16S}{3a_{\max}}} \tag{2}$$

According to equation (1), the displacement, velocity, acceleration and acceleration of polynomial modified trapezoidal acceleration and deceleration algorithm are obtained, as shown in Fig.4.

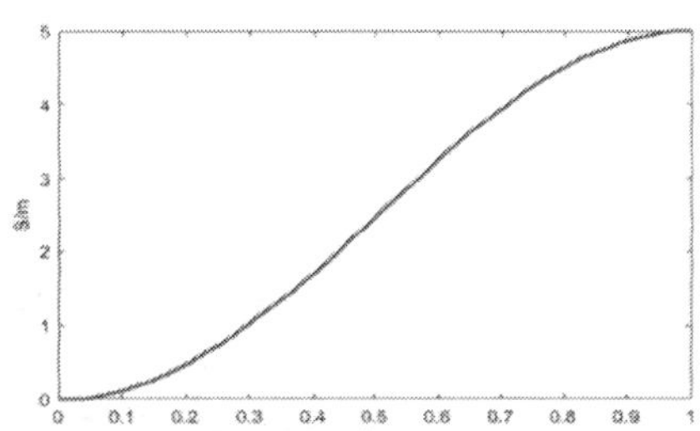

(a)

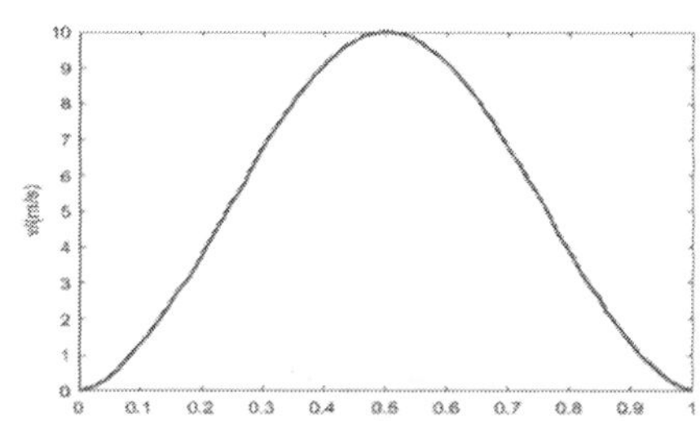

(b)

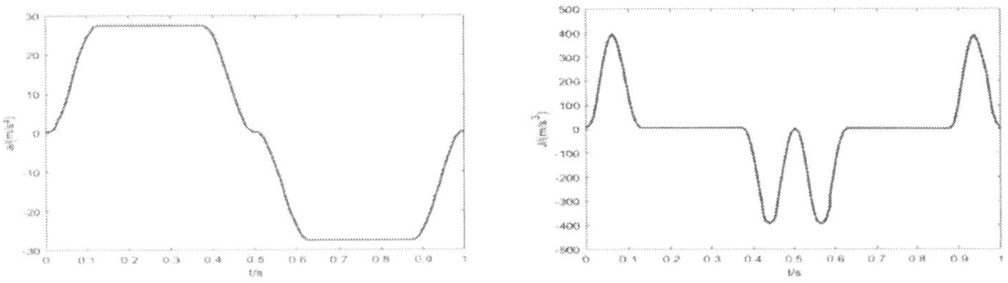

(c) (d)

Fig.4 Polynomial modified trapezoidal motion diagram

It can be seen from Fig.4(d) that the change of acceleration is stable, which can prevent jitter caused by acceleration mutation.

2.2. Overall sorting path optimization

Nowadays, many studies focus on single trajectory optimization without focusing on overall sorting path planning[7]. Random sequential sorting and fixed vertical sorting are mostly adopted, which is relatively single. Now, overall sorting path is optimized to improve efficiency. Seeking the shortest path of sorting, this problem is similar to the traveling salesman (TSP) problem[8]. The artifacts and placed on the conveyor belt for the two big urban agglomeration, the end of the actuator for the traveling salesman, urban agglomeration in the dynamic change of workpiece, traveling salesman, alternately in the two big urban agglomeration. A path planning method for dynamic workpiece sorting based on improved genetic algorithm is proposed. The specific process is as follows:

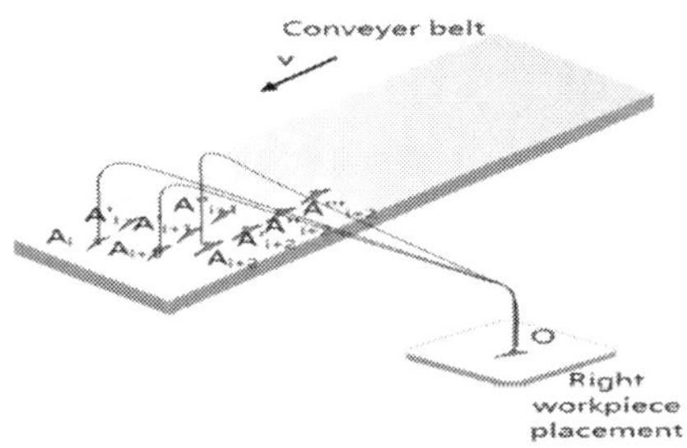

Fig.5 Overall sorting flow chart

As shown in Fig.5, point O is the starting position of the end-effector, and the grasping sequence is $O \rightarrow A_i \rightarrow A_{i+}1 \rightarrow A_{i+2}$, As can be seen, each grasping stroke is different. In the process of the end-effector grasping, the conveyor belt also maintains a uniform motion state. Under ideal conditions, the end-effector and the workpiece to be captured reach the same position at the same time. Starting when the end of the actuator in O, at this time, the artifacts i in A'_1, the artifacts i + 1 in A'_{i+1}, when the artifacts i is captured, the end-effector and the artifacts i should arrive at A_1 at the same time, when the artifacts i grab completed and placed it, the artifacts i + 1 position from A''_{i+1} to A'_{i+1}, in the end-effector motion process, The artifacts changes from A'_{i+1} to A_{i+1}. The motion time of the artifacts shall be equal to the motion time of the end-effector, i.e. $t_{A'iAi}=t_{OAi}$, Among them:

$$S_{OA_i} = \frac{3a_{\max}}{16} T^2{}_{OA_i} \qquad (3)$$

$S_{A'iAi}$ represents the distance of the workpiece from A'_i to A_i, and S_{OAi} represents the distance of the end-effector from point O to point A_i.

12 times per beat as a sorting task, the total travel is $\sum 2S_{OA_i}$.

2.3. Improved genetic algorithm path optimization

2.3.1. Algorithm overview.

It's principle is the use of biological "survival of the fittest" evolutionary law. The parameters of the problem to be solved were encoded into chromosomes, and the information of chromosomes in the population was exchanged by means of selective crossover and mutation, and the chromosomes satisfying the optimization goal were generated[9]. The workflow is shown in Fig.6.

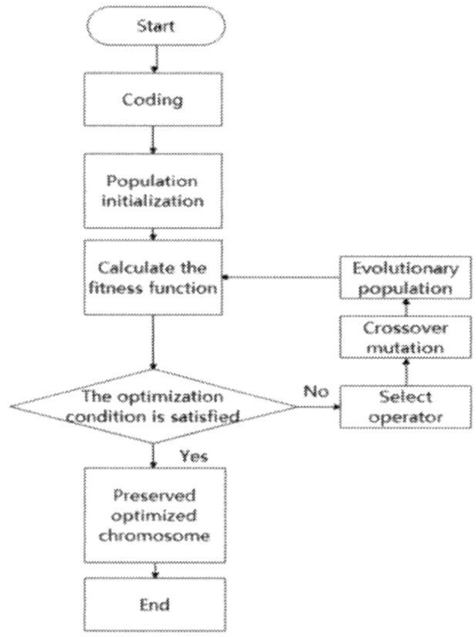

Fig.6 Genetic algorithm workflow

2.3.2.Coding improvement.

The point where the workpiece is placed is numbered O, will stay on the conveyor belt sorting number (1 - n), the part with the position of the workpiece was fetching as (A_1-A_n), chromosome is defined as A linear sequence, random one path coding ($O \to A_1 \to O \to A_2 \to \ldots \to A_i \to O$), will appear alternately in the code number O and A use of chromosomes, According to odd and even split into two codes, respectively ($A_1 \to A_2 \to \ldots \to A_i$) and O, coding O as the fixed point, only need to operate on the former.

2.3.3.Fitness function.

Fitness function value is the basis for evaluating the quality of a chromosome, this problem is equivalent to the TSP problem that has been grouped, directly making each group the shortest path[10], and the fitness function can be obtained as follows:

$$F = \frac{1}{\sum 2S_{OA_i}} \tag{4}$$

2.3.4.Improvement of genetic operator.

The crossover variation of genetic operators directly affects the efficiency and convergence speed of genetic algorithm. According to the size of individual fitness value, the operation of survival of the fittest is adopted to generate a new population, and the fitness of each chromosome is divided by the sum of fitness, as the selection probability of each chromosome; The crossover operation determines the global search ability of genetic algorithm. Mutation operation determines the local search ability of genetic algorithm. Under the condition of real-time performance of each beat calculation, a predetermined number of iterations is adopted as the termination condition

3. Experimental Verification

Using the trajectory of the end-effector of the improved parallel robot and according to the sorting time sequence optimized by the improved genetic algorithm, the workpiece grabbing experiment was carried out. The conveyor belt speed is set as 50mm/s, the number of sorting per beat is 12, the initial population number is 100, and the number of iterations is 200. If optimization is feasible, the total path through the sorting will be shorter than the traditional sorting path. Simulated as shown in Fig.7.

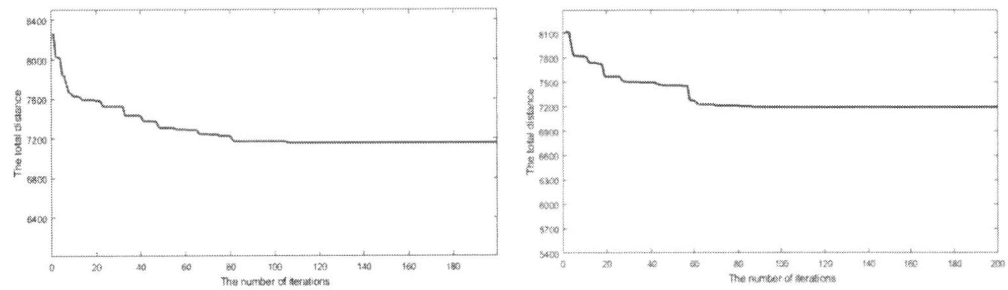

(a) (b)

Fig.7 Sorting optimization curve

In order to verify the efficiency of using the improved genetic algorithm, the total distance of the traditional random sorting and fixed vertical sorting was compared with the sorting optimized by the improved genetic algorithm, and the shortening conditions were calculated respectively, and average calculated by formula $\left(\frac{x_1+x_2+\cdots+x_n}{n}\right)$, Specific data reference table 1.

Table 1. Sorting path optimization results

The serial number	Total sorting stroke /mm			Shorten percentage /%	
	Random sorting	Fixed longitudinal sorting	Optimization of sorting		
1	8081.80	7605.32	7204.15	10.86%	5.27%
2	7954.29	7696.57	7167.35	9.89%	6.88%
3	8102.14	7712.10	7295.16	9.96%	5.41%
4	8013.65	7596.73	7108.72	11.29%	6.42%
5	7996.72	7554.12	7301.94	8.69%	3.34%
6	8003.20	7656.37	7257.83	9.31%	5.21%
7	8094.19	7628.96	7279.49	10.07%	4.58%
8	8112.94	7580.23	7186.92	11.41%	5.19%
9	7974.27	7598.29	7149.30	10.35%	5.91%
10	8061.85	7636.68	7219.28	10.45%	5.47%
The average	8039.51	7626.54	7217.01	10.23%	5.37%

By comparison, the optimized sorting path is better, shortening 10.23% on average compared with random sorting path and 5.37% compared with fixed vertical sorting path.

4. Conclusion

In this paper, a sorting path optimization method based on improved genetic algorithm is proposed to solve the sorting path optimization problem of randomly distributed workpiece on a conveyor belt. In tests, the optimized sorting path more shorter than random fixed longitudinal sorting and sorting, increased by 10.23% and 5.37% respectively, and proves that the algorithm can effectively improve the efficiency of the sorting of the Delta parallel robot, a single sorting path to prevent the jitter caused by acceleration mutation, has high accuracy and stability, and can be applied in practical engineering.

References

[1] Wu, X.Q., Huang, J.C., Zhou, L., Lin, J. (2019) Design of intelligent sorting system for parallel robot. J. Mechanical and electrical engineering., 2: 224-228.
[2] Yang, L., Chen, L.S., Xie, Y.C., (2021) Design of intelligent chip sorting Production Line based on Robotstudlo. J. Computer Measurement and Control., 8: 137-141.
[3] Zheng, C.H., Ye, Z.Q., Fu, H.X., (2015) Design and implementation of intelligent dense cabinet based on upper and lower computer structure. J. Journal of qiqihar university: natural science edition., 5: 28-32.
[4] Yin, G.L., Bai, R.L., Wang, Y.J., Li, X., (2015) A time-optimal trajectory planning method for parallel robots. J. Computer engineering., 10: 192-198.
[5] Li, X.H., (2016) Design and Research of Visual Sorting and Tracking System for Delta Parallel Robot. D. South China University of Technology., Guang Zhou.
[6] Wang, Y.J., Bai, R.L., Ji, F., (2016) A time optimal trajectory planning method for Delta robot. J. Computer engineering., 12: 295-301.
[7] Feng, L.H., Zhang, W.G., Gong, Z.Y., Lin, G.S., Liang, D.K., (2014) Research progress and status quo of Delta series parallel robots. J. Robot., 3: 375-384.
[8] Wang, Z., Liu, R.M., Zhu, Y.G., Wang, X., (2019) An improved genetic algorithm for solving TSP problem. J. Electronic measurement techniques., 23: 91-96.
[9] Xu, G.S., Xu, Z.Y., Zhou, J.J., Zhang, W.J., Shao, Z.P., (2021) Research on Path Planning Algorithm of Inspection Robot Based on Genetic Algorithm. J. Mechanical Design and Manufacturing Engineering., 06: 93-98.
[10] Zhou, R., Ma, H.W., (2021) Laser cutting collaborative operation path planning based on improved genetic algorithm. J. Logistics technology., 10: 50-55.

Sensitivity study of ring-shaped electrostatic sensor based on nonparametric estimation

Zhe Kan*, Yuanzhe Li

School of Information and Control Engineering

Liaoning Petrochemical University,Fushun,China

*Corresponding author email: kanzhe@lnpu.edu.cn

Abstract. In this paper, aiming at the problem of the electrostatic sensor signal satisfying the gaussian distribution, the non-parametric kernel estimation method is introduced, and the electrode induction model of the electrostatic sensor is finally fitted by combining the goodness of fit and the simulation data samples. This model satisfies the gaussian distribution and the electrostatic signal satisfying the gaussian distribution is given in the theory. Maxwell simulation software was used to simulate the theoretical sensitivity of the electrostatic sensor and the axial and radial spatial sensitivity characteristics of different sensor parameters were obtained. Within a certain range, the relative permittivity of the insulating tube is also discussed. Finally, an insulating tube with a relative permittivity of 3 is selected as the material of the insulating tube. Finally, the experiment is carried out on the experimental equipment and the conclusions obtained in the articleare confirmed.

1. Introduction

Pneumatic conveying process widely exists in chemical process, and a series of researches have been carried out on the measurement of gas-solid two-phase flow parameters. Many scholars in China have also studied electrostatic sensors, and a large number of discussions have been carried out by changing the model structure and other conditions. An electrostatic sensor model is studied with an array of bar electrodes and a square insulated pipe based on electrostatics, as well as the mirror-charge method establishes a mathematical model[1]. The actual results obtained through the experimental device are basically consistent with the simulation results, which indicate that the established analytical mathematical model is correct and effective[2]. In order to overcome the uneven sensor sensitivity distribution, an electrostatic sensor with mesh electrode is designed by COMSOL software. Higher sensitivity is achieved and the sensitivity distribution is more uniform. In 2012, the different flow patterns of two-phase flow is analyzed[3-7], and it is concluded that the uneven sensitivity distribution of the electrostatic sensor is caused by the non-linearity of the electrostatic sensor itself and the design structure model[8]. Therefore, the matrix electrode electrostatic sensor is designed, and the sensitivity distribution is consistent with the reality through simulation, which also provides a direction for the realization of the model optimization of the electrostatic sensor. Electrostatic sensor has a simple structure model, and the price is relatively low. The equipment is simple and convenient to install, and it is suitable for some places with bad environment and other advantages. At present, the main research institutions for electrostatic sensors are Manchester University in the United Kingdom; Greenwich University; Tsinghua University; Southeast University[8-11]; Tianjin University[12]. ABB, PCME and other companies in the UK are developing electrostatic sensors, among which ABB and

Journal of Physics: Conference Series **2121** (2021) 012019 doi:10.1088/1742-6596/2121/1/012019

the University of Teesside have jointly developed a plug-in electrostatic load detector, which can be used normally in thermal power plants.

When particles collide and rub in the pipe, a large amount of static charge will be generated on the particles and the pipe. Currently, sensors for electrostatic charge measurement can be divided into intrusive and inductive types according to the installation methods. Among them, the flow of intrusive particles is easy to cause electrode wear, which requires frequent electrode replacement to avoid affecting the accuracy of measurement results. While the induction type has no effect on particle flow, and the flow of particles in the pipeline will not cause electrode wear, which mainly includes the ring electrostatic sensor.

2. Electrostatic sensor principle

In the process of pneumatic conveying, there is a wide range of gas-solid two-phase flow state, and the fluid is in a fast flow or jitter. Vibration and other motion state, Collision and separation between solid particles and pipelines make particles carry a considerable amount of static electricity. The electrostatic sensor's schematic diagram is shown in Figure 1.

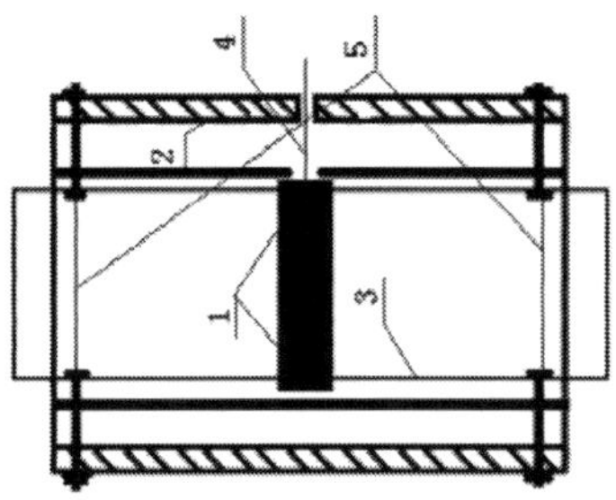

Fig.1 electrostatic sensor

In Figure 1, the annular electrostatic sensor consists of a PVC tube, a metal shield, and a copper electrode ring. Label 1 is the ring electrode, Label 2 is the shielding cover, Label 3 is the insulating tube, Label 4 is the electrostatic signal lead line of the electrode, and Label 5 is the fixing bolt. A metal ring of fixed width b is used as an electrostatic electrode on the outer wall of the insulating tube.

The amplification factor and the electrostatic signal are amplified to 1-5V for the collector to collect. The collector chooses NI acquisition card, and the receiving software of the upper computer adopts LabVIEW software. The laboratory measurement system is shown in Figure 2.

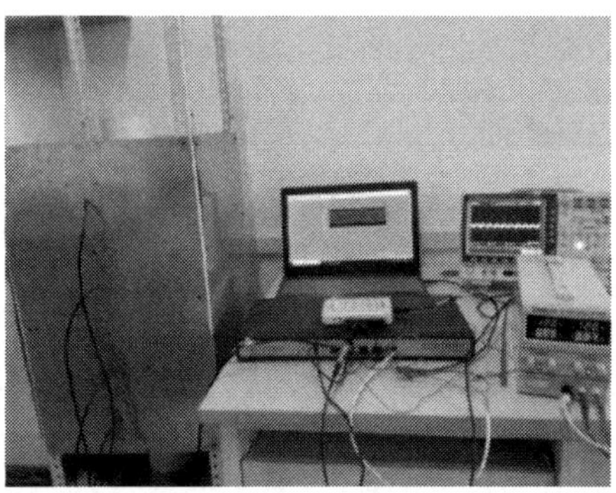

Fig.2 Sensor experimental system

3. Theoretical model of electrostatic sensor

3.1 Theoretical model

$$\begin{cases} \nabla \bullet \nabla \phi(r,\theta,z) = -\dfrac{\rho(r,\theta,z)}{\varepsilon(r,\theta,z)} \\ \phi(r,\theta,z)\mid_{(r,\theta,z)\in\Gamma_s} = 0 \\ \phi(r,\theta,z)\mid_{(r,\theta,z)\in\Gamma_d} = Cons \end{cases} \quad (1)$$

Where, $\phi(r,\theta,z)$ is as electric potential distribution in the field of function. $\rho(r,\theta,z)$ is charge density of the body for the space, and $\varepsilon(r,\theta,z)$ of the dielectric is for the space distribution of the dielectric constant. Γ_s and Γ_d are respectively electrode of the space position of shielded enclosure, Cons is a constant.

3.2 Model building

The three-dimensional static sensor model established by Maxwell software is shown in Figure 3. The middle ring of the pipe is the electrode of the static sensor. In the model, Z axis represents the electrode and the axial direction of the pipe, X axis represents the radial direction, and Y axis represents the tangential direction.

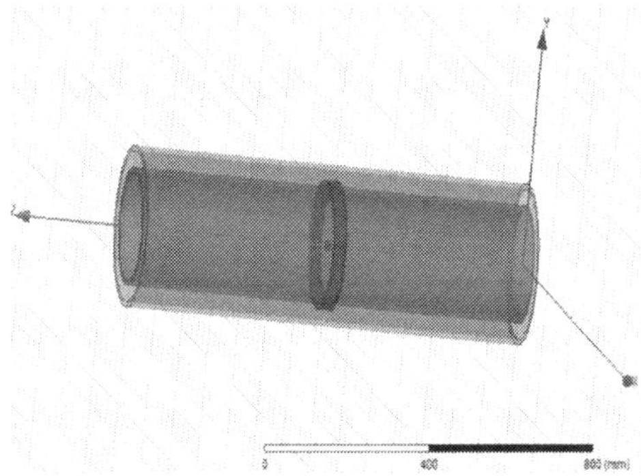

Fig. 3 Solid model of electrostatic sensor

Maxwell software is used to draw the electric field intensity vector and cloud maps at different spatial positions in 2D. The electric field vector and cloud maps at the radial space centre position, and adopted the default XY Cartesian coordinate system is as shown in Fig. 4.

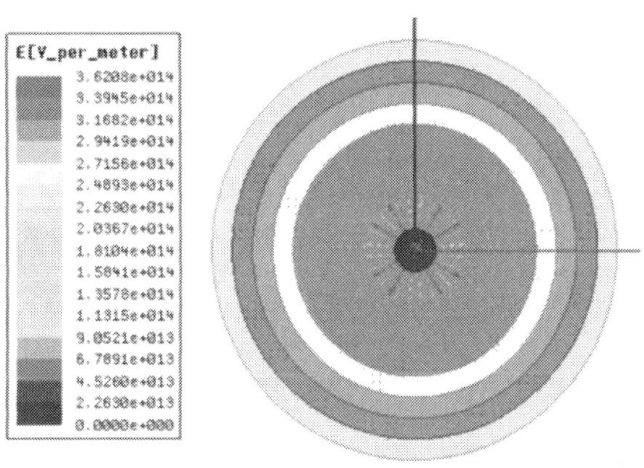

Fig. 4 Vector diagram and cloud diagram of the electric field

In the post-processing process of Maxwell3D, the field calculator is needed to calculate the induced charge on the electrode surface, and the spatial sensitivity characteristics of the electrostatic sensor can be obtained according to the following equation

$$S = \left| \frac{q(r, \varepsilon, \theta, z)}{Q} \right| \tag{2}$$

Therefore, $q(r, \varepsilon, \theta, z)$ is electrostatic sensor sensing electrode charge, Q is charged particles take charge of its own.

3.3 Non-parametric estimation mathematical model

The advantage of non-parametric function estimation method is that there is no specific assumption on the function model and it avoids the risk of model selection. However, the expression is complex, difficult to explain, and a large amount of calculation is a big problem of non-parameter[13].

The kernel density estimation method is suitable for small and medium-sized data sets and can quickly generate an asymptotic unbiased density estimation. Assuming x1, x2, ..., xn is the sample of random variable x, and the probability density function of x is f(x). Given a kernel function K(x) and a positive number h is called "window width," At $n \to \infty, h \to 0$, the variable x of kernel density estimation can be expressed as:

$$\hat{f}(x) = \frac{1}{nh} \sum_{i=1}^{n} K\left(\frac{x - x_i}{h} \right) \tag{3}$$

Relative to the kernel function, the choice of the bandwidth on the result of data fitting is bigger. When the h value, the greater the kernel density estimation function $\hat{f}(x)$ curve is smooth, but it ignores the more details.

$$Q(z) = \frac{\sum_{i=1}^{19} [K_h(z - z_i) * Q_i]}{\sum_{i=1}^{19} K_h(z - z_i)} \tag{4}$$

Especially, $K_h(z - z_i) = \frac{1}{\sqrt{2\pi} h} e^{-\frac{(z - z_i)^2}{2h^2}}$

Based on the theory of nonparametric estimation, the formula of induced charge of electrostatic sensor is determined:

$$Q(r, \varepsilon, z) = A(r, \varepsilon) e^{-\frac{z^2}{B(r, \varepsilon)}} \tag{5}$$

Among them, $Q(r, \varepsilon, z)$ is in the relative dielectric constant of ε insulation pipe and charged particles in the radial position z electrode, and $A(r, \varepsilon)$ and $B(r, \varepsilon)$ is a function of r. $A(r, \varepsilon)$ is decision, and $Q(r, \varepsilon, z)$ is amplitude. $B(r, \varepsilon)$ is decision, and $Q(r, \varepsilon, z)$ is shape. The first-stage charge amplification principle in Figure 3 is used to convert the amount of electrode charge into voltage, and the measurement principle satisfies the expression (2). The electrode charge signal can be converted to 1-5V output through the filtering amplifier circuit, which can be used for collection and analysis by the acquisition card. Finally, the sensitivity formula of electrostatic sensor is determined as follows:

$$S = \left| \frac{Q(r, \varepsilon, z)}{Q} \right| = \frac{A(r, \varepsilon) e^{-\frac{z^2}{B(r, \varepsilon)}}}{Q} \tag{6}$$

4. Simulation and experiment

Sensitivity is one of the important parameters, and the evaluation of electrostatic sensor is operated characteristics based on the definition of spatial sensitivity. The finite element simulation and calculation of the field calculator can be found that the charged particles lie in the space position of PVC pipes, such as the size of the sensor electrode are related to spatial sensitivity characteristics of electrostatic sensor. Therefore, in order to understand whether the electrostatic sensor has good operating characteristics, it is important to calculate the space the sensitivity parameters.

4.1 Simulation experiment

The electrode width b of the electrostatic sensor is set as 30mm, 40mm, 50mm and 60mm respectively, and the diameter of the pipe D is as 100mm. And the field calculator is used to calculate the amount of induced charge in different axial and radial positions, and it is obtained the sensitivity of the corresponding electrostatic sensor. The simulation figure is as shown in Fig. 5.

From figure 5, it can be seen within a certain range that W/D is the greater the relative sensitivity is higher. Electrostatic sensor is width b the growth, and the probe sensitivity is reduced. And in the axial centre of basic graphics presented, the normal distribution can be seen in the centre of the radial position on either side. When the W/D is 0.5 and 0.6, its change is in surge. When W/D is 0.3, it can be seen that the sensitivity is low than 0.4 W/D. So W/D 0.4 relative is the most appropriate, it is adopted in this paper pipe. Its diameter is 100 mm width of 40 mm electrode.

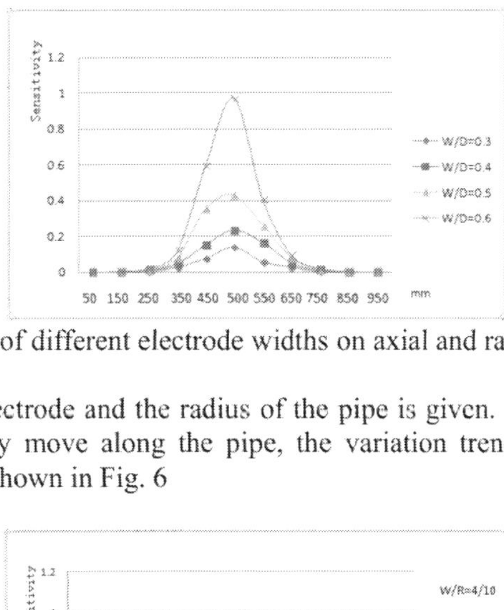

Fig. 5 Effects of different electrode widths on axial and radial sensitivity

The axial width of the electrode and the radius of the pipe is given. W/R is as fixed values, when the charged particles radially move along the pipe, the variation trend of the charged particles at different spatial positions is shown in Fig. 6

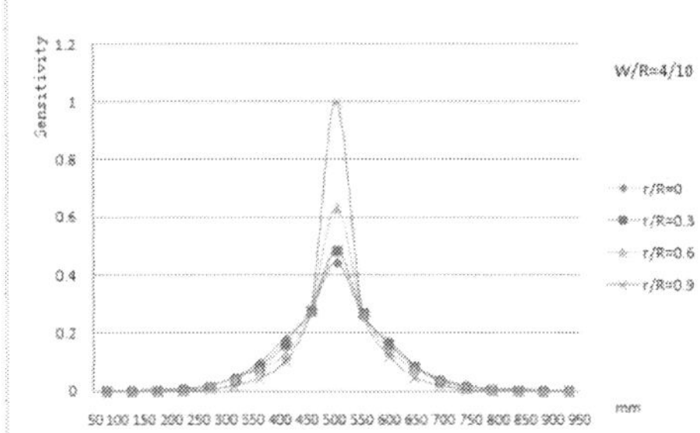

Fig.6 Simulation results of a particle at the different spatial positions

One conclusion can be drawn from Figure 6. The particles are in the axial position $450<Z<550$. The larger the radial coordinate is, the larger the r/R is. And the electrode sensitivity is the greater in the axial position $50<Z<450$ and $550<Z<950$. With the increase of radial coordinates, the sensitivity of the electrode gradually decreases, which also reflects the inhomogeneity of the spatial sensitivity of the electrostatic sensor.

4.2 Experient

The experimental equipment mainly includes electrostatic sensor module, signal processing module, signal acquisition and signal display module. The sensor has four signal output lines, and representing the width of the induction electrode is 5mm, 10mm, 20mm and 40mm respectively from bottom to top. The plastic ball is allowed to fall from the same height. The measurement result is shown in Fig. 7.

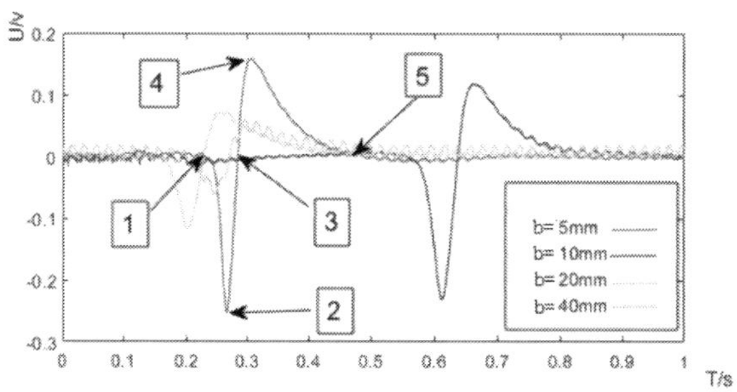

Fig.7 Voltage values of a particle under the different electrode widths

From the figure 7 it can also be found, electrode width is wider, and the measured voltage signal amplitude is smaller. Because the balls are at the same speed, the change of the signal amplitude can reflect the uniformity of the pipe inner space sensitivity. Sensitivity is uniform, so the voltage signal size change is slow. The amplitude is the less, and it is consistent with the results of the simulation.

5. Conclusion

The principle of the electrostatic sensor is simply introduced. By a non-parametric kernel estimation method commonly used in statistics, the estimated value of the induced charge at each axial position can be calculated with R programming language. And then the goodness of fit is introduced to judge the effect of the fit, and the results show the advantage that is the kernel function using Gaussian kernel function for the estimated value of the induced charge. The original data reached 0.95 between the goodness of fit. The results provide a certain basis for the sensitivity formula using Gaussian distribution theoretical basis. The greater the electrostatic sensor sensitivity is relatively higher. Within a certain range, the relative dielectric constant insulating tube is larger, and the electrostatic sensor sensitivity is higher. But with the increase of the relative dielectric constant, the sensitivity of the probe is reduced.

Finally, the experimental equipment and experimental result analysis are briefly introduced, and the corresponding relationship of Q-U graph is explained. The experimental results match the results of Maxwell simulation.

Acknowledgments

This work was supported by the General Project of the Liaoning Province Education Department (L2020019), China.

References

[1] S. Zhang, Y. Yan, et al. Modeling and experimental verification of a square electrostatic sensor array for gas-solid two-phase flow measurement [J]. Journal of Central South University (Science and Technology), 2018, 49 (3): 618-625.

[2] W. Zhang, B. Yang, et al. Measurement of particle velocity and concentration at the elbow of reticulated electrostatic sensor [J].Journal of Instrumentation, 2016,37 (12): 2790-2796.

[3] J. Li, Chuanglong Xu, et al. Spatial sensitivity of electrostatic sensor matrix [J]. Journal of Engineering Thermophysics, 2012,33(12):2108-2111
[4] Z. Kan, Fuqun. Shao, DING L. Measurement of re-lative velocity based on electrostatic sensor [J]. Journal of Shenyang University of Technology, 2010, 32(1):90-94+109
[5] Z. Kan, Fuqun Shao. Measurement of mass flow rate of gas/solid two-phase flow based on electrostatic sensor [J].Journal of Northeastern University (Natural Science), 2010,31 (4): 473-476.
[6] Z. Kan, Fuqun SHAO. Discussion on Accuracy of Speed Measurement Based on Electrostatic Sensor [J]. Control and Instruments in Chemical Industry, 2010,37(3): 69-72.
[7] Z. Kan, SHAO F Q, LI Q H. Measurement of velocity by spatial filtering of plug-in electrostatic sensor [J]. Chinese Journal of Sensors and Actuators, 2012, 25(5):609-612
[8] B. Zhou, D. Y. Yang, et al. Analysis on dynamic sensitivity of ring-shaped electrostatic sensor [J]. Proceedings of CSEE, 2008,28(14):44-49
[9] Y. B. Feng, H. M. Gao, et al. Study on the Space Sensitive Characteristics of Stereo Mesh Static Sensor [J]. Chinese Journal of Scientific Instrument, 2015,36 (9): 1953-1960.
[10] H. M. Gao, Q. Chang, et al. Research on the superficial gas velocity measurement method of dense-phase pneumatic conveying based on electrostatic sensor array [J]. Chinese Journal of Scientific Instrument, 2015, 36(4):787-794.
[11] Chuanglong Xu, J. Li, et al. A spatial filtering velocimeter for solid particle velocity measurement based on linear electrostatic sensor array [J]. Flow Measurement and Instrumentation,2012,26:68-78
[12] Chao Wang, et al. Gas-solid two-phase flow measurement and research apparatus based on electrostatic sensing [J]. Journal of Electronic Measurement and Instrument, 2011,25(1):1-9
[13] Y. Chen. Nonparametric statistical research on tail distribution of non-life insurance loss data [D]. Yun nan University of Finance and Economic, 2010

Research on voltage stability and control strategy of power system considering grid connected charging of electric vehicles

Weidong Fang, Hao Lv*, Yiting Jiang, Lingzhi Li

Fujian Key Laboratory of Automotive Electronics and Electric Drive, Fujian University of Technology, Fuzhou 350118, China

*Corresponding author's e-mail: 2191905051@smail.fjut.edu.cn

Abstract. The development of new energy vehicles is an important measure to deal with the growing energy demand and climate change. Especially in recent years, with the support of national policies and the maturity of electric vehicles(EVs) related technologies, the number of EVs has increased explosively, and the situation is very good. However, it also means that a large number of charging loads will be connected to the power grid, which will put great pressure on the safe and stable operation of the power grid. Although there have been many studies on the impact of EVs integration into modern power grid, most of the EVs load models are based on probability function and lack accuracy. Therefore, starting with the actual operation data of EVs charging station, this paper studies the influence of a large number of EVs charging loads on the static voltage stability of power grid. It is found that the charging load of large-scale EVs is added to the power grid, which significantly reduces the stability of power grid voltage, especially at the place connected to the EVs load and far away from the balance node. In addition, when the charging station adopts the time-of-use(TOU) price strategy, it can effectively improve the voltage stability of the whole network.

1. Introduction

In recent years, the rapid development of society has also brought a series of problems, such as the shortage of fossil energy, global warming and excessive emission of pollutants, which has brought great challenges to the development of modern society [1]. As a promising element of sustainable development and environment-friendly means of transportation, EVs have attracted extensive attention [2]. However, the rapid development of EVs has also brought various challenges to the growing charging demand of the power grid [3]. Nowadays, power systems are forced to operate closer to safety limits [4-6]. As a mobile random load, grid connected charging of large-scale EVs will increase the burden of the power grid and pose a potential threat to the voltage stability of the power grid, especially cluster charging, which may cause regional power grid collapse in serious cases [7-8]. Voltage is one of the important indexes to evaluate the stability of power system [9-10]. It is very important to evaluate the voltage stability of power system to ensure the safe operation of the system. In this paper, the existing actual charging station data are processed, and the influence of EVs charging load on power grid voltage stability is studied by comparing and analyzing the value of voltage stability index L of each node of power grid under the three conditions of only basic load, disorderly charging of EVs and TUO price strategy.

2. Data processing of EVs charging station

The charging of EVs connected to the power grid will impose relatively large load demand, and the load characteristics of EVs are different from other traditional system loads. The start and duration of charging and the actual power consumption of EVs load can not be simply predicted [11]. Therefore, starting with the real EVs charging data, the experiment is closer to the actual situation. The data comes from a charging station in Fuzhou. The station is a comprehensive charging station, with a total of five charging piles, each pile has two charging ports, all of which are DC fast charging. We analyzed the data of two months when the charging station adopted different pricing strategies, namely December 2020 and April 2021. The pricing strategy adopted in December 2020 is 0.8 ¥/kW·h for a whole day; In April 2021, the time-sharing pricing strategy was adopted from 6:00 to 23:00, it was 1.23 ¥/kW·h, and from 23:00 to 6:00, it was 0.88 ¥/kW·h..

We analyze and process the original message data, and extract the field information such as time, charging stake number, charging socket number.The time interval of each data is 7 seconds. Here, we take the voltage and current data appearing for the first time in each minute as the data at that minute, and multiply the voltage and current to be the power at that time. We then add the power of each charging port of each charging pile at the same time to obtain the overall power of the charging station for each minute. As shown in table 1, the power of the charging station for four minutes is given.

Table 1. Total power per minute of charging station.

Time	Power(W)
2021-04-02 00:00:00	144360.72
2021-04-02 00:01:00	84066.20
2021-04-02 00:02:00	135400.46
2021-04-02 00:03:00	135863.94
...	...

In this way, the daily charging load curve of the charging station can be drawn. Considering the worst situation, we select the day with the largest peak valley difference in a month as the representative data of that month. They are the daily peak valley difference of load in December 2020 and April 2021, respectively. It is obvious that December 20, 2020 and April 13, 2021 are the days with the largest load peak valley difference in two months respectively, so we select the data of these two days for the experiment, as shown in figure 1 and figure 2.

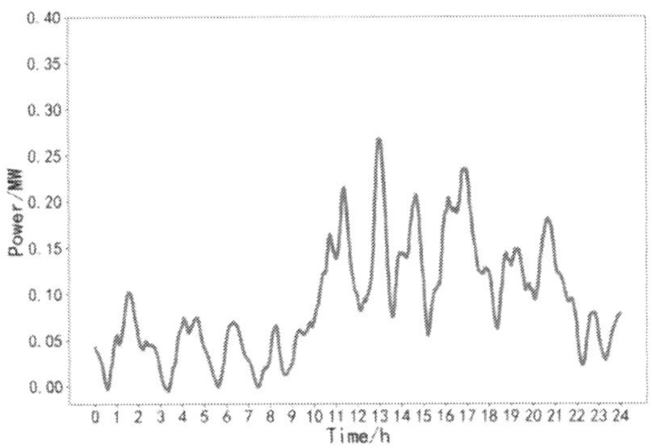

Figure 1. Daily load curve of charging station on December 20, 2020.

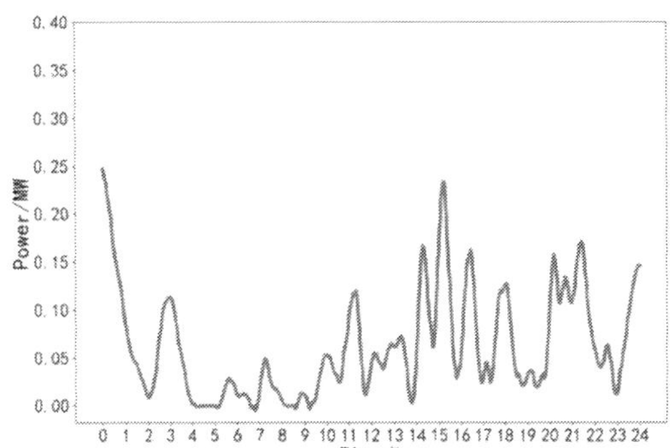

Figure 2. Daily load curve of charging station on April 13, 2021.

3. Simulation example calculation

3.1. Calculation of voltage stability index

In [12], the formula of load bus voltage stability load index is given by using voltage equation. This technique uses the measured values of bus voltage phasor and no-load voltage to calculate the voltage stability index. This index gives the distance from the bus to the voltage stability limit. The voltage stability index is given by the following equation:

$$L = \frac{4[V_0 V_1 \ \cos(\theta_0 - \theta_1) - V_1^2 \cos(\theta_0 - \theta_1)]}{V_0^2} \quad (1)$$

Where V_0 and θ_0 is the no-load bus voltage and phase angle respectively. V_1 and θ_1 is the voltage and phase angle of the bus after the EVs is connected to the power grid and the superposition of the original load. When the value of L at every load bus in the system is less than 1.0, the system is voltage stable. As the value of L approaches 1.0 at any bus, the system approaches its stability limit and becomes unstable when L exceeds 1.0 at the referred bus.

3.2. Power flow calculation

PSASP software is selected for the simulation platform, and WSCC-9 node system is used for the test. As shown in figure 3, the reference capacity is 100mva, bus 1 is a balance node with unlimited output, bus 2 and bus 3 are PV nodes, and other buses are PQ nodes.

Combined with the typical daily load curve given in reference [13], it is taken as the basic load at bus 6. In addition, the charging loads of EVs under the guidance of disordered and time of use pricing strategies are added respectively. The EVs load power factor is set to 0.95, Suppose there are 100 similar charging stations in the region. Other buses are the original conventional load of the system. The voltage and phase angle of each bus are calculated by power flow. As shown in table 2, the voltage amplitude and phase angle of each bus under no-load condition. Then calculate the voltage amplitude and phase angle under other loads, and bring it into equation (1) to obtain the value of voltage stability index L of each bus under different loads.

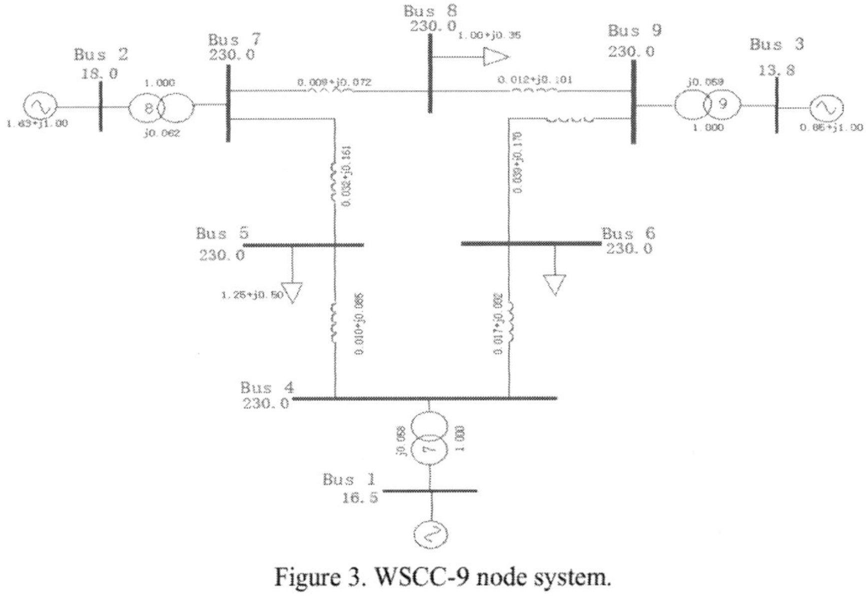

Figure 3. WSCC-9 node system.

Table 2. Voltage amplitude and phase angle of each bus under no-load condition.

Bus name	Voltage amplitude	Voltage phase angle
Bus 1	1.04	0
Bus 2	1.02506	30.64294
Bus 3	1.02502	25.43784
Bus 4	1.02559	7.27665
Bus 5	1.02918	13.36107
Bus 6	1.03812	12.57936
Bus 7	1.04583	25.18871
Bus 8	1.05453	24.13246
Bus 9	1.04822	22.78013

4. Results and discussion

As shown in figure 4 and figure 5, the L index of each bus when the time-of-use tariff strategy is adopted by the charging station and the charging station respectively. We can see that since bus 4 is closest to the balance node and the L index at bus 4 remains at a very small value, the voltage stability at bus 4 closest to the balance node is the highest. Bus 8 is far away from the balance node, its L value is large and its voltage stability is low. The load at bus 6 is the heaviest, and the L value at bus 6 is also high, with large fluctuation range and unstable voltage. Next, we compare the L index values of buses 6 and 8 under the conditions of only basic load, disordered charging and TOU price, as shown in figure 6 and figure 7.

It can be seen that after the EVs charging load is connected, the L index value at the bus increases obviously and the voltage stability decreases. Compared with the situation after disordered charging and charging stations adopt TOU price strategy, generally speaking, the L index value decreases and the voltage stability improves. However, with the time of use price strategy, the L index value increases significantly in the valley price period, resulting in the reduction of voltage stability in this period.

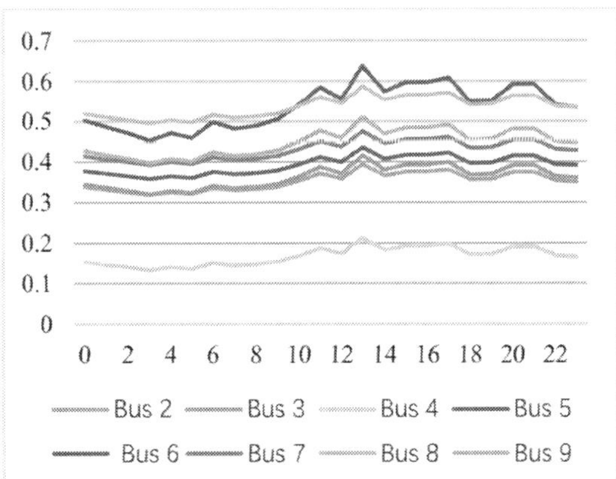

Figure 4. L index of each bus during disordered charging.

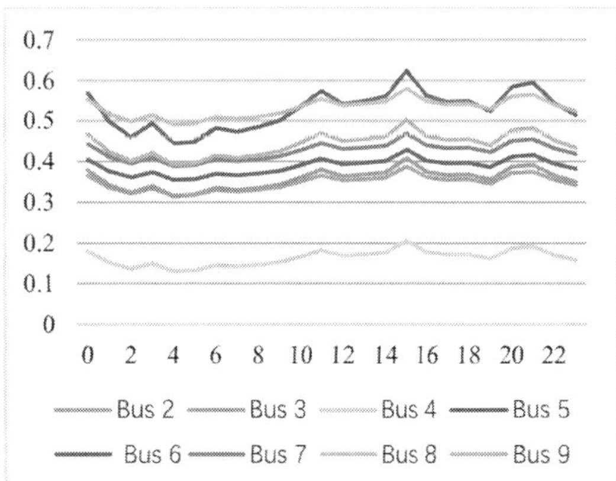

Figure 5. L index of each bus during TOU price charging.

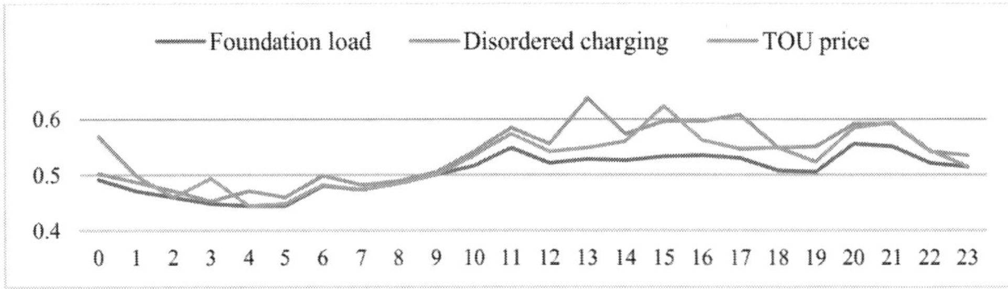

Figure 6. L index of three cases at bus 6.

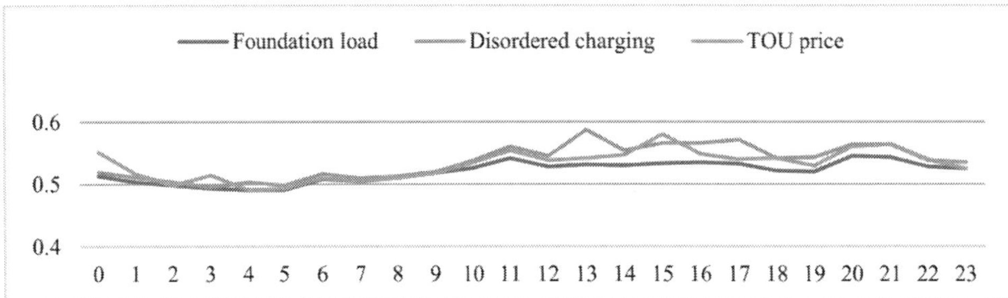

Figure 7. L index of three cases at bus 8.

5. Conclusion

Research shows:

- When the electric vehicle charging load is connected to the power grid, it will reduce the stability of the power grid voltage, especially at the bus added to the electric vehicle load and the bus away from the balance node.
- The time of use pricing strategy can effectively improve the voltage stability of the power grid at the whole time, but it will reduce the voltage stability during the valley price period. Therefore, it is necessary to formulate the charging price strategy in a planned way according to the local load characteristics, so that the power grid can operate within a safe range without transforming the existing lines.

Acknowledgments

Portions of this research were funded through the support of the Technical Consultation on the Integration of Electric Vehicles into Modern Power Grids(GY-H-17041,GY-H-20115),the Development and Demonstration Application of New Energy Vehicle Supervision Service System in Fujian Province(GY-Z160141), Fujian New Energy Vehicle Supervision Service Platform (Gulou District Science and Technology Bureau)(GY-Z160142).

References

[1] Lv J B, Song H., Liu Y, et al. Study on the influence of EVs charging on the voltage quality of distribution network [J]. Electrical measurement and instrumentation, 2018, 55 (22): 33-40

[2] Notice of the State Council on printing and distributing the development plan of energy saving and new energy vehicle industry (2012-2020) [J]. Bulletin of the State Council of the people's Republic of China, 2012 (20): 26-31

[3] Kadurek P, Ioakimidis CS, et al. EVs and their impact to the electric grid in S. Miguel. In: Proceedings of international Conference on power engin.

[4] Chendur Kumaran R, Venkatesh TG, Swarup KS. Voltage stability – case study of saddle node bifurcation with stochastic load dynamics. Int J Electr Power Energy Syst 2011;33:1384–8.

[5] Tiwari R, Niazi KR, Gupta V. Line collapse proximity index for prediction of voltage collapse in power systems. Int J Electr Power Energy Syst 2012;41:105–11.

[6] Khani D, Sadeghi Yazdankhah A, Madadi Kojabadi H. Impacts of distributed generations on power system transient and voltage stability. Int J Electr Power Energy Syst 2012;43:488–500.
[7] Wu X, Hu X, et al. Stochastic control of smart energy management with plug-in EVs battery energy storage and photovoltaic array. J. Power Sour 2016;333:203e12.
[8] Dickerman L, Harrison J. A new car, a new grid. IEEE Power Energy Mag 2010;8(2):55e61.
[9] Wang Z, Fan S, et al. Coordinated charging strategy of plug-in EVs for maximizing the distributed energy based on time and location. J Eng 2017;13:1740e4. 2017.
[10] Burchett S, Douglas D, et al. An optimal thevenin equivalent estimation method and its application to the voltage stability analysis of a wind hub. IEEE Trans Power Syst 2018;33(4):3644e52.
[11] Impact of EVs fast charging on power system voltage stability[J]. International Journal of Electrical Power & Energy Systems, 2014.
[12] T. K. Rahman, G. B. Jasmon, "A New Technique for Voltage Stability Analysis in a Power System and Improved Loadflow Algorithm for Distribution Network", Proceedings of 1995 International Conference on Energy Management and Power Delivery, 1995. EMPD '95, Vol. 2, pp. 71 4 – 719.
[13] Li Meng, Cheng Haozhong, Yang Zonglin, et al. Improved forecasting method of typical daily load curve using fractal interpolation [J]. Journal of power system and automation, 2015

Design of the Power System and Remote Control System for the Unmanned Ship

Hao Zhang*, **Zeyu Zhang**

Wuhan Institute of Marine Electric Propulsion, Wuhan, China.

*Corresponding author's e-mail: 18995619539@163.com

Abstract. In lakes and reservoirs with high environmental protection requirements, it is necessary to adopt pure electric intelligent water quality monitoring ships, adopt pure electric power systems and intelligent remote control systems. This article specifically completed the design of the power system and remote control system of the smart ship, and elaborated the design method in detail. The system has the characteristics of high reliability, miniaturization, and modularization. The successful development of the system has greatly improved the intelligence of water quality monitoring ships, and has a positive effect on small intelligent ships in lakes and reservoirs.

1. Introduction

With the continuous improvement of emission requirements in inland rivers, lakes, ports and other fields, some environmental protection water areas have unmanned boats with electric boats for water quality monitoring and cleaning operations. As a new type of power ship with silent, environmental protection and zero emission, electric boats are gradually receiving industry attention. The use of electric boats has better quietness, environmental protection, is more environmentally friendly, and has a more comfortable experience.

The power system of the unmanned ship includes both the propulsion system and the remote control system. The power system is composed of the power supply system, the propulsion system, the remote control system, and the auxiliary system. The remote control system includes the propulsion control system of the ship and the remote control equipment on the mother ship. The propulsion control box distributes power to the load, and receives instructions from the mother ship to control the normal operation of the propulsion system and feed back the status of the propulsion system and power distribution system to the mother ship.

2. The Design Of Power System

In normal navigation conditions, the power supply system consists of two sets of 576V lithium battery packs (including BMS) and two power boxes. The lithium battery pack has a capacity of 100AH and a rated voltage of 576V. The lithium battery pack is centrally monitored and protected by BMS. The two power boxes take power from lithium batteries respectively, and provide power and protection for the ship through switches, DC/DC switching power supplies, fuses, etc., to meet the power supply of the branch control system. In shore power working conditions, the power supply system is provided by the mother ship with two power sources. At this time, the shore power is only used by the heater. During charging conditions, the charging interface box connect to the charging power supply cabinet on the mother ship to charge the lithium battery pack, which will be fully charged within 4 hours. The propulsion system includes a frequency converter (including the inverter controller CU320), a permanent magnet motor, and the propulsion system's power and control electricity are taken from the power box.

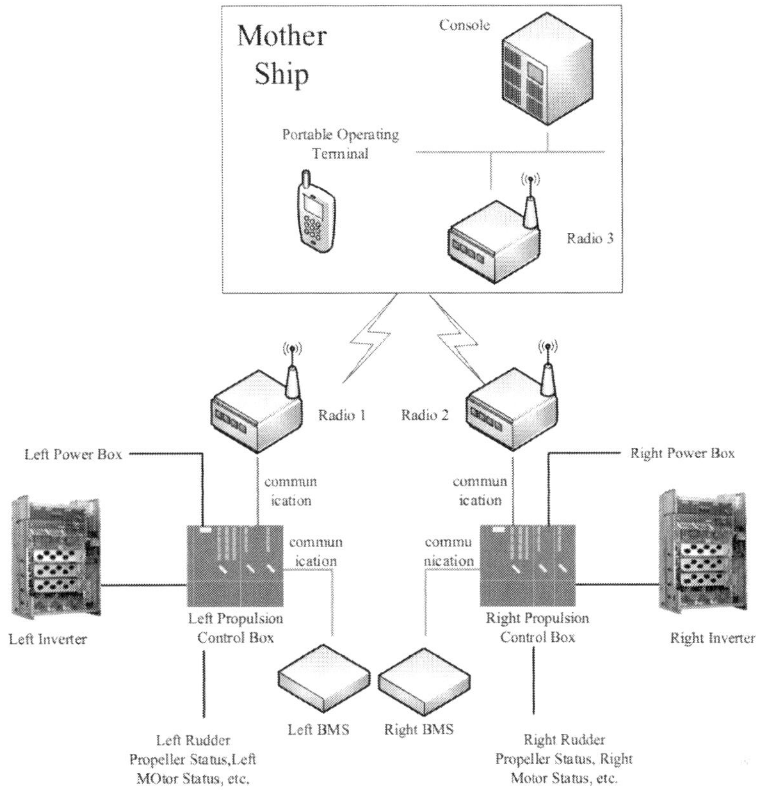

Figure 1 Block diagram of intelligent controller system

2.1 The main functions of the power system

The main functions of the power system are described as follows:

a) Power supply function

The power of the whole ship is provided by the direct current provided by two sets of battery packs for the load of the whole ship. Part of the DC power supplies power to the two sets of power propulsion branches, and part of it is inverted into AC power through the power box to supply power to the ship's daily loads. The power supply of the two branches of the cooling water pump, the data transmission station and the control system are independent of each other, and the failure of one power supply branch does not affect the normal operation of the other power supply branch equipment.

b) Propulsion function

The output power of the propulsion motor is limited by the inverter to ensure that the propulsion motor is not overloaded. The propulsion system also monitors the battery current and automatically reduces the propulsion speed when the battery current exceeds the set upper limit. In the speed control mode of this system, the frequency converter can adjust the speed of the motor according to the speed setting signal given by the control system. When a set of propulsion branches fails, the remaining set of propulsion branches can operate normally. The propulsion also has protection functions such as short-circuit, over-current, DC bus under-voltage, and motor overheating to ensure its own safety and reliability.

2.2 Design of the Battery System

The power system consists of a lithium battery system (including a lithium battery pack and BMS) and a power box. The power of the whole ship is provided by the DC power provided by two battery packs for the loads of the whole ship. One part of the DC power supplies power to the two sets of electric propulsion branches, and the other part is inverted into AC power through the power box to power the daily loads of the whole ship. The power supply is independent of each other, and the failure of one power supply branch does not affect the normal operation of the equipment of the other power supply branch. As the power source of the whole ship, the lithium battery system is composed of lithium battery pack and BMS[1].

At present, the development of lithium batteries, lithium iron phosphate batteries and ternary lithium batteries are two more important development directions, in terms of cycle times, lithium iron phosphate batteries and ternary lithium batteries are significantly better than other types of lithium batteries. The type of lithium battery is classified according to the battery's cathode material. Ternary lithium battery refers to the type of lithium battery composed of nickel, cobalt, and manganese as the cathode material of the battery[2]. Compared with lithium iron phosphate as the cathode material The battery has the advantages of higher energy density, higher battery cell nominal voltage, longer battery life, etc. The energy density of ternary lithium batteries is significantly better than that of lithium iron phosphate batteries. However, the thermal stability of the ternary lithium battery is quite different from that of the lithium iron phosphate battery, and the stability of the working environment is very high. After comprehensive comparison and consideration of the working environment of marine conditions, the lithium iron phosphate type lithium battery is selected, and the structure of the lithium iron phosphate battery has two different types, mainly cylindrical and square. The two installation types of lithium iron phosphate batteries form multiple boxes according to the requirements of the ship's capacity to meet the installation requirements.

3. Design of remote control system

3.1 Main functions of the remote control system

The left propulsion controller and the right propulsion controller are independent of each other, respectively monitoring the status of the two propulsion branches, and the main controller controls the power box switch status and radio communication.

The ship contains two digital transmission radio stations, which are independent of each other, and when one of them fails, the normal operation of the other one will not be affected. The mother ship contains a set of portable terminals and a set of radio stations and control stations. The operator of the mother ship can control the unmanned ship through the portable terminal or the control station. Operation location conversion: The location switch button on the control station can be used to switch between different operation locations. The priority of the control station is higher than that of the portable operation terminal. Multiple operation modes: support standby mode, joint operation mode, independent operation mode, etc. Different operation modes can be selected on the panel. Real-time monitoring information: The ship's relevant information can be monitored in real time on the display screen during operation, including BMS-related information, motor speed, motor power, bus voltage, equipment temperature, alarm information, etc., with power-off storage function, mother ship control station Historical faults can be recalled at any time, and no less than 1000 faults can be stored. The control station panel contains control buttons and indication buttons, which can indicate the real-time status of the system. Communication failure protection: When a communication failure occurs, the propulsion system will automatically stop and wait for the corresponding mother station operation instruction, which will not cause an out of control state. Hierarchical fault function: The unmanned ship control system is divided into two fault handling modes according to different fault levels, namely first-level fault (serious) and second-level fault (general). Different faults adopt different fault handling methods. The first level fault (serious) reports the fault and stops directly, the second level fault (general) reports the fault and limits the propulsion power. Manual emergency stop function: Both the control station and the portable operating terminal contain emergency stop buttons, allowing the operator to adopt the emergency stop mode.

3.2 Topology of the remote control system

The control system consists of a propulsion control box, a frequency hopping radio station, a mother ship control station and a portable operating terminal. The mother ship operating station transmits control information to the propulsion control box through the frequency hopping radio. The propulsion control box accepts control commands to control the start, stop, speed increase, and deceleration of the propulsion system. At the same time, the propulsion control box accepts BMS, frequency converter, motor, etc. The status information is communicated to the mother ship control station and the portable operation terminal through the radio. The portable operation terminal can be operated in the full deck area of the mother ship. The control station and the portable operation terminal support standby mode, joint operation mode, independent operation mode, etc., and different operation modes can be selected on the panel[3]. When a communication failure occurs, the propulsion system will automatically shut down and wait for the corresponding mother station operation instructions, without causing an out-of-control state. Both the control station and the portable operating terminal contain emergency stop buttons, allowing the operator to adopt the emergency stop mode.

3.3 Design of remote control system

3.3.1. Control station (mother ship)

The control station and the radio station realize data interaction through RS485 or RS422 communication, and have the control and monitoring functions of the unmanned ship, which specifically includes the following functions:

Operation location conversion: The location switch button on the control station can be used to switch between different operating locations. The priority of the control station is higher than that of the portable operation terminal; Multiple operation modes: support standby mode, joint operation mode, independent operation mode, etc., different operation modes can be selected on the panel; Real-time monitoring information: The ship's relevant information can be monitored in real time on the display screen during operation, including BMS-related information, motor speed, motor power, bus voltage, equipment temperature, alarm information, etc., with power-off storage function, which can be used at any time Recall historical faults, and store no less than 1000 faults. The operation station panel contains control buttons and indication buttons, which can indicate the real-time status of the system. Communication failure protection: When a communication failure occurs, the propulsion system will automatically shut down and wait for the corresponding mother station operation instruction, which will not cause an out of control state. Hierarchical fault function: The unmanned ship control system is divided into two fault handling modes according to different fault levels, namely first-level fault (severe) and second-level fault (general). Different faults adopt different fault handling methods. , The first level fault (serious) reports the fault and stops directly, the second level fault (general) reports the fault and limits the propulsion power. Manual emergency stop function: Both the control station and the portable operating terminal contain emergency stop buttons, allowing the operator to adopt the emergency stop mode.

3.3.2. Portable Terminal

The portable terminal can communicate with the radio by wireless communication, and can also exchange data with the radio in the form of RS485 or RS422 bus. It has the control and monitoring functions of the unmanned ship, which specifically includes the following functions:

Operation location conversion: After the operation authority of the control station is transferred to the portable terminal, the activation button on the portable terminal activates the operation authority of the portable terminal; Multiple operation modes: support standby mode, joint operation mode, independent operation mode, etc., different operation modes can be selected on the portable operation terminal; Real-time monitoring information: The ship's relevant information can be monitored in real time on the LCD screen during operation, including BMS-related information, motor speed, motor power, bus voltage, equipment temperature, alarm information, etc. It does not have the power-down storage function. Among them, the portable terminal is provided with key information indicators such as a fault alarm indicator, a running indicator, and a ready indicator[4];Communication failure protection: When a communication failure occurs, the propulsion system will automatically stop and wait for the corresponding mother station operation instruction, which will not cause an out of control state. Hierarchical fault function: The unmanned ship control system is divided into two fault handling modes according to different fault levels, namely first-level fault (severe) and second-level fault (general), and different fault handling methods are adopted for different faults , The first level fault (serious) reports the fault and stops directly, the second level fault (general) reports the fault and limits the propulsion power; Manual emergency stop function: Both the portable operation terminal and the control station contain emergency stop buttons, allowing the operator to adopt the emergency stop mode.

4. Conclusion

This paper designs the power system and remote control system of the intelligent water quality monitoring ship in lakes and reservoirs. The power system has the characteristics of high integration, intelligence and simulation. The remote control system greatly improves the intelligent level of the power system. . The design can be extended to other small cruise ships, official ships and other pure electric ships, which has a positive effect on the development of smart ships.

References

[1] Zhao Xilong, (2018) Multifunctional automatic aquaculture operation ship energy management system. [Master's degree thesis]. Jiangsu: Jiangsu University.

[2] Fu Xiang, Wang Honglei, Huang Bin, Yang Yuqiang, (2018) Analysis of BYD e5 Electric Vehicle Power Battery and Its Control Technology"[J]. Automobile and Driving Maintenance (Maintenance Edition), 8:156-157.

[3] Liu Yingjie. Power and control system selection of unmanned rescue boats, (2020). Ship Electric Technology, 3 :1-4.

[4] He Hao, (2013) Research and design of energy-saving vacuum suction and exhaust control system. [Master's thesis]. Hunan: Central South University.

Research and Application of Automatic Measurement of Resistance Value Based on a Thermal Control Device

Tao Feng [1], Lei Huang[1], Mingyu Guo[1], Zhen Xu[1], Guanghui Fu[1], Yue Yu[1]

Institute of Satellite Environmental Engineering in Beijing

Email: fengtao@cast.casc

Abstract: The resistance measurement system of thermal control device is mainly composed of four parts: fixture, interface circuit board, switches module and data acquisition system. The main contents of the system design are as follows: To make the system realize the automatic resistance measurement function of thermocouple, heating belt and heating plate, and improves the measurement speed and robustness of the software; To achieve the friendliness of human-machine interface;

At room temperature, it can collect 40 output voltage signals of thermocouple, and judge whether the temperature collection of thermocouple is accurate or not.

1. Introduction

This paper briefly describes the development background and the main requirements of the system, and summarizes the function completion of the system design, if the main technical indicators are consistent with technical agreements and some technical problems encountered in the process of system design, solutions and results.

2. Overall Design for the System

2.1 System Overview

The system design requires the ability to measure multiple components under test at one time. Generally, the number of input channels of a single resistance measurement module cannot meet the requirements of the technical specifications. Multiple resistance measurement modules can meet the design requirements of the number of channels, but multi-card control requires more control resources, and there is a high complexity on the measurement and control. Therefore, this system design uses a resistance measurement module, with a group of 48 thermocouple measurement equipment and a multichannel matrix switch card. When the resistance (thermocouple) is measured, the computer serial port of the measurement and control is used to send the measurement instruction to the resistance measurement module (thermocouple measurement equipment).The instruction is to specify the measurement information such as the measurement channel. The measurement module analyzes the measurement instruction, controls the corresponding detection channel through the DO switch matrix, establishes the measurement loop of the specified channel. Then the channel resistance value Rx is measured by the resistance measurement module, and the measured values are fed back to the measurement and control computer through RS232 (USB) communication interface to complete the measurement of a channel. The measurement and control computer issues the measurement and control instructions in turn according to the condition of the measured parts until all the measurement channels complete the measurement.

2.2 System Composition
The resistance measurement system of thermal control device is mainly composed of four parts: fixture, interface circuit board, switches module and data acquisition system. The main contents of the system design are as follows:

1). Optimize and improve the fixture structure of the existing automatic test system for thermistor acceptance, reduce the complexity of operation, increase the clamping function of heating belt and heating plate, and improve the clamping reliability of the fixture;

2). Optimize and improve the measurement software of the existing automatic test system for thermistor acceptance, and increase the thermocouple, heating belt, heating plate resistance measurement function through the optimization of software measurement algorithm to improve the measurement speed and robustness of the software.

3). Optimize and improve the man-machine interface of the existing measuring tools, and improve the user-friendly interface by improving the integration and reducing the weight of the tools and the number of parts;

4). Add measurement function for thermocouple. At room temperature, it can collect 40 output voltage signals of thermocouple, and judge whether the temperature collection of thermocouple is accurate or not.

2.3 Structural Design
The part of system structure is a design of integration, which integrates the measurement and control computer and its accessories, fixture, power module and resistance measurement module. The overall structure of the system is shown as follows:

Figure 1: Overall Structure of the System (PANSINO SOLUTIONS/ measuring system for thermal control device organization)

2.3.1 Thermistor and Thermocouple Fixture Design

The thermistor and thermocouple are connected by two scattered wires. The size of the scattered wires is about 3mm to 5mm. The system designs a special fixture for this connection.

2.3.2 Design of Heating Belt Fixture

The heating belt is different in size due to its mechanical size, and the measuring connection has three ways: two, four and six wire. The use of a thermistor fixture in section 5.2.2.1 is not convenient for the installation of the test piece. Therefore, the front panel of the system is designed with a six-core aviation socket and a six-core cable. [1]The other end of the cable has six flat-mouth clips, three in black and three in red and all of them are to be paired for use. The renderings and mechanical dimensions of flat mouth clamps are shown as follows:

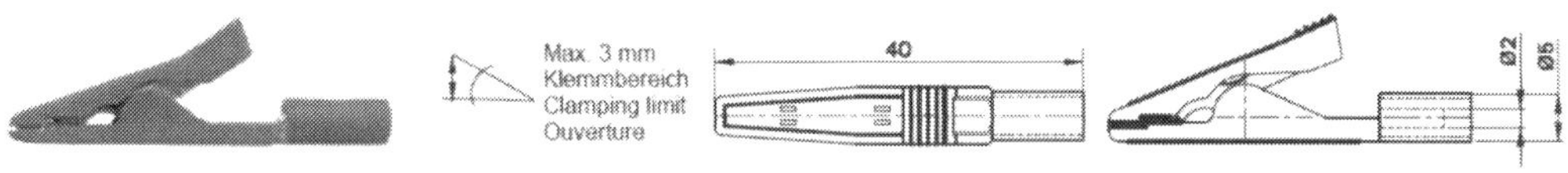

Figure 2: Effect Diagram and Mechanical Dimensions of Flat Mouth Clamp

When measuring, connect two ends of one heating belt to a pair of flat mouth clamps (black and red) respectively. [2] If there are three ways, connect three pairs of measuring cables respectively

2.4 Module Design for Resistance Test

The measured part and the measuring device are connected by cable. In order to accurately measure the resistance value of the measured part, the factors affecting the measurement accuracy such as the wire resistance value and the fixture contact resistance must be eliminated as far as possible. The system design uses four-wire to measure resistance, which can reduce the influence of the measurement route resistance on the measurement accuracy. For the lead resistance between the test piece and the test device, the lead terminal can be short-circuited to measure the lead resistance before measurement, and the lead resistance of each test piece can be deducted after the measurement.

2.4.1 Resistance Measurement

The schematic diagram of four-wire resistance measurement is as follows:

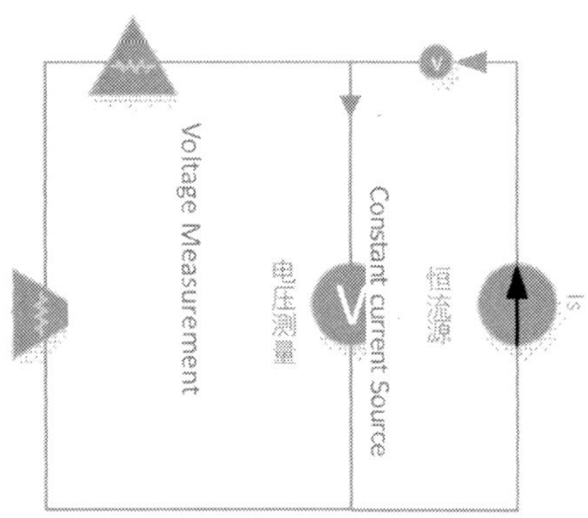

Figure 3: Schematic Diagram of Four-Wire Resistance Measurement

The constant current source Is outputs the excitation current, the current through Rx produces a voltage drop Ux, the voltmeter measures the voltage drop Ux at both ends of Rx, and then according to ohm's law Rx=Ux/Is. Since the RL is much smaller than the voltmeter input resistance, the voltage drop UL on the RL is negligible. The constant current source module outputs constant current excitation Is, and the output of Is varies slightly with the ambient temperature or the resistance value of the measured component. In order to accurately obtain the constant current excitation size in a single measurement, the system design measures the constant current excitation size of the constant current source output each time. [4]In this way, the influence of constant current output instability caused by environment and other factors on the measurement results can be avoided.

Based on the previous design experience of resistance testing hardware, it is estimated that the resistance measurement accuracy of the system is between ± 0.5%, which can meet the resistance measurement accuracy requirements of the Technical Specifications.

2.4.2 Design of Current Source

The constant current source circuit design is a voltage control current source based on dual operational amplifier structure of Howland current source, and as shown in the figure:

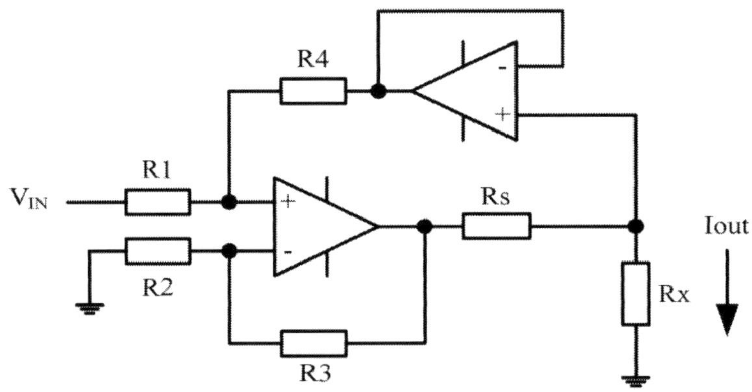

Figure 4: Schematic Diagram of Constant Current Source Circuit

When R1*R3=R2*R4, Iout=R4* V_{IN} /R1*Rs.

Stable test current output can be achieved by setting appropriate resistance value and selecting operational amplifier with excellent performance. Before the resistance test, the system measures the output current of the constant current source, and then the voltage at both ends of the bridge of the tested item is collected for several times, and some digital filtering is carried out. [3] Finally, the bridge resistance of the tested item is calculated according to ohm's law.

The system design chooses to use the high-precision DAC output voltage as Vin, which is used to generate different current excitation. The system control chip controls the DAC output through the SPI interface. According to the resistance value of the measured object, the control chip controls the DAC output matching voltage, and then generates matching excitation current.

2.4.3 Protection Circuit

The design idea is as follows: the current value applied to the measured resistance is monitored, and the high precision resistance Rs is used to convert the current signal into voltage signal. The voltage signal is fed into a high-speed voltage comparator with a pre-set reference voltage Vref. The output signal of the comparator is formed through the self-locking control circuit to form the relay control signal, which controls the on-off of the resistance test loop and feedback the control state signal to the upward machine.

When the test current on the resistance being measured exceeds the set value, the voltage on Rs increases and exceeds the reference voltage Vref. At this time, the output signal of the voltage comparator changes from low level to high level, and the high level signal triggers the operation of the self-locking control circuit. On the one hand, the self-locking control circuit locks the alarm state, and feedback the alarm signal to the MCU related IO; On the other hand, the output control signal is given to the reed relay on the test loop. The relay disconnects the test loop and the test current on the measured resistance is cut off.

The voltage comparator for protection in the current scheme selects high-speed AD8561R with the maximum delay time of 7ns, which triggers the relay for fast operation. Pickring's quick response reed relay 111P-1-A-5/1D was used as the relay.

2.4.4 Automatic Resistance Measuring Range

The resistance measurement module is used to measure the resistance of heating belt and thermistor. The resistance of the heating belt is in the range of about 100 ω, and the resistance of the thermistor varies according to the model at room temperature. The resistance measurement module needs to determine the appropriate measurement range according to the resistance value of the component under test.

The range selection method of resistance measurement: the resistance measurement module uses a small excitation current output to roughly measure the resistance value, select the appropriate measurement range according to the measured resistance value, and then accurately measure to get the final resistance value. The measuring range of resistance can be 0~100 ω, 100 ω ~1K ω, 1K ω ~100K ω, and 100K ω ~1M ω.

2.4.5 Zero calibration of resistance measurement

After the system is powered on and the testing software is started, the system automatically calibrates the zero point of resistance measurement.

Zero calibration method: the system design a standard reference channel CHref. When the equipment is inspected, the loop resistance Rref of CHref is steadily measured, which is used as the reference. [5]The relative difference Δrx between loop resistance Rx and CHref in all other channels was measured and recorded.

The loop resistance of each channel is mainly affected by the temperature, and the relative difference between each channel is basically unchanged. Therefore, the resistance Rref of the reference channel Chref is only checked once before each measurement, and then the loop resistance Rx of each channel is calculated from Δrx and is subtracted from the resistance values measured later.

2.5 Module Design for Thermocouple Measurement

The system uses NI 9213 thermocouple measurement kit, which includes a 16-channel thermocouple measurement module and a 4-slot CompactDAQ case, to complete thermocouple measurement. The measurement kit communicates with the measurement and control computer through the USB interface of the chassis to exchange measurement instructions and data.

The main technical specifications of the kit are described as follows:

- Differential input channels: 16 channels
- Resolution: 24bit
- Maximum voltage range: -78.125mV~78.125mV
- Accuracy of voltage measurement：38μV
- Sample frequency：78Hz

Except for the number of channels, other main performance indicators of the kit meet the requirements of system design and technical indicators. System requirements require that the number of system design channels is not less than 40. Therefore, three sets of thermocouple measurement channels with a total of 48 channels are selected in this scheme to meet the system design requirements.

2.5.1 Line-connecting Method.

The NI 9213 can be connected to the thermocouple input signal. The positive thermocouple is connected to the TC+ terminal, and the negative thermocouple is connected to the TC- terminal. It connects the shield layer to the COM end when using a shielded thermocouple. The shielded end is connected to the common mode reference voltage of the thermocouple. Common-mode reference voltage refers to the voltage within the range of the thermocouple common-mode voltage ±1.2V. When the floating thermocouple or the potential difference between the thermocouple and ground is within ±1.2V, ground the COM terminal and the shield terminal.

2.5.2 Cold End Compensation and Measurement Error

The NI 9213 measurement module multiplexes 16 thermocouple input channels, 1 cold end compensation (CJC) channel and 1 automatic zeroing channel to ADC.

Devices or heating devices near the NI9213 will increase the terminal temperature of the NI9213, that means the terminal temperature will be different from the temperature of the cold end compensation sensor, which will affect the measurement error of the thermocouple. Terminal thermal gradient can lead to terminal temperature differences. In this case, the measurement result not only has the determining measurement error, but also has the relative accuracy error between channels. Therefore, it is necessary to minimize the influence of thermal gradient on the measurement results. The system mainly includes the following points:

1). The system needs to be preheated for 15 minutes before measurement;

2). Try not to measure the thermocouple in an overly cold or overheated environment;

3). Minimize the flow of heat or air around the terminals and keep the ambient temperature as stable as possible;

4). Use a smaller diameter cable to complete the thermocouple conductor.

2.5.3 Thermocouple Qualification Judgment

The system thermocouple measurement is performed at room temperature. Multiple temperature points cannot be measured. It is impossible to determine whether the thermocouple is working on its temperature curve. In order to initially judge whether the thermocouple works normally, the system chooses to use DS18B20 temperature measurement module to measure the room temperature as the standard temperature value T0. Then the output voltage of the thermocouple is measured and converted to the corresponding temperature Ttc. Comparing T0 with Ttc, the thermocouple is considered to work normally if the following relationship is satisfied:

$$|(Ttc - T0) \div T0 \times 100\%| \leq 10\% \quad \textbf{OR} \quad |Ttc - T0| \leq 2^{\circ}C \qquad (1)$$

2.5.4 Multi-way Switchover

The system needs to be able to collect 40 thermal control devices, including thermistor, heating plate, heating band resistance and thermocouple output voltage values

The thermistor, heating plate and heating belt are resistive devices. Their resistance value under normal temperature conditions needs to be measure only. The system design uses custom resistance test cards for measurement; The output voltage system of thermocouple is measured by NI 9213 thermocouple test module. Since the output of thermistor and thermocouple will not be measured at the same time, the system design uses a set of fixtures. According to the different measurement objects, the data acquisition equipment corresponding to the measurement objects is selected through the multi-way switch.

The tested components are connected to the system through fixtures. According to the different types of the tested objects, the software controls the switch matrix to switch and select, and the tested components are connected to the corresponding acquisition equipment. In the figure above Thermocouple Measurement Module consists of 3 NI 9213 thermocouple measurement cards, and each card has 16 channels, so there is a total of 48 channels. 48 thermocouple output voltages can be measured simultaneously. Resistance Measurement Card is a custom resistance test card and multi-way switching card composition. The resistance test card consists of a resistance measurement path and a 12 - way multi - way switching unit and can measure up to 48 channel resistors through four sets of 12 switch switches each.

3. Software Design

3.1 Software Function

The system software mainly includes four modules: measurement configuration, data acquisition, data management and file management. Each module separately contains several functional modules.

3.2 System Software Functions
The system software mainly consists of four parts. The following sections describe in detail the functions of each component, organizational unit and the detailed design scheme.

3.2.1 Measurement Configuration
The software measurement configuration function consists of two functions: parameter import and export and parameter configuration. The system configuration parameters are as follows:

- Model code: satellite model code;
- Product code: code of the product under test;
- Measurement time: start time based on the system time;
- Location of survey: Record the location of survey;
- Surveyor: The name or code of the person who performs the measurement;
- Qualification criteria: resistance measurement qualification criteria and thermocouple qualification criteria;
- File storage path setting: includes configuration parameter path, test data storage path, and report generation path.

Parameter Import and Export Function

Parameter import: the user can edit and save the configuration file in Excel. The software reads the specified configuration file, loads the configuration parameters into the system. And it only needs to select the specified model code and product code when measuring.

Parameter export: the user can edit parameters on the parameter configuration interface. After editing, click "Save". The software will update or add the configuration parameters to the configuration file for the next import.

Parameter Configuration

Software design for parameter configuration interface is used to import parameters or add parameters to edit. This interface has the following functions:

- Parameter loading: The user selects the specified code product, loads the existing configuration data, and can edit it;
- New parameters: The user can add and edit a new set of configuration parameters for new product measurements;
- Parameter storage: After editing the parameters, save the new parameters to the configuration file and update the software configuration data

3.2.2 Data Collection
Data acquisition module is mainly composed of: thermocouple measurement, resistance measurement (thermistor and heating band) measurement, matrix control, measurement display, qualification judgment, sound and light alarm and data storage and other functional units.

Thermocouple Measurement

The thermocouple measurement unit mainly performs the following functions

- NI 9213 self-test, zero calibration;
- Measure the thermocouple temperature of the specified channel according to the measurement configuration;
- Collect the RTD temperature data;
- Judge whether the thermocouple is normal according to the measurement parameters, highlight the state of the product, and make sound and light alarm when defective products occur;
- Store measurement data.

Resistance Measurement

The resistance measurement unit mainly performs the following functions:

- Resistance measurement hardware self - check, zero calibration;
- Measures the resistance value of the resistance of the specified channel according to the measurement configuration,;
- The resistance measurement process is divided into three main steps:
  - Output minimum test current, preliminary measurement resistance value, and adjust the measurement module gear according to the resistance value;
  - Output suitable excitation current, delay appropriate time according to the different gear selection, and wait for the measurement circuit voltage stability;
  - Collect the resistance value of the measured resistance frequently, and the sampling results are processed to obtain the stable and effective resistance value
- Judge whether the resistance value is normal according to the measurement parameters, highlight the state of the product, and make sound and light alarm when defective products occur;
- Store measurement data

Matrix Control

According to the measurement selection, the software sends a switch command to the matrix switch module to switch the measurement signal to the thermocouple measurement hardware or the resistance measurement hardware.

Qualified Judgement

The qualification is judged according to the measurement parameter configuration for each measurement of the thermocouple temperature value or resistance value of a channel, and output the indication information whether the status is qualified or not.

Sound-light Alarm

If the qualified Judgement determines that a product meets the parameter requirements, it is a defective product. The software controls the acousto-optic alarm device to output acousto-optic indication signal, and flashes the measured data of the channel on the interface.

Data Storage

After the measurement is complete, the measurement data is saved to the specified file. If no file is specified, the file is saved with a specific string as the name. The data is also recorded in the database.

3.2.3 Data Management

Database Management

During the software installation, the system database and corresponding data tables are automatically created. The software provides a database management module, where the user can create new tables or delete existing tables.

Data Query

The users can query the test data of a specific product through the data query interface. Or they can specify search conditions to query test data that meet the search conditions.

3.2.4 File Management

System Log

During the running of the software, the system automatically records work logs in the background, including measuring basic information and system running status. When a fault occurs, the system can be queried and located. The system log file is in "*. Log" format and can be opened using an application program such as Notepad.

Configuration Files

Users can edit and input parameters through the parameter configuration function unit, update the existing parameters or add new parameters, and update the configuration file.

Data Report

The Software designs the generation interface for data report. The user can generate the measurement data report of the specified product through this interface. After the measurement report is completed, the software stores it in the specified path and stores it as "a *. xls" file with the specified name or customized name.

4. Test Application

After testing, the system design function is complete and is in line with the functional requirements of the project. Thermistor measurement, thermocouple measurement, heating sheet measurement and room temperature measurement accuracy meet the technical requirements.

No.	Path	Standard Resistance Value R0(kΩ)	Measure Resistance Rx(kΩ)	Error =(Rx-R0)/R0*100%
1	1	1	1.01	1%
2	2	1	1.01	1%
3	3	1	1.01	1%
Note: If all measurement errors meet the Technical Specifications, the test passes.				

Figure 5: the functional requirements of the project

5. Conclusion

At present, the research and application of this set of system fill the blank in the field of thermal control device resistance value on batch automatic measurement and increase the accuracy of resistance value measurement, and save the time and labor cost of batch measurement. There is a broader market prospect at the aspect of application.

References

[1] Ying X. (2008). Design of Hardware Circuit for Measuring Temperature Characteristic Curve of Thermistor, Journal of Modern Electronics Technology., 16:10.

[2] Xi H.l., Zhao M.F. (2010). A Brief Description of Temperature Measurement method using Thermistor, Journal of Xi Science and Technology., 5:4.

[3] Lin X.L., Huang M.X., Zhang X.W.(2004).Research on thermistor Measurement method, Journal of Reliability and Environmental Testing of Electronic Products., 1:12.

[4] Fan X.F., Guan C.B., Zhang L., Li H.J., Liu Q.H., Sun L.Z. (2021).Development of the Supply unit of the Active Potential Control Test Device, Journal of Spacecraft Environmental Engineering., 1:83.

[5] Liu R., Chen J.G.(2021).Design and Implementation of Low Power Target Detection System Based on Wireless Sensor Network, Spacecraft Environmental Engineering.,2:218

Research on Key Technology of Tile T/R Module Design

Yu Zhang, Hailong Zhao*, Xuefeng Song and Zenglu Li

The 13th Research Institute, CETC, Shijiazhuang, Hebei, 050051, China

*Corresponding author's e-mail: zhaohl@cetc13.cn

Abstract. Tile T/R module has the characteristics of miniaturization, lightweight, high integration and high performance, which are very difficult to develop. This paper studies the key technologies involved in the development of T/R module, and focuses on the key technologies such as high integrated multifunction chip technology, high performance Radio Frequency (RF) 3D interconnection technology, high integrated multichannel RF design technology and thermal design technology. Through research and analysis, a design scheme of tile T/R module with miniaturization, high performance and high reliability is proposed.

1. Introduction

T/R module is the core part of millimetre wave system. It is widely used in electronic communication systems such as active phased array radar, satellite navigation communication and electronic countermeasures. The performance of T/R module will directly affect the technical indicator of the whole system. Improving the performance of T/R modules and reducing volume, weight and manufacturing cost are the continuous goals of T/R module technology research [1-2].

In the development of millimetre wave circuits and systems, T/R module shows the development trend of miniaturization, high integration, high performance and multichannel. The tile T/R module developed on the basis of 3D high density integration technology [3-4], has incomparable advantages in reducing module cost, volume and weight compared with the current block T/R module, and is easier to array integration.

The development of tile T/R module involves many key technologies, which is very difficult. Fortunately, in recent years, some tile T/R modules realized by 3D packaging technology have appeared in China. In 2019, Liu Weiqiang of the 20th Research Institute of China Electronics Technology Group Corporation presented a multichannel, low cost and high density broadband tile T/R module integration scheme [5]. In 2020, Yu Lei of Science and Technology on Electronic Information Control Laboratory introduced the design of a 3D broadband tile T/R module that can be applied to phased array system [6]. In 2021, Wang Qingyuan of the 13th Research Institute of China Electronics Technology Group Corporation designed a four channel tile type 3D integrated T/R module for radar by using MEMS bulk silicon 3D heterogeneous integration technology [7].

The design of tile T/R module needs to meet the requirements of miniaturization, high performance, good heat dissipation and good electromagnetic compatibility, etc. Therefore, the key technologies of module design lie in the aspects of high integrated multifunction chip technology, high performance Radio Frequency (RF) 3D interconnection technology, high integrated multichannel RF design technology and thermal design technology, etc. as shown in Figure 1.

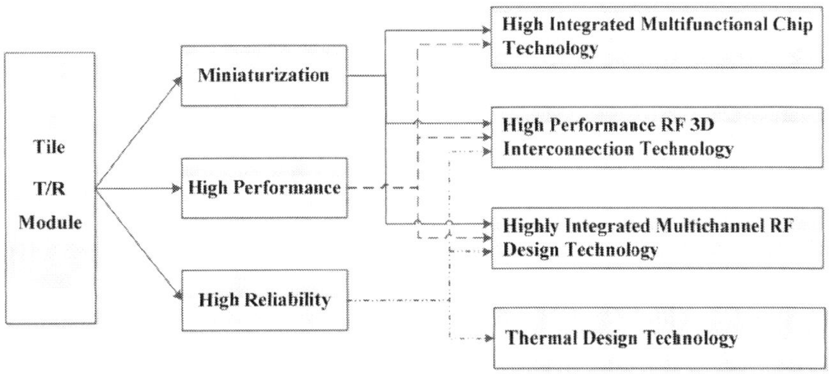

Figure 1. Key technology of tile T/R module design

2. High integrated multifunctional chip technology

2.1. Design of four channel multifunctional chip

The difficulty in the design of four channel multifunctional chip lies in the design of high precision amplitude and phase indicator under high integration. Relying on the multifunctional RF chip design, process, detection and reliability platform for T/R, high precision amplitude and phase control circuit, low gain temperature variation technology, power amplification and other technologies are studied. At the same time, separate external power modulation circuits and serial-parallel conversion circuits which are usually required in T/R channels are integrated to complete the product development. It greatly reduced the production and manufacturing costs of traditional analog T/R modules, and greatly saved the volume of modules, as shown in Figure 2.

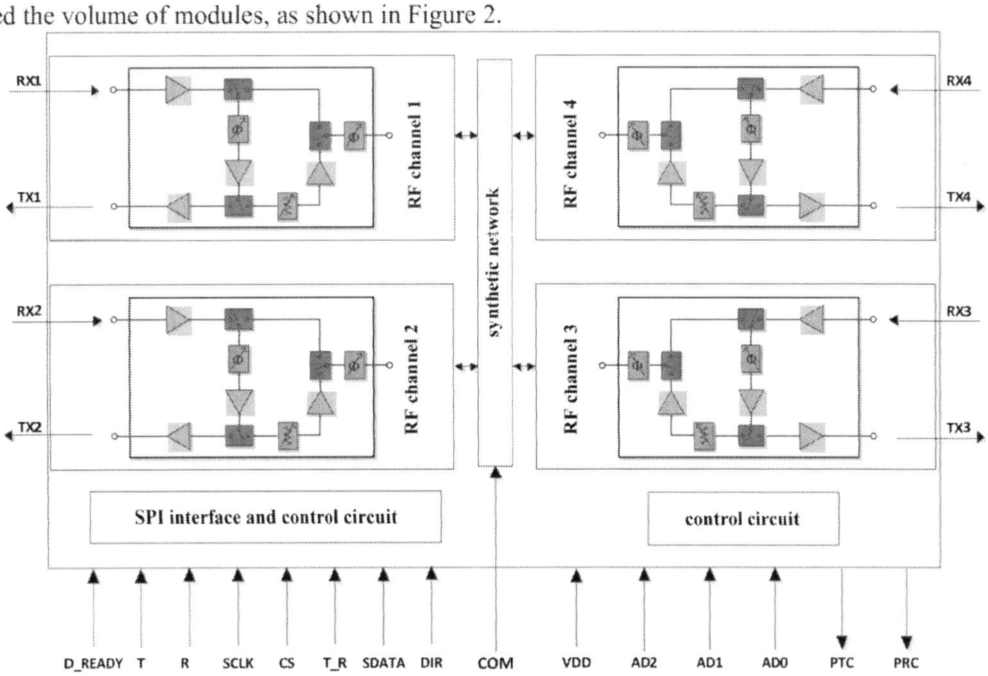

Figure 2. Schematic diagram of four channel chip

2.2. Integrated design of multifunctional chip

The integrated design of multifunctional chip includes low noise amplifier and switch matching technology and switch and power amplifier large signal matching technology.

The application of high integrated multifunction chip effectively saves space, which is a necessary means for the development of tile T/R module.

3. High performance RF 3D interconnection technology

The core of tile T/R is the high density assembly and interconnection of chips and components. The electromagnetic field structure presents 3D characteristics and the size is reduced. At present, the assembly density of 2D Multi Chip Module (MCM) has reached the theoretical limit, and the packaging density of 3D-MCM based on 2D-MCM has been greatly improved. The advantages of 3D-MCM mainly include: ① The interconnection between components is shorter, which can obtain lower digital delay and smaller parasitic parameters; ② MCMs are stacked and interconnected along the vertical direction, which can obtain greater assembly density and realize the miniaturization of the system; ③ It can reduce the system noise, reduce the overall power consumption and improve the transmission rate.

For 3D integrated microwave circuits, the key to forming 3D structure is how to realize the vertical interconnection between planar microwave circuits. The main microwave vertical interconnection methods used in this design include RF connector interconnection and new vertical interconnection between boards, which has the advantages of small space, low insertion loss and easy installation.

(1) RF connector

Compared with fuzz buttons and other currently developed vertical interconnection methods, RF connectors occupy less space, have better performance in high frequency and broadband modules, have low cost and easy assembly, and provide convenience for repair. The traditional vertical interconnection structure is that the RF connector is sintered on the box body, and the pin of the connector is connected with the strip line by soldering tin, bonding gold wire or gold belt. The amount of solder is difficult to control during welding, and the amount of solder directly affects the output stationary wave of the cascade. Therefore, the mode of bonding gold belt is used in this paper.

(2) Vertical interconnection between boards

For 3D integrated microwave circuits, the key to forming 3D structure is how to realize the vertical interconnection between planar microwave circuits. Vertical interconnection refers to the interconnection required for the design of power supply, grounding and inter-layer signals in 3D modules. Interconnection should not only ensure the integrity of microwave signal, but also have the characteristics of simple structure. The 3D vertical interconnection technology adopted by this module is substrate vertical interconnection. The simulation and results of microwave substrate 3D vertical interconnection are shown in Figure 3 and Figure 4.

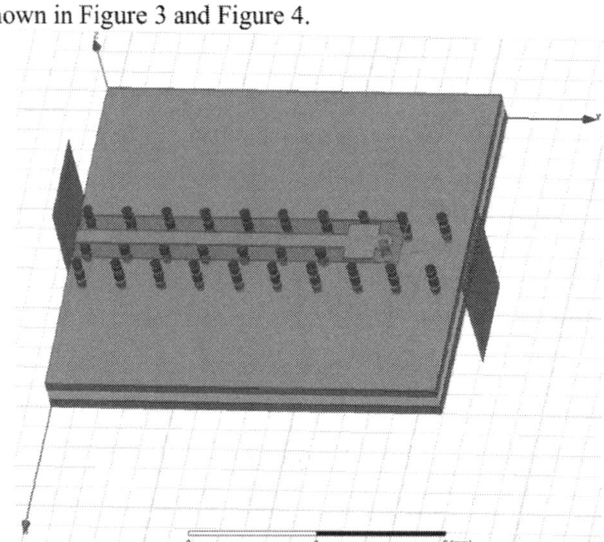

Figure 3. Simulation of 3D vertical interconnection of microwave substrate

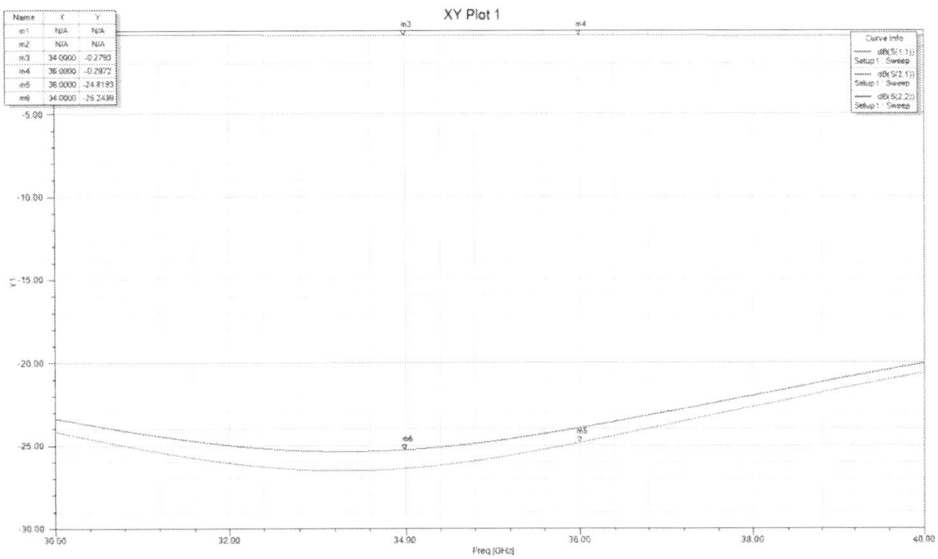

Figure 4. Simulation results of 3D vertical interconnection of microwave substrate

The upper and lower cavity microwave interconnection adopts the way of serial holes on the back of the microwave board and interconnection between plates. Although coplanar waveguide is used, there are still some risks in microwave broadband performance. The local circuit is modelled and simulated for microwave discontinuity. The simulation results are shown in Figure 5 and Figure 6, which can meet the performance requirements of modules.

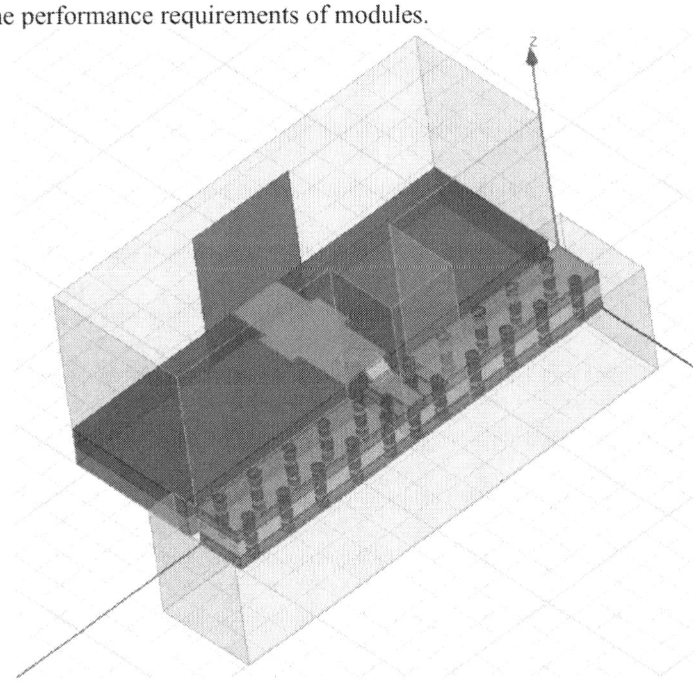

Figure 5. Microwave discontinuity simulation

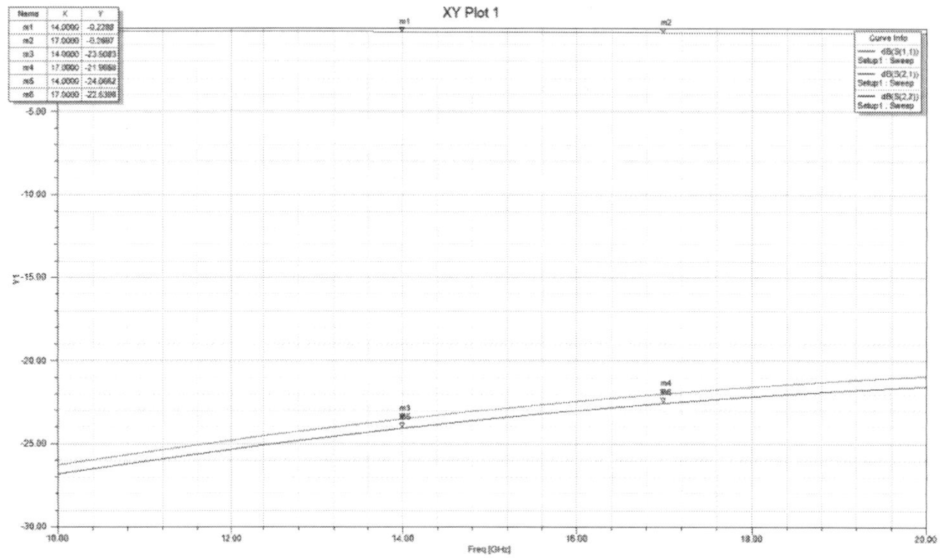

Figure 6. Microwave discontinuity simulation results

4. High integrated multichannel RF design technology

The solution of miniaturized multichannel RF Design of high density integration is to adopt multilayer interconnection substrate technology. The key technologies involved include MCM multilayer wiring technology, high density assembly technology and multichannel isolation technology.

4.1. MCM multilayer wiring technology

Multilayer board and multilayer wiring is the basic way to realize miniaturization. We must start with design and simulation to optimize circuit design and reduce parasitic effect and crosstalk between circuits. For the process manufacturing, the process parameters should be strictly controlled to improve the consistency of the performance of multilayer microstrip circuits, so as to facilitate the correction and modification of the design.

4.2. High density assembly technology

At present, it has a variety of micro assembly processes: thin film, thick film, rack bonding, bonding, mounting, welding, spot welding, welding, etc. The assembly process is directly related to the reliability of the module, so the process route suitable for the module should be formulated on the basis of the existing mature process technology.

4.3. Multichannel isolation technology

The isolation of module circuit design is realized through reasonable layout. In the layout of the overall circuit, the digital circuit, analog circuit and RF circuit can be partitioned to reduce the coupling of electric field and magnetic field and achieve the effect of circuit isolation. Adding ground wall isolation between RF modules to isolate the electric field coupling path between modules can significantly reduce the coupling between circuits. When the common power supply of multiple circuit modules is grounded, the static current in the circuit will produce a voltage drop on the power line. For RF circuits, the dynamic current on the power supply will also form a dynamic voltage drop, resulting in crosstalk between modules. In this design, each amplifier is equipped with a separate power modulation circuit, and the in-chip filter capacitor is used to achieve the purpose of power isolation between amplifiers.

In the module structure design, the partition wall is designed in the key circuit with high gain in the cavity, which can effectively avoid signal crosstalk. There shall be no gap between the height of partition wall and box cover to prevent the formation of resonant cavity.

5. Thermal design technology

The thermal design of tile T/R module is an important part of tile module design. When designing heat dissipation, it is necessary to consider how to further improve the transmission efficiency of T/R modules and optimize the heat dissipation performance of the module while meeting the requirements of output power. The chips and devices in the T/R module have the highest withstand temperature. When the temperature of some chips increases by 10℃, the failure probability will increase by an order of magnitude. When the temperature of the chip is higher than the maximum temperature required for the chip to work normally, the chip may be permanently damaged and the whole T/R module may fail.

The heat dissipation mode of T/R module is usually comprehensively selected by measuring the power per unit area of each chip in the module. By comparing the power consumption and working state of components, the device can meet the heat dissipation requirements while working and make the structure compact. The thermal design is mainly carried out in the following ways:

(1) Optimize the power chip, reduce the thermal resistance of the chip, and improve the output power and efficiency;

(2) Select the metal aluminium alloy shell material with high thermal conductivity, and weld the power single chip to the molybdenum copper carrier as the heat sink to effectively transfer heat;

(3) Evenly arranged the heating devices through reasonable circuit layout and process design;

(4) By studying the heat dissipation measures and heat dissipation paths, ensure the metal under the heating device is continuous and complete, so that the heat can be transmitted in time;

(5) Verified by thermal analysis software. Build the thermal analysis model by finite element analysis software and perform thermal simulation. By reducing the junction temperature of the semiconductor chip through the thermal design, the reliability of the circuit can be better improved.

6. Conclusion

In order to meet the development requirements of miniaturization and high integration of active phased array radar, tile T/R module needs to have the characteristics of miniaturization, high performance and high reliability, etc. The development process involves a variety of key technologies. Through high integrated multifunction chip technology, high performance RF 3D interconnection technology and high integrated multichannel RF design technology, the integration of T/R module is improved, the volume and weight of T/R module are reduced, and the performance of T/R module is improved. Through high performance RF 3D interconnection technology, high integrated multichannel RF design technology and thermal design technology, the structure, layout and packaging process of T/R modules are optimized and the reliability of modules is improved.

References

[1] Zhang G Y, Zhao Y J. (2006) Phased Array Radar Technology. Publishing House of Electronics Industry, Beijing.
[2] Hu C M, Zhou Z P, Yan W. (2010) Technology of T/R Module for Phased Array Radar. National Defense Industry Press, Beijing.
[3] Hauhe M S, Wooldrige J J. (1997) High Density Packaging of X-Band Active Array Modules. IEEE Trans on Components, Packaging, and Manufacturing Technology, 20(3): 279-291.
[4] Schreiner M, Leier H, Menzel W, et al. (2003) Architecture and Interconnect Technologies for a Novel Conformal Active Phased Array Radar Module. IEEE MTT-S Microwave Symposium Digest, pp: 567-570.
[5] Liu W Q, Ju J C, Guo C. (2019) Design on Multichannel Tile-Type T/R Module of Ku-Band. Modern Navigation, (2): 136-141.
[6] Yu L, Wu C Y, Jie H, et al. (2020) The Design and Implementation of Wideband Tile T/R Module. Electronic Information Warfare Technology, 35(1): 68-71.
[7] Wang Q Y, Wu H J, Zhao Y, et al. (2021) A 3D Integrated T/R Module with Bulk Silicon MEMS Technology. Semiconductor Technology, 46(4): 300-336.

Research on Pump Product Intelligent Fault Diagnosis System Based on Dig Data

Yimei Xu[1,*], Xiaoqing Xu[2]

[12] School of Information Engineering, Xi'an Eurasia University, Xi'an, Shaanxi, 710065, China

[*]Corresponding author's e-mail: xuyimei@eurasia.edu

Abstract. With the integration of information technology and manufacturing industry and the improvement of equipment monitoring data and computer computing ability, equipment fault diagnosis has entered the era of "big data". Using big data analysis method for fault diagnosis, the fault diagnosis model can learn its own characteristics and complete fault identification, so that the fault diagnosis is more intelligent and automatic on the basis of high identification accuracy. This paper mainly studies the application of fault diagnosis method and big data analysis method in pump product fault diagnosis.

1. Introduction

Today, the world has set off a new round of industrial transformation with the transformation and upgrading of manufacturing industry as the primary task. The deep integration of the new generation of information technology and manufacturing industry will promote the service transformation and product upgrading in the industrial field and reshape new competitive advantages[1].

As an important part of water conservancy and hydropower system, pump products have higher and higher requirements for safety and reliability. Most of the relevant national maintenance standards are very conservative. With the improvement of equipment design and manufacturing technology, material technology and operation scheduling technology, the relevant specifications and standards seriously do not adapt to the actual situation of the project. The implementation of the specifications leads to a large number of "over maintenance" problems in the industry. Frequent unnecessary disassembly and assembly not only destroys various gaps between dynamic and static running in of the unit, Moreover, most of the maintenance and disassembly are not as professional and standard as the construction and installation, resulting in inadequate reassembly and gap guarantee, forming potential safety hazards, affecting the service life of the unit, unplanned shutdown, resulting in major economic losses and even casualties[2].

Relying on big data and integrating high-quality upstream and downstream resources of the industrial chain, study fault analysis and diagnosis methods based on big data, and strive to establish an intelligent fault diagnosis system with independent learning ability, which is an effective means to prevent accidents and unplanned downtime, the development direction of equipment maintenance, an important part of enterprise intelligent operation and maintenance, and the implementation of "Made in China 2025" Development planning.

Journal of Physics: Conference Series **2121** (2021) 012024 doi:10.1088/1742-6596/2121/1/012024

2. Failure diagnosis method

2.1. Knowledge driven fault diagnosis method
Knowledge-driven fault diagnosis method does not need quantitative mathematical model. It is based on mechanical principle knowledge and human experience knowledge. It can not only introduce more information about the diagnosis object, but also make full use of the experience knowledge of human experts. Therefore, this method is suitable for fault diagnosis of non-linear systems and complex large-scale systems. Depending on the structure, it can be divided into qualitative model-based approach and symptom-based approach. Methods based on qualitative model include sign digraph method, fault tree method and qualitative simulation. Symptom-based methods include expert system method, fuzzy theory method, etc[3].

Knowledge-driven fault diagnosis method requires a great deal of mechanical principle knowledge and expert knowledge as support, and most of them combine reasoning mechanism formed by expert system with parameter estimation and state estimation. However, there are obvious shortcomings:

- It is difficult to update the knowledge base.
- Inference mechanism has poor flexibility and is difficult to adapt to the updated knowledge base.
- Manual operation is required and human-machine interaction is complicated.

2.2. Data Driven Fault Diagnosis Method
Because the I/O signal model of the system has certain connection with the fault source in terms of amplitude, phase, frequency and correlation, when fault occurs, some signal processing methods and feature extraction methods (such as various spectrum analysis methods, correlation analysis, time series feature extraction methods, adaptive signal processing methods and information fusion) can be applied for fault detection and separation. It does not need to establish an accurate model of the object, avoids the difficulties in establishing the model, so it has strong adaptability.

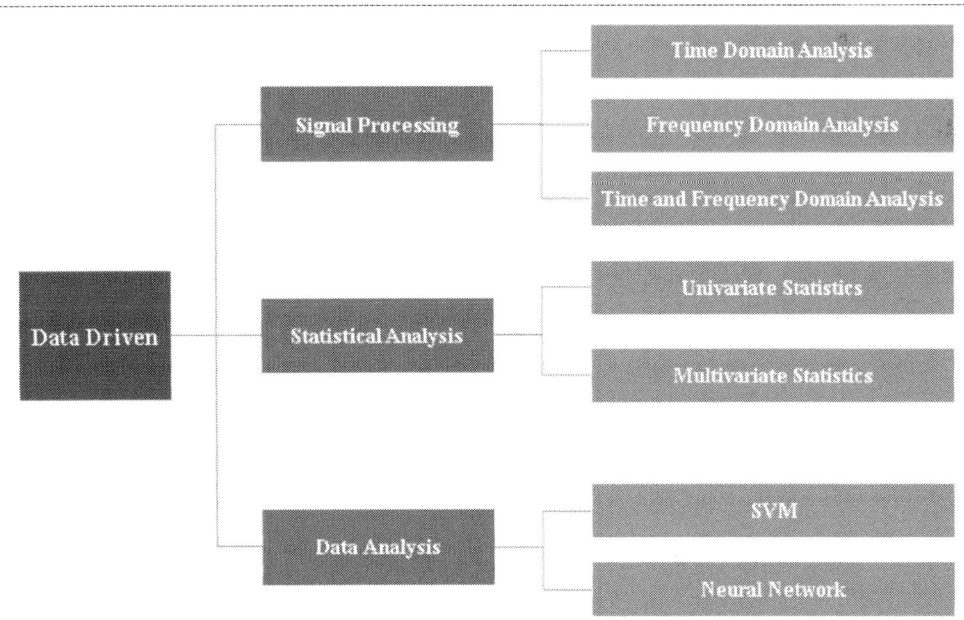

Figure 1. Classification of data driven fault detection methods

2.3. Value-driven fault diagnosis method
With the continuous increase of industrial monitoring data and the continuous improvement of computer

computing ability. In the development of industrial 4.0, the deep learning method has been well inherited and developed in the field of fault diagnosis. Many scholars attempt to use the deep learning model to evaluate the operation status, diagnose faults and forecast the life of industrial equipment.

The method of equipment fault diagnosis based on deep learning mainly trains the deep learning model in the way of supervised learning according to the data monitored in different states during the equipment working process, and then uses the trained model to diagnose the fault. The general process includes data acquisition, data preprocessing, deep learning model building, deep learning model training and fault diagnosis using the model. During data acquisition, one-dimensional data such as vibration, temperature and other data can be collected, as well as two-dimensional data in the form of pictures. Data preprocessing mainly includes data cleaning, data integration, data protocol and data conversion. In order to obtain sufficient data training model, data enhancement may also be included. In the process of constructing deep learning model, network structures such as NN, RNN and ANN are widely used at present. In the process of model training, select the appropriate optimization function and adjust the corresponding data parameters to get an accurate diagnosis model[4].

3. Fault diagnosis system based on big data

Knowledge driven and data-driven fault diagnosis methods mainly rely on the richness of domain knowledge, the accuracy of diagnosis model and the completeness of data samples. These methods have the advantages of simplicity, interpretability and easy development, but they are easy to be disturbed by the environment. In the face of large-scale nonlinear and non-stationary complex systems, their effect is not satisfactory, and the interaction between data signals of complex coupled systems is variable. At the same time, with the exponential growth of monitoring data, the research on fault diagnosis is facing the great challenge of processing industrial big data.

The fault diagnosis method based on value driven industrial big data background can reduce the use of expert experience and reduce the cost of manual participation in the diagnosis process, improve intelligence and enhance the accuracy of fault diagnosis.

3.1. System architecture design

This topic builds an end-to-end fault detection system. The framework structure of pump product fault diagnosis system is shown in Figure 2.

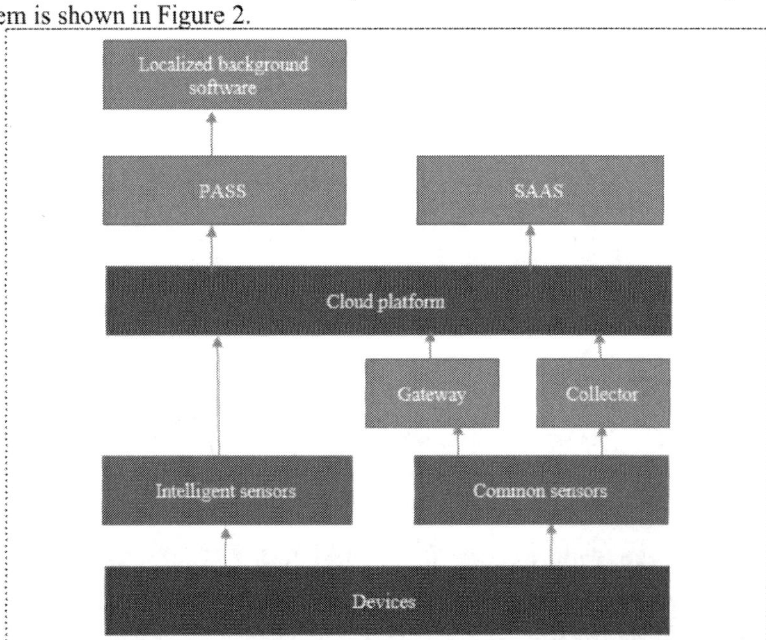

Figure 2. Framework structure of pump product fault diagnosis system

3.2. Design of fault diagnosis module

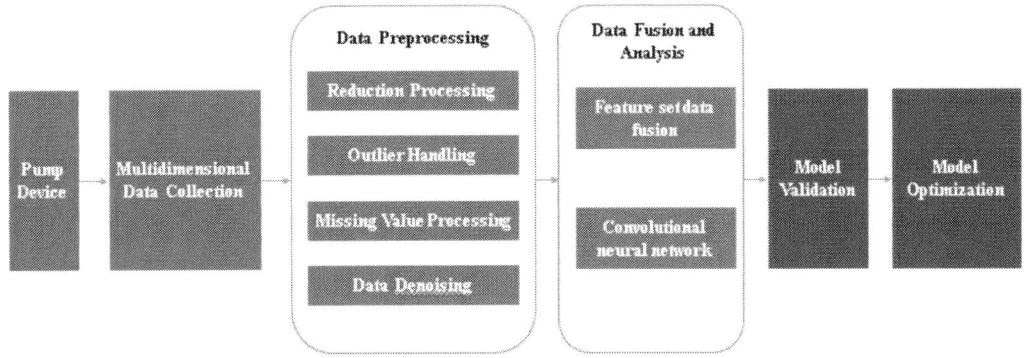

Figure 3. Fault data processing module architecture

Using the data integration and fusion technology and deep learning technology in big data analysis technology, find the relationship between multidimensional data by analyzing one-dimensional time domain, two-dimensional time domain, operation environment and inherent attribute data, so as to judge the type, location and severity of fault, find the cause of fault, provide remote guidance for equipment maintenance for maintenance parties, and shorten the maintenance time, Reduce maintenance cost; Before the failure, predict the remaining life of the product, the parts and devices that may fail in the short term, and do preventive maintenance to reduce the production delay and other losses caused by the failure[5][6].

1. Data preprocessing technology

Data sources are diverse, including operation environment data, inherent attribute data, real-time operation status data, and data types and formats are also different, including structured, semi-structured and unstructured data; Historical discrete data and real-time continuous data acquisition; The data quality caused by sensor failure or network transmission is poor, which requires preprocessing such as de duplication, replenishment, de abnormality and so on.

2. Data analysis and modeling

Data screening: in the face of a large number of data, select the data that can characterize the key factors. On this basis, take the average, maximum, minimum, variance, range, median, etc. to generate more feature quantities. Reduce the dimension by PCA principal component analysis. The data is divided into training set samples and test set samples, and establish the training classifier model.

3. Deep learning algorithm

Industrial big data analysis methods include traditional statistical analysis algorithms, general machine learning algorithms, vibration analysis algorithms for rotating equipment, time series algorithms for time series data, text mining algorithms for unstructured data, etc.

The system adopts the two-dimensional convolution neural network fault diagnosis model of signal. For the sampled samples, the video image is obtained through continuous wavelet transform, and then the time-frequency image is compressed as the input of the two-dimensional neural network model for training in the way of supervised learning. Avoid that only one-dimensional time domain will lack valuable information.

4. Conclusion

The data fusion technology and deep learning technology of big data analysis technology are applied to the fault diagnosis and analysis of industrial equipment, which can not only judge the type, location and severity of the fault, find the fault cause, provide remote guidance for equipment maintenance for the maintenance party, but also predict the fault and reduce accidents and unplanned downtime, And reduce the huge economic losses and casualties. The research results of this subject can help pump enterprises establish pump equipment fault diagnosis and prediction system, realize intelligent upgrading of pump equipment, improve product added value, improve service level and user experience of pump enterprises, and reshape new competitive advantages.

Acknowledgments
Sponsor: School Scientific Research Project of Xi'an Eurasia University: Research on Fault Diagnosis Technology Based on Big Data (2021XJZK02)

References
[1] Wei X.B., Lu X.J., Li T.B. (2021) Fault Analysis System Design of Real-time Monitoring of Large Pump Motor State, Electrotechnical Application, 09:43-47.
[2] Song G.D. (2021) Research on Fault Diagnosis and Preventive Management Method of Coal Mine Drainage Pump. Automation & Instrumentation, 06: 105-105.
[3] LI F., WANG Q., Hu J.X, Tian Y. (2021) Combined Data-driven and Knowledge-driven Methodology Research Advances and Its Applied Prospect in Power Systems, Proceedings of the CSEE,41: 4377-4390
[4] Dai J.Y., Tang J., Huang S.Z. (2018) Signal-based intelligent hydraulic fault diagnosis methods: Review and prospects. Mechanical Engineering, 32:1-22
[5] SIANO D, PANZA M A.(2018) Diagnostic method by using vibration analysis for pump fault detection, Energy Procedia,148:10-17.
[6] Chao Q., Tao J.F., Wei X.L. (2020) Identification of cavitation intensity for high-speed aviation hydraulic pumps using 2D convolutional neural networks with an input of{RGB}-based vibration data. Measurement Science and Technology, 31:105102

X-ray Image Recognition Method for Crimping Defects of Strain Clamp Based on OpenCV

Pengwu LI[1a*] **, Ronghai LIU**[1]

Power Science Research Institute of Yunnan Power Grid Co. , Ltd. , Kunming 650217, China

a*2192224044@ncepu.edu.cn

Abstract:The crimp quality of the tension clamp of the line affects the safety of the power grid. At present, the measurement of the quality of the tension clamp has certain limitations. An X-ray image detection method for strain clamp based on image processing technology is proposed. Firstly, the X-ray image of the strain clamp is taken and the image is preprocessed. Secondly, by selecting the defect part image of the typical defect sample, using some statements of OpenCV Python Library in Python to quickly identify and detect other X-ray pictures, and mark the defect part of other pictures. Finally, whether it is a defect is determined according to the cumulative value image of the gray value distribution of the marked defect area. Through the research in this paper, the rapid identification of the X-ray image of the crimping position of the tension clamp can be realized, which has important reference value for the engineering application of the crimping quality inspection of the tension clamp.

1. Introduction

As an important part of the transmission line, the tension clamp is composed of aluminum casing and steel anchor [1]. The aluminum pipe and the steel anchor part are pressed, and formed to combine the wire with the tension resistance clamp, so that it can bear all the tension of the connecting wire and turn on the current. Due to the high safety factor of the line design, accidents will not occur immediately after construction even if there is a quality defect of crimping [2-3]. However, in the case of dancing and long-term breeze vibration after putting into operation, the wire clamp is prone to fracture, thus causing power accidents and endangering social safety.

In practical engineering practice, the front-line workers often use vernier calipers and straight connecting pipes to complete the inspection of the appearance and size of the tension fixture and its connecting line when disassembling it. Before construction, experimental samples are usually made for force analysis according to the requirements of engineering indicators, and manual methods are used to check the crimping quality of tension-resistant wire clips [4]. This manual inspection method requires the construction personnel to have a high level of skills and practice, and it needs to conform to the specification. In addition to the operation space is high above the ground, it not only has potential safety hazards, but also increases the difficulty of operation and is not easy to supervise. Therefore, it is necessary to improve this manual inspection method [5-7].

At present, there are many researches on the quality of tension - resistant wire clamp. Wang Yi Saitama [8] et al. first extracted edge information from X-ray images of tension-resistant line clips, then obtained inflection points by calculating the maximum local curvature on image edge contour, and finally realized automatic detection of tension-resistant line clips' crimping quality by calculating and comparing coordinate differences of each detection position. Li Junhua et al. calculated the depth value based on the sound pressure curve of finite element simulation and the depth value read by phased array imaging, which to a certain extent solved the problem that the quality of internal crimping of tension-resistant wire clips in traditional transmission lines could not be judged [9]. Based on the basic theory of electromagnetic field, Ouyang Kejian et al. used large-scale universal finite element software to conduct simulation calculation on the anti-electromagnetic interference of the detection device, and combined with the simulation calculation to optimize the design of live detection device, so as to complete the live detection of the quality of the tension line clamp of the transmission line .

All of the above researches can detect the quality of tension clamp, but there are some problems such as high cost and low efficiency. In this paper, the image processing method is used to quickly locate and identify the defects in X-ray images, which has important reference value for the engineering application of the quality inspection of tensioning wire clamp.

2. Preprocessing of tension clip X - ray images

In the process of image shooting, the unpredictability of the environment often leads to the difference of the final image brightness, size, resolution and other information. In order to reduce the influence of unnecessary factors on the algorithm, it is necessary to carry out some image preprocessing work before the implementation of the main process of the algorithm, such as filtering and denoising, image enhancement, etc.

2.1. Filtering and denoising

Noise has a great impact on the acquisition, input, processing and output results of image processing [10]. Therefore, image denoising must be carried out before image processing. At present, the image denoising technology has been developed more mature, there are many effective image denoising methods, this paper uses gaussian filtering.

Gaussian filtering adopts convolution kernel to filter the image through Gaussian function. Suppose the original image is f(u, v), g(u, v) after filtering, and the dimension of The Gaussian kernel matrix H(u, v) is (2k+1)×(2k+1), then:

$$H(u,v) = \frac{1}{2\pi\sigma^2}\exp\left[\frac{(u-k-1)^2+(v-k-1^2)}{2\sigma^2}\right] \quad (1)$$

$$g(u,v) = f(u,v) \times H(u,v) \quad (2)$$

In Equation (1), σ represents variance;Equation (2) represents convolution filtering. The effect of gaussian filtering before and after is shown in Figure 1.

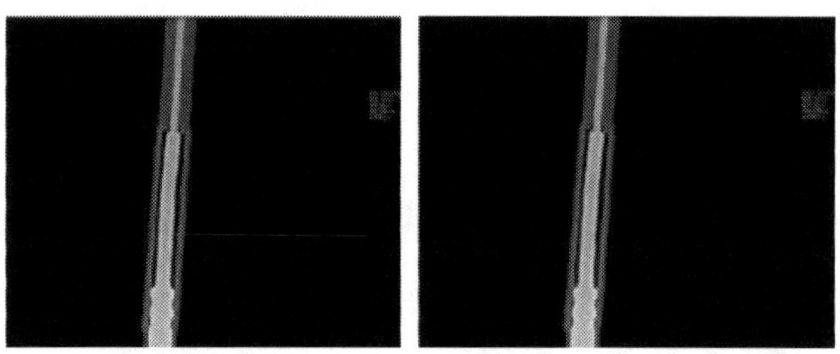

(a)filter before (b) after filtering

Fig. 1 Effect picture before and after Gaussian filtering

2.2. Image enhancement

After image enhancement processing, the recognition of X-ray image can be enhanced and the difficulty of defect location identification can be reduced. In this paper, the image contrast enhancement algorithm is used to process the acquired tensioning line clip X-ray images.

Image histogram equalization applies image histogram to adjust contrast in the field of image processing. This method is usually used to increase the local contrast of many images, especially when the contrast of the region to be processed in the image is quite close[11] . By this method, the gray value of the target image can be evenly distributed on the histogram. Histogram equalization achieves this function by effectively extending commonly used gray values, thus ensuring that local contrast enhancement is achieved without affecting the overall contrast. The effect before and after mean enhancement of gray histogram is shown in Figure 2.

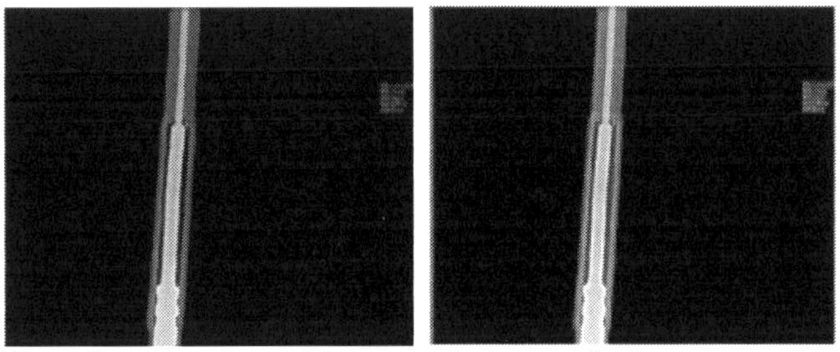

Fig. 2 Effect diagram of gray histogram before and after mean enhancement

3. Defect Location

This paper adopts template matching algorithm to locate defects. Template matching algorithm is a technique to find the best matching (similar) part in one image with another template image. After the template image is selected, the matching degree between it and the pixel value of the selected template image is calculated from left to right and top to bottom on the image to be detected. The greater the matching degree, the greater the similarity between the template image and the detection region of the image to be detected. In this paper, three of these algorithms are used to locate the defect location, and the final defect location is obtained by overlapping the three images. In this paper, the implementation image is 1063×664 PNG image, and the development environment is Pycharm. OpenCV and NumPY libraries are required. The selected template image is shown in Figure 3.

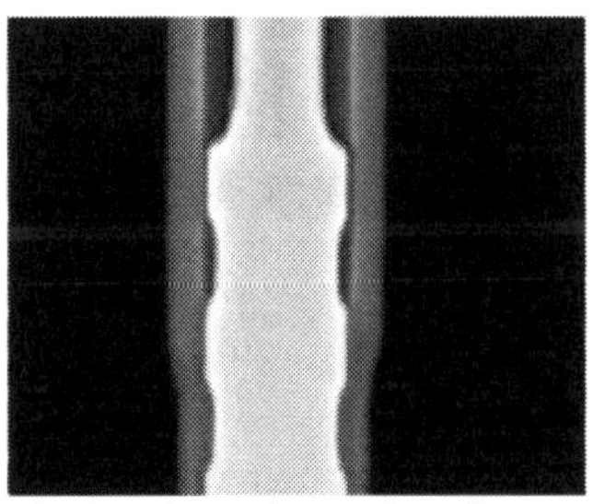

Fig 3 selected template image

There are three methods used and the results are output respectively. The three methods are normalized square variance matching, normalized correlation matching and normalized correlation coefficient matching.

The normalized square variance matching method normalized the correlation matching method, making the algorithm more adaptable to the change of light and gray linear. The smaller the matching value R(x, y) is, the higher the matching degree between the template image and the detected image region is. Therefore, when the matching degree between the template image and the detected image region is the highest, the matching valueR(x, y) is 0. Conversely, when the matching degree is lower, the matching value R(x, y) is larger. Matching value R(x, y) function expression is:

$$R\ (x,y)=\frac{\sum_{x',y'}(T(x'y')-I(x+x',y+y'))^2}{\sqrt{\sum_{x',y'}T(x',y,)^2\cdot\sum_{x',y'}I(x+x',y+y')^2}} \qquad (3)$$

The effect diagram of the normalized square error matching method is shown in Figure 4.

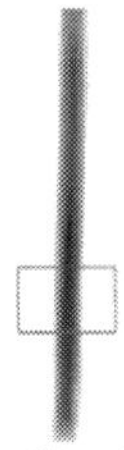

Fig. 4 Effect drawing of normalized square error matching method

The effect diagram of the normalized square error matching method is shown in Figure 4. The normalized correlation matching method normalized the correlation matching method, and used the pixel multiplication operation between template and target image to judge the correlation between the two. Therefore, the larger the result is, the greater the matching degree between template and detected image region is, and the smaller the matching degree between template and detected image region is. Its function expression is:

$$R\,(x,y)=\frac{\sum_{x',y'}(T(x',y')\cdot I(x+x',y+y'))}{\sqrt{\sum_{x',y'}T(x',y')^2\cdot\sum_{x',y'}I(x+x',y+y')^2}} \tag{4}$$

The effect diagram of normalized correlation matching method is shown in Figure 5.

FIG. 5 Effect drawing of normalized correlation matching method

The normalized correlation coefficient matching method subtracts the average value of the detected image and the template image, and then divides the variance of each image, so that the target image and the template are standardized, which can ensure that the image and the template will not affect the calculation result even if the brightness of the image and the template changes. The calculated correlation coefficient is between -1 and 1. 1 means that the template is exactly the same as the detected image region, -1 means that the brightness of the template is opposite to the detected image region, and 0 means that there is no linear relationship between the template and the detected image. The functional expression of the normalized correlation coefficient matching method is:

$$R\,(x,y)=\frac{\sum_{x',y'}(T(x',y')-I'(x+x',y+y'))}{\sqrt{\sum_{x',y'}T(x',y')^2\cdot\sum_{x',y'}I'(x+x',y+y')^2}} \tag{5}$$

The effect diagram of the normalized correlation coefficient matching method is shown in Figure 6.

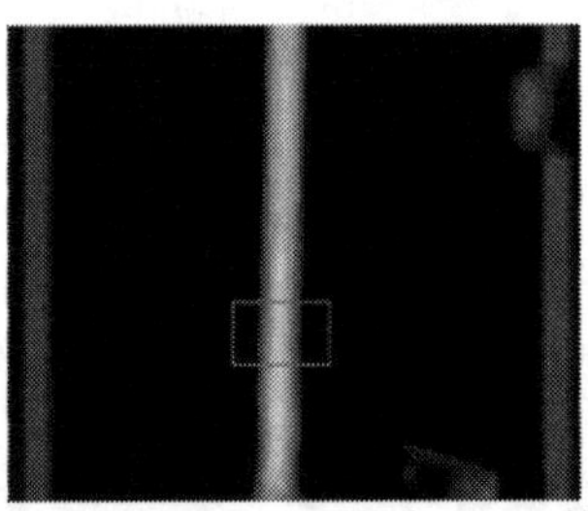

FIG. 6 Effect picture of normalized correlation coefficient matching method

The images obtained by these three algorithms are superimposed, and the one with the largest number of circled positions can be determined as the defect image. The final position of the crimp defect is shown in Figure 7.

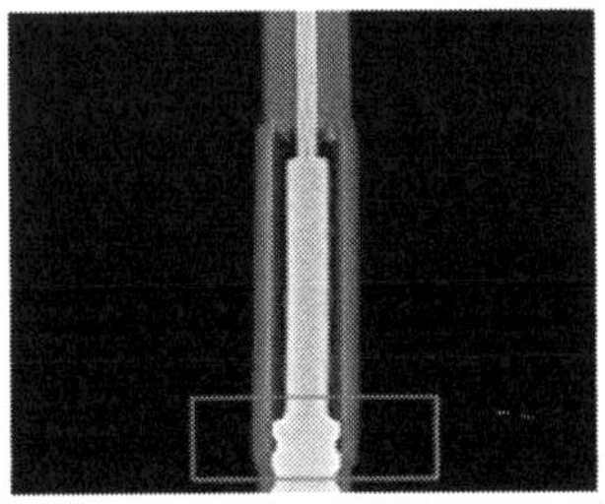

FIG. 7 The final location of the crimping defect

4. Detection of defective parts

The position image of wire crimping defect is taken out separately for defect judgment. Since the gray histogram in image processing reflects the corresponding function between the number of pixels and the occurrence frequency, it can effectively reflect the distribution of gray value in the image, so it can be used to find the jump discontinuities of the change of gray value, so as to automatically calculate whether there are crimping defects at the crimping place. As shown in Figure 8, the whole image is scanned line by line, and the median value is taken and recorded according to the change of gray value.

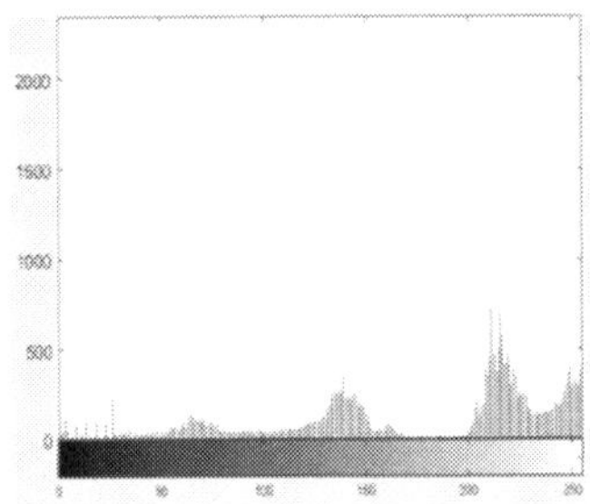

FIG. 8 Histogram of grayscale image

Initialization window pixels for A pixel, the window along the horizontal line, statistics of gray level distribution of the total value, change for the first time, drop point to point A, continue to move the sliding window, when the first appeared on the jump discontinuity point, record center position as point B, read through the AB two pressure welding wire insertion depth, The gray accumulated value images with defects and without defects are shown in Figure 9 and 10 respectively.

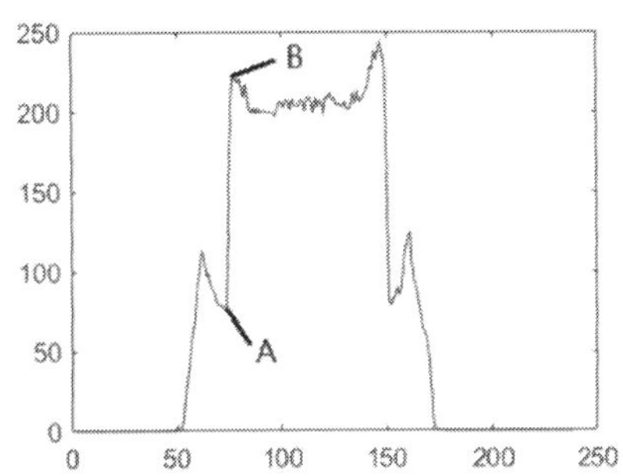

FIG. 9 Defect grayscale cumulative value image

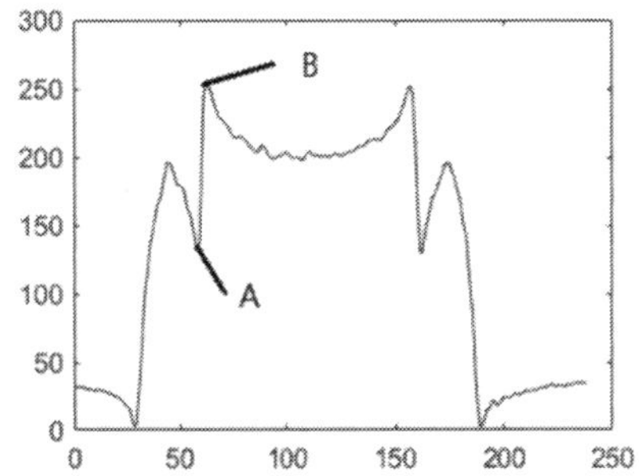

FIG. 10 Image of non-defective grayscale cumulative value

Cumulative gray distribution diagram of defect location image with inadequate crimping Defect Distribution image Because there is no vacant part during steel anchor crimping, there is no such large cumulative gray difference between AB points. According to the above processing results, according to the function relationship, there is a corresponding relationship between the size in the detected image and the actual size, which is approximately a linear relationship. Therefore, dimension information such as the depth of tension line clamp can be calculated according to the length of the detected image.

Set the threshold value as 150, that is, the difference of Y value between two points AB is 150 gray units. If it is lower than the threshold value, no defect will be judged; otherwise, insufficient crimping defect will be judged. According to the proposed method, X-ray image recognition of tension-resistant wire clip was carried out. Partial detection results are shown in Table 1, and all detection results are shown in Table 2.

Table 1 Part of the test results

Image category	Gray scale accumulation diagram	AB difference	Is it a defect	Judgment situation
Defect	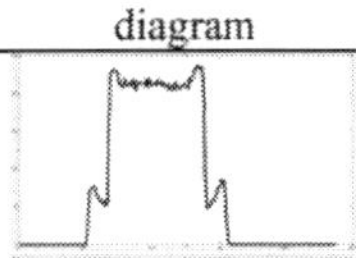	182	Is	Correct
Defect	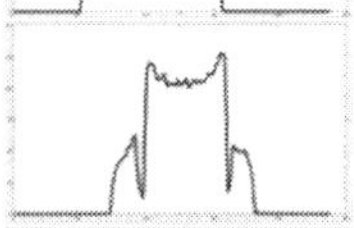	211	Is	Correct
Defect	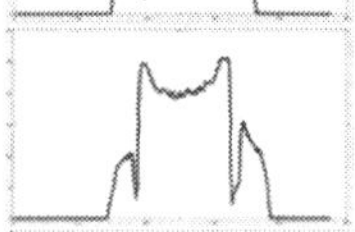	214	Is	Correct
No Defect	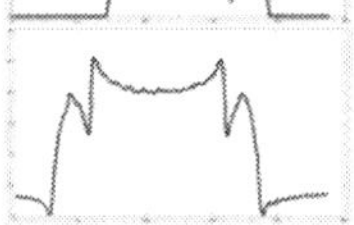	123	No	Correct
No Defect	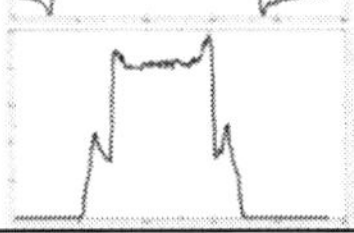	143	No	Correct

Table 2 Statistics of all test results

Image category	No Defect	defectiveness
Amount	500	300
Judged as no defect quantity	453	26
Judged as the number of defects	47	274
precision rate	90. 6%	91. 3%

5. Conclusion

In this paper, the X-ray images taken by the tension clip are preprocessed. Secondly, by selecting partial images of typical defect samples and using template matching method, other X-ray images can be quickly recognized and detected, and the defect parts of other images can be marked. Finally, the defect can be determined by the image of the gray value distribution of the marked defect area. Through the research of this paper, it can realize the rapid identification of X-ray images of tensioning wire clamp, which has important reference value for the engineering application of tensioning wire clamp quality detection.

References

[1] ZHOU Lixian, SUN Tao, GU Jian . Evaluation of Strain Clamp Crimping Property for La rge Cross-section Conductor [J]Proceedings of the CSEE , 2020, 40(S1):149-154.
[2] WANG Ruo-ming, ZHAN Ma-ji, JI Kun, YAN Bo, WANG Fu-cheng, DU Xiao-dong. Fail ure Reason of Tension Clamp for High Voltage Electicity Transmission[J]MATERIAL S FOR MECHANICAL ENGINEERING, 2017, 41(03):112-116.
[3] LIU Chun, CHEN Hongdong, OUYANG Kejian, TANGYuanfu, HE Pengfei, LI Wenbo, G ONG Jing. Analysis on the Strain Clamp Fracture of ACCC[J]. ELECTRIC POWER, 2015, 48(10):97-100+106.
[4] SUN Hebin, ZHOU Zhiyi, LI Jun, ZHANG Li. Quality Inspection and Defect Analysis of Tension Clamp[J]. Electric Power Science and Engineering, 2020, 36(02):48-54.
[5] LU Zhan-jie, WANG Jin, TANG Gui-ji, WU Zhang-qin, HAN Guo-dong, WAN Shu-ting. Research on the Relationship between Steel Core Crimped Size and Tension of Ove rhead Wires Based on X-ray[J]. Science Technology and Engineering, 2014, 14(34):18 5-189.
[6] GUO Taotao, WANG Dada, YU Hong, GUO Tieqiao, WU Zhangqin. Selection of Focal L ength Parameter in the GIS Equipment's X-Ray Visualization Testing[J]. Southern Po wer System Technology, 2012, 6(03):111-114.
[7] LIU Sanw ei, XIE Yi, ZHANG Jun, DUAN Jianjia, HUANG Fuyong, DUAN Xiaoli, ZE NG Zeyu. High VoltageCable X-ray Digital Image Depth Processing and Technology of Buffer Layer Defect Intelligent Recognition[J]. Southern Power System Technolog y, 2020, 14(12):66-70.
[8] WANG Wei, YUAN Qi, GU Junjie, et al. Application of X-ray nondestructive flaw dete ction technology in transmission lines press fittings [J] . Journal of Shanghai Jiaotong University, 2018, 52(10) : 1189 — 1194.
[9] LIU Ronghai, YANG Yingchun, GENG Leizhao, ZHENG Xin, TANG Faqing. Application of X-ray Image Recognition Technology in GIS Defect Diagnosis[J]. HighVoltage Ap paratus, 2019, 55(06):62-69.
[10] A. Vallan and F. Molinari. A vision-based technique for lay length measurement of meta llic wire ropes[J]. IEEE Trans. Instrum. Meas. , 2009, 58(5):1756-1762.
[11] R. C. Gonzaletz, R. E. Woods. Digital Image Processing[M]. 3rd ed. London, U. K. : Pe arson, 2008.

Research on main insulation defect of cable fusion joint based on IDIP-DR detection technology

Sanwe LIUi[1,*], Fuyong HUANG[1], Chao QIU[2], Xiaoli DUAN[1], ChaowenFANG[1], Jianjia DUAN[1], Zeyu Zeng [1]

[1]State grid Hunan electric power co, LTD. Electric power research institute, 41000, Changsha China

[2]State grid Hunan electric power co, LTD. Zhuzhou Power supply company, 41500, Zhuzhou China

[3]State grid Hunan electric power co, LTD. Xiangtan Power supply company, 41500, Xiangtan China

[*]Liu sanwei: 604208086@qq.com

Abstract. The cable connector is the weak link of the electrical part of the cable. Most of the cable breakdown accidents occur at the connector, which will have a huge impact on the safe power supply of the city. In recent years, with the development and maturity of 10 kv cable fusion joint technology, high-voltage cable fusion joint technology is gradually applied in the field, but the key process problems in the production process lack of effective control means. In order to solve such problems, this paper use IDIP - DR digital radiographic testing technology. The key technology of high voltage cable melting joint detection, effectively detect the main insulation problems bubble construction technology. Through the anatomical, proved the accuracy of the test results, for the subsequent company to carry out the weld joint process control provides an effective means of. The causes of defects were analyzed and the corresponding process control suggestions were put forward.

1. INTRODUCTION

In recent years, the scale of high voltage cable lines of 110 kV and above has been increasing with an annual growth rate of about 12% [1]. Due to the influence of cable length and construction, it is necessary to make cable joints every 500m or so for connection. Cable joints are the weak link of the electrical part of the cable, and faults often occur here [2-3]. Traditional prefabricated mechanical joints have some process risks [4-5] :

Due to fasteners loosening, node corrosion and other reasons, the electrical conductivity of mechanical pressure connection resistance will change with the growth of time. If the pit is deformed after mechanical crimping, local electric field distortion will be caused, which is easy to cause the serious result of joint breakdown.

At the point of mechanical connection, the tensile strength is greatly reduced. Due to the tension of the cable line itself and other reasons, the connection point will become the weak link of the tension of the wire. Under long-term stress, it is easy to cause the loosening of the connection point and affect the electrical connection of the joint.

The connection point of the cable connector is the weak place to withstand the impact of large current. There is a resistance difference between the connection point and the conductor.

As a kind of cable fusion joint technology which is fundamentally different from the traditional prefabricated mechanical joint, the joint manufacturing is in. Therefore, in view of the key manufacturing process control of the weld joint of high voltage cable, this paper proposes to use idiP-DR digital radiographic detection technology to fill the technical gap of the field control of the weld joint of cable.

2. HIGH-VOLTAGE CABLE FUSION JOINT TECHNOLOGY

Cable fusion splicing technology is to use strong aluminum, active in the flame ignited reagents to produce high temperature in an instant, through the displacement reaction, will came out of the copper oxide copper displacement, reaction process will release a lot of heat on both ends of the conductor and the displacement of copper melting together, finish on both ends of the conductor of welding, and gradually restore cable structure of each layer, as shown in figure 1.

Fig.1. Cable weld joint

Compared with traditional prefabricated mechanical joints, high-voltage cable fusion joints have the following characteristics in field applications:

1. The joint is not affected by the instantaneous high current. When the short-circuit current is affected, the melting speed of the welding joint is slower than that of the general electric conductor, so it is not easy to be damaged.

2. Strong corrosion resistance and integrity. The welding at high temperature belongs to the intermolecular connection, and there is no mechanical stress. After welding, the joint at both ends of the conductor becomes a complete part with strong integrity.

3. The resistance of the fusion point is low. After the same metal is used for welding, the connection points are of the same material as the conductors at both ends of the cable body, and the resistance value is small.

Compared with traditional prefabricated mechanical joints, welded joints have certain advantages. However, because welded joints are made on site and affected by the production environment, their construction technology is prone to problems. So far, there is no effective process control means. Therefore, the following digital detection method based on IDIP-DR is proposed in this paper to strengthen the process control of welding joint of high voltage cable in the field production.

3. IDIP-DR digital radiographic detection technology

Because of the obvious density difference between conductor and main insulation after recovering by fusion technology, X-ray digital imaging technology can be used to detect and distinguish between conductor and main insulation.

Figure 2 shows the schematic diagram of the testing equipment in this paper, in which 1-X-ray machine; 2- plate detector; 3- Portable power module; 4- Wireless communication components; 5-X-ray machine control system; 6- Cable grayscale image receiver; 7- Cable grayscale image display tablet computer; 8- Special fixed bracket; 9- Cable body.

X-ray that penetrates the inspected workpiece is received by flat plate detector, and then the crystal circuit inside the flat plate detector converts it into current signal according to the dose intensity of X-ray, and finally presents it on the terminal computer in the form of digital image [6]..

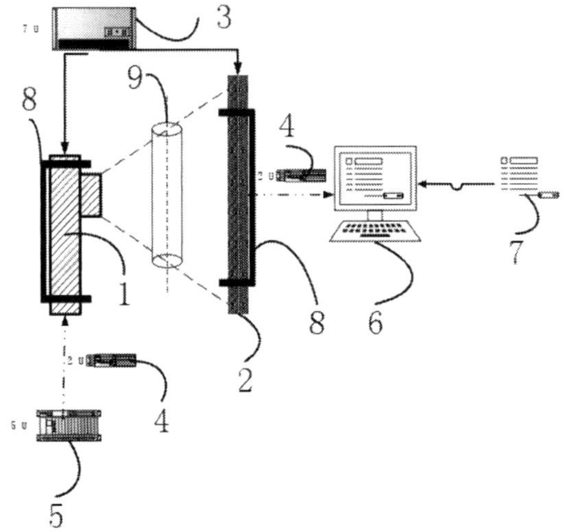

Fig.2 The fundamentals of digital X-ray imaging

As the gray scale of the DR device reaches 16bit=65536 gray scale, and the difference between the defects and the surrounding images is small in gray scale, the human eye in the original image cannot clearly identify the defects, as shown in FIG. 3(a).

The gray value range of the gray image is compressed to the gray range that can be recognized by the human eye, and the redundant components in the image are removed, so that the gray value of the image is more concentrated in the recognition range of the human eye. In this paper, sparse transformation and projection measurement are implemented for images based on compressed sensing principle [7-8].

The image is sparse and reconstructed. After image in-depth processing [9], the gap between the defect and the surrounding images increases in gray scale and can be clearly identified, as shown in Fig. 3(b).

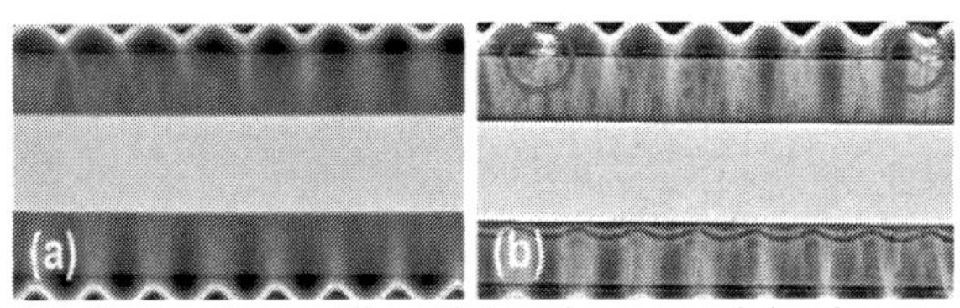

Fig.3 Cable X - ray image like original image and depth - processed image

4. Application Case analysis

This paper has tested and applied the welding joint of 110 kV high voltage cable in a city, as shown in Figure 4 and 6 Idip-dr testing was performed after the cable was restored to primary insulation. Figure 4 and 6 are the original images of detection. Without in-depth image processing, the human eye cannot directly recognize the defects. After the advanced treatment proposed in this paper, the bubbles on the left and right interfaces of the main insulation are clearly displayed, as Figure 5 and 7 show.

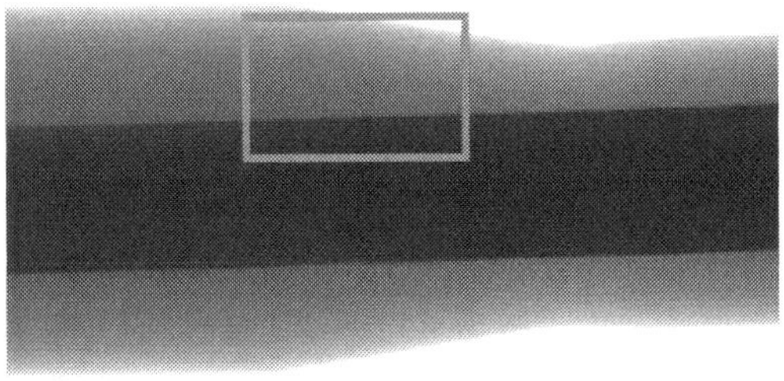

Fig.4 Check original picture on the right side of weld joint

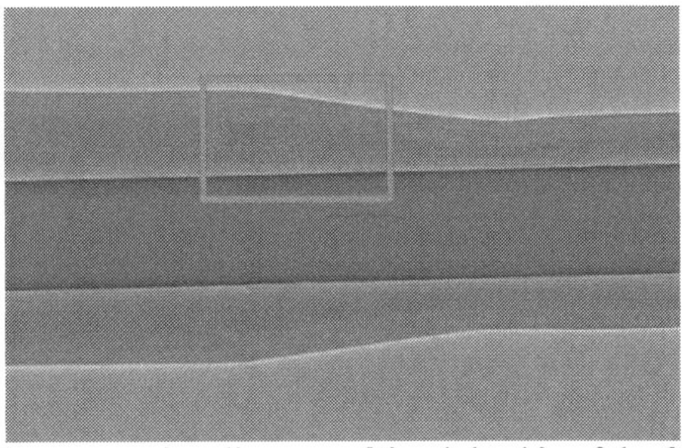

Fig.5 Inspection processing diagram of the right side of the fusion joint

Fig.6 Check original picture on the left side of weld joint

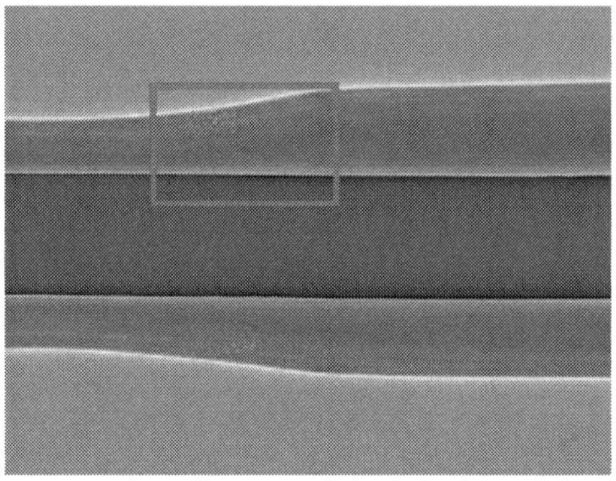

Fig.7 Inspection processing diagram of the left side of the fusion joint

In order to further verify the accuracy of the test results, after the position calibration of the above detection points, the section was cut parallel to the axial direction for dissection. As shown in Figure 8 Zeyuand 9, the dissected cable main insulation has different sizes of smile bubbles with diameters from 0.5 mm to 2.6mm.

Fig.8 Physical view of cable after transverse cutting and dissection

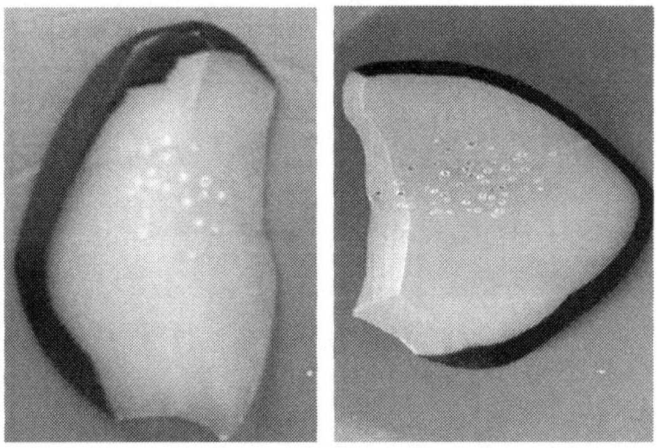

Fig.9 After transverse cutting and dissection, the cable was sectioned

In this paper, the causes of bubbles are analyzed: for the production of ultra-high voltage XLX cable, the nitrogen pressure in the pipeline should not be less than 1.2mpa. To maintain such a high pressure, the main reason is to reduce the number and diameter of microholes in the insulating layer and reduce the local discharge value; The temperature of the cross-linked pipe should match the nitrogen pressure. When the pipe temperature is high and the nitrogen pressure is low, micro-pores and obvious bubbles will be generated [10]. In the process of XLPE crosslinking, the pressure and temperature inside the mold have a direct influence on the number and size of microholes inside the insulating layer [11]. During the cross-linking process of the main insulation at the production site, the power was cut off for 15 minutes due to the unstable power supply of the generator. As a result, the pressure and temperature in the grinding tool did not meet the cross-linking conditions, resulting in the bubble defect mentioned above.

5. Conclusions and Suggestions

In this paper, the IDIP-DR digital radiographic detection technology was used to detect the key manufacturing process of the weld joint of high voltage cable, and the bubble construction process problem of the main insulation was effectively detected. Through on-site dissection, the accuracy of the detection results was proved, which provided an effective means for the subsequent companies to carry out the control of the weld joint manufacturing process. It provides a new control method for the weld joint technology of high voltage cable in engineering application. In addition, in the process of restoring the main insulation, it is necessary to ensure the reliability of the whole power supply and voltage supply, otherwise similar defects are likely to occur.

Acknowledgement

This paper is supported by Hunan Electric Power Co., LTD scientific project *Research and application of cable Line defect diagnosis and fault location technology* (5216A521001M).

References

[1] Power transmission Specialty of State Grid Corporation, December 2020 [R]. Beijing: State Grid Corporation, 2020.
[2] Liu CHANGming. Statistics and analysis of High voltage cable fault in Shimen Power Plant [J]. Hunan Electric Power, 2001, 32(05):49-49.
[3] Duan Jianjia; Chen fine jade; Wei Wenbang. Failure Analysis and Experimental Research of 110kV Cable Joint [J]. Hunan Electric Power, 2017, 37(01):63-65.
[4] PENG Bo, HUANG Fuyong, WANG Cheng, et al. Fault Location and Analysis of 110kV Power Cable [J]. Hunan Electric Power, 2012, 112(02):24-26.
[5] LUO Jihui, XU Mingliang, LI Chunyang, et al. Defect Analysis and Preventive Measures of 110 kV Cable Prefabricated Intermediate Joint [J]. Science & technology & innovation, 2017,25 (12):111-114.
[6] Zhou Yihuan, Wang Jianchen, Wang Yanan, et al. Research on X-ray detection of foreign body defects in GIS equipment [J]. High Voltage Electrical Apparatus, 2019, 55(1):41-46.
[7] Wang Yan, Lian Qiusheng, Li Kai et al. Optical Technique, 2010, 36(3): 350-355. (in Chinese)
[8]Duarte MF, Davenport MA, Takhar D, Sun T, Kelly KF, Baranuik RG. Signal-pixel imaging via compressive sampling [J]. IEEE Signal Processing Magazine, 2008, 25(2): 83-91.
[9]Gan H, Cheng Z, Yang S, et al. Circulant and toeplitz chaotic matrices in compressed sensing [J]. Journal of Computational Information Systems, 2015, 11 (4) : 1231-1238.
[10] [LU N. Analysis of manufacturing factors and effective control of high-voltage XLPE cable breakdown. Power Engineering Technology, 2017, 36(3):43-48.
[11] Luo Fengkui, Zhao Dandan, Huang Fuhuan. Reduce the insulation defects in the production process of 10kV XLX cable. Wire and cable, DOI: 10.16105 / j.carol carroll nki DXDL. 2017.01.012.

Research on the Architecture and Key Technology of Data Two-way Interactive Distribution Automation System

Yinghua Song[1,2*], Gang Wang[3]

[1] NARI Group Corporation (State Grid Electric Power Research Institute), Nanjing, 211106, China

[2] NARI Technology Co., Ltd, Nanjing, 211106, China

[3] State Grid Liaoning Electric Power Supply Co., Ltd, Shenyang, 110004, China

*Corresponding author's e-mail: ssyyhh786@163.com

Abstract. On the basis of analyzing and summarizing the current situation of the construction and use of distribution automation systems at home and abroad, according to the requirements of the use of distribution automation systems, a data two-way interactive distribution automation system architecture design is proposed. In response to the requirements for multi-source data collection in the distribution network, a distribution automation system architecture with partitioned collection and bidirectional data flow is proposed to improve the perception of distribution network equipment. Propose the self-recognition technology of power distribution system and terminal based on IEC61968 and IEC61850 model mapping to realize the fast access of power distribution terminal. Propose a grid integration modeling and global decision analysis method based on model splicing technology to realize multi-dimensional data fusion and analysis applications.

1. Introduction

Since the construction of smart grids has gradually accelerated in 2009, the construction of distribution automation systems has been carried out in nearly 160 cities in China, and the main domestic urban agglomerations have been basically covered. The distribution automation systems are all in operation monitoring, fault handling, and information interaction. It has achieved good construction results and played an important role in the operation and dispatch of the distribution network [1,2,3].

2. Situation Analysis

The power distribution automation system still has the following shortcomings in terms of technology:

(1) Insufficient flexibility of system architecture [4,5]

Internationally, the distribution automation systems of companies such as ABB and Siemens generally include FES, SCADA, DMS and other parts. The FES part is responsible for terminal data channel management and data access, and the SCADA part is responsible for the operation, monitoring and control of the distribution network. The DMS part is responsible for the realization of fault handling, operation analysis and operation optimization of the distribution network.

(2) The efficiency of power distribution terminal access and debugging is low [6,7]

The important characteristics of the distribution network are complex lines and a large amount of equipment. For a distribution network with more than 400 10kV lines, after the construction of the distribution automation system, the number of power distribution terminals that need to be connected is more than 1,600. The 104 protocol transmission between the electric terminal and the automatic system only solves the problem of data transmission, the meaning of the data is unclear, the necessary association relationship between the data is lacking, the terminal equipment lacks the self-description function, and the terminal access debugging efficiency is low.

(3) The extended application software has insufficient practicability and cannot adapt to the integrated analysis application of multi-application data[8,9,10]

The distribution automation system covers all aspects of distribution network production, operation and service. The integration and sharing of data and information in each link, and then the realization of integrated analysis and decision-making, is an important trend in the development of current power distribution applications. The existing power distribution automation system does not support the unified modeling of multiple application data, nor does it support the modeling of new equipment such as distributed power sources and electric vehicles. The application software does not support the integration of multi-source data forms, state estimation, network reconstruction, etc. Insufficient practicality of extended application software.

3. Overall design

3.1. Design Principles

The design of the data two-way interactive distribution automation system follows the following design principles:

(1) Provide a broad range of support for distribution network modeling

Combined with the development trend at home and abroad, the design of the data two-way interactive distribution automation system fully considers the scale of the distribution network and the needs of business system integration. According to the law of distribution network operation, the distribution network system requires main distribution and configuration integration, that is, the distribution network system has a construction covering the entire network (500kV, 220kV and above main network, 110kV, 35kV and 10kV distribution network, and 380V low-voltage users). It also supports modeling of distributed power sources, micro-grids, energy storage, and electric vehicle charging and swapping stations.

The scale of the system fully considers the planning level of the city's substations, distribution lines, distribution transformers, and urban population, meets the development needs of at least 5 years, fully considers the needs of static graphics and dynamic real-time data modeling, and supports big data processing.

(2) Support cross-regional multi-source data collection interactive mode, and build a generalized distribution network automation system architecture

The data two-way interactive power distribution automation system consists of two major subsystems deployed in the production control area and the information management area. The production control area collects the wired communication terminal. The information management area collects the wireless communication terminal, and the two subsystems exchange data in two directions. The data two-way interactive power distribution automation system support platform can support the district application mode, and the support platform supports the C/S and B/S hybrid architecture.

(3) Comprehensively optimize and enhance the analysis and application functions of the distribution network

In view of the technological development trend of multi-application data information integration, the distribution network analysis application software should fully consider the actual situation of the distribution network application data is not strong, and comprehensively use the real-time data collected by measurement, quasi-real-time data and manual data to match the distribution network. The network data is supplemented and comprehensively analyzed to realize the effective analysis of the data of the operation status of the distribution network, and realize the integrated modeling and global decision analysis of the distribution network.

3.2. Architecture design

Following the above design principles, the data two-way interactive power distribution automation system DMS-3220 realizes the architecture system of "one supporting platform, multiple application subsystems". Functionally, the system is divided into three major subsystems: distribution network data collection, operation monitoring, and analysis application. Among them, the distribution network operation monitoring and distribution network analysis application subsystems belong to real-time applications. They are deployed in the safety zone I and can be used for III. District applications provide service support. Distribution network data collection is deployed in District I and District III to meet the requirements of multi-source data partition collection, and two-way data interaction can be realized between the two regions.

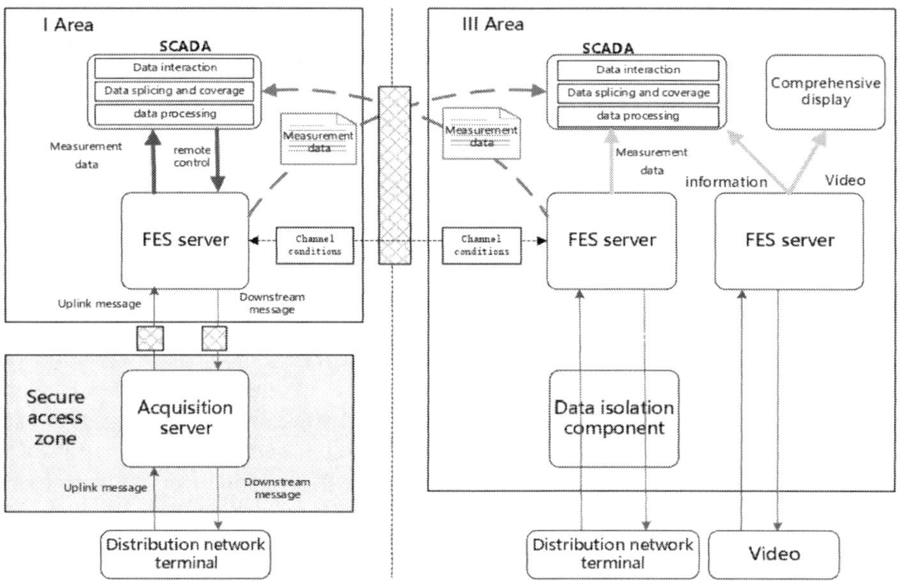

Fig.1 system architecture diagram

(1) One supporting platform

Following the principles of standard form, development, scalability, advancement, security, etc., build a standard support platform to provide general technical support for the development, operation and management of various applications of the system, and provide unified exchange services, model management, Data management, graphics management, and standardized information interfaces meet the needs of various real-time, quasi-real-time and production management services of distribution network dispatching, establish a complete and rigorous distribution network model, and support distribution network data collection, operation monitoring and analysis applications .

(2) Three application subsystems

Based on a unified support platform, it meets the requirements of dispatching operation control and production business respectively. Data collection is the basis for the perception of power distribution terminals, which realizes data collection and interaction of different types of terminals in multiple regions, and provides data support for distribution network analysis applications. Operation monitoring applications are the basis of the real-time dispatching business of the distribution network, and use information to realize the real-time monitoring of the power grid, fault judgment and processing and other applications. Analytical applications complete the effective analysis of the operating status of the distribution network, use the operating data of the distribution network and the result data provided by other application software to analyze and evaluate the operation of the distribution network, and realize the optimized operation of the distribution network.

Journal of Physics: Conference Series **2121** (2021) 012027 doi:10.1088/1742-6596/2121/1/012027

4. Key technology

4.1. Multi-application integrated architecture design

The data two-way interactive power distribution automation system provides technical support for the flexible combination of multiple application modes including the integration of main distribution, integration of distribution and grabbing, and integration of prefecture and county from the three levels of support platform, data processing, and distribution network application. The design of multi-application integrated system is shown in the figure below.

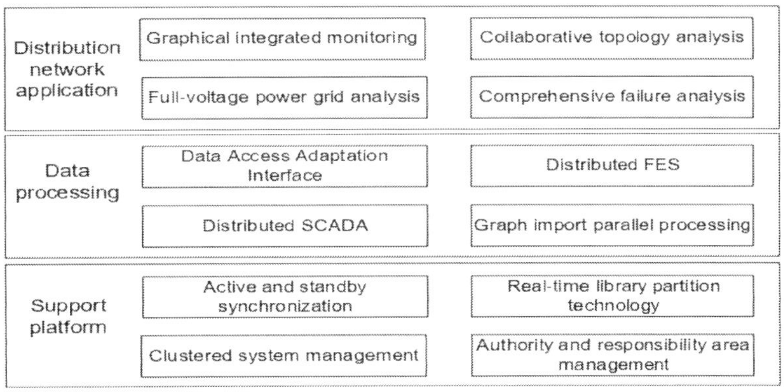

Fig.2 Multi-application integrated architecture design

At the support platform level, it focuses on providing unified service and system management and power grid modeling support, realizing real-time database partitioning, master-distribution integrated responsibility area and authority management, and supporting multi-application clustered system management. Dynamically set the real-time library partition range according to the regional setting, and support the synchronization and switching management of master and backup.

At the data processing level, the focus is on the performance and efficiency of multi-application integrated systems. Distribution automation system partition data acquisition is realized through distributed front-end, and distributed SCADA calculation realizes data processing and grid analysis calculations downloaded to the real-time database partitioned power grid.

At the distribution network application level, it mainly realizes the integration of application and display, and realizes the integrated graphical monitoring application of the grid based on the integration of the operation and distribution and the whole grid analysis.

4.2. Plug and play of power distribution terminal based on IEC 61850 and IEC61968 model mapping technology

IEC61850 provides a unified standard for substation automation, and provides a foundation for realizing interoperability between different intelligent devices. The DMS-3220 system applies the technology and methods of IEC61850 to distribution network automation, and combines the widely used IEC 60870-5-104 standard to solve the communication problems between power distribution terminals and between power distribution terminals and automation systems. Finally, the plug and play of the power distribution terminal is realized, which improves the efficiency of on-site construction and maintenance of the power distribution terminal.

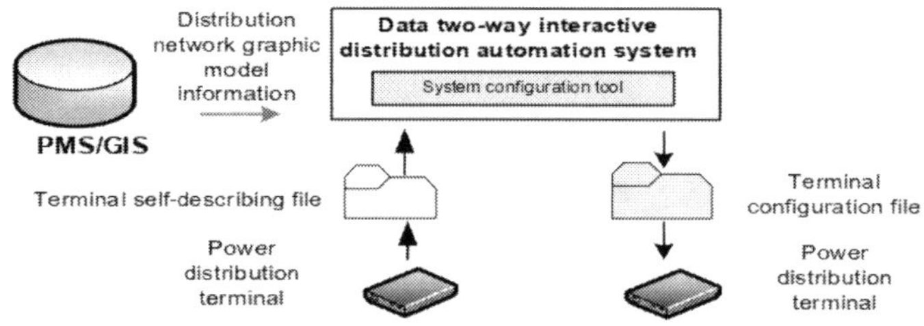

Fig.3 Principle of Plug and Play Terminal

The power distribution terminal transmits the equipment self-description information and related information such as the service capabilities of the power distribution terminal to the automation system, and the system can select/subscribe to the services it needs from the service items supported by the terminal. According to the distribution network model information of the system, the automatic mapping function from the IEC61850 model sent by the distribution terminal to the IEC61968 model of the distribution automation system is realized, and the system configuration file of the terminal is generated and sent to the distribution terminal. The power distribution terminal generates the configuration information of the terminal itself according to the received system configuration file, realizes the automatic identification of the power distribution terminal equipment.

4.3. Multi-application data integration whole-grid modeling and analysis
The data two-way interactive distribution automation system supports integrated modeling and model import of high, medium and low voltage distribution networks. Contains high-voltage diagram (110kV/35kV station information, including 10kV outlet switch and load), 10kV feeder part (from 10kV outlet switch to distribution transformer), and low-voltage part (from 10kV distribution transformer to low-voltage user). The modeling principle is shown in the figure below.

4.3.1. Modeling of high-voltage distribution network
Mainly refers to the equipment and lines in the 10kV and above substations. The modeling completely adopts the existing dispatching automation system modeling method and follows the IEC61970 standard for modeling. The plant is used as the modeling container, and the high-voltage distribution network and the medium-voltage distribution network are demarcated. Outgoing switch at 10kV.

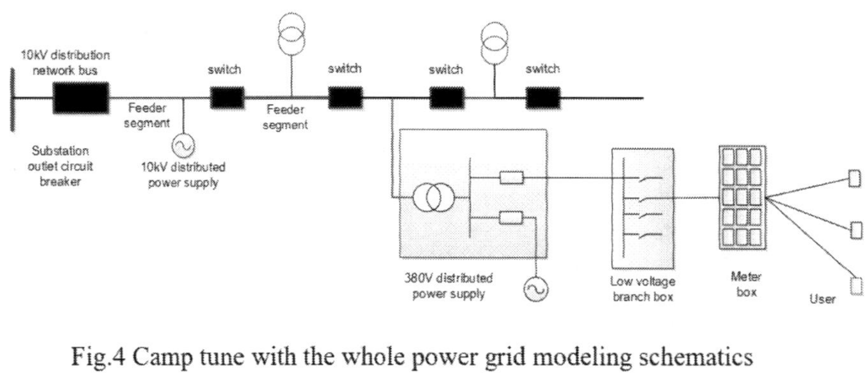

Fig.4 Camp tune with the whole power grid modeling schematics

4.3.2. Medium voltage distribution network modeling
Mainly refers to the 10kV/20kV/6kV feeder part, which is modeled according to the IEC 61968 standard. The boundary between the medium-voltage distribution network and the high-voltage

distribution network is the 10kV/20kV/6kV outlet switch of the substation, and the boundary with the low-voltage distribution network is the 10kV/20kV/6kV distribution transformer. The medium voltage distribution network model is used to manage the distribution network feeder and station model outside the substation. The equipment container in the distribution network includes two types: feeder and switch station.

4.3.3. Low-voltage distribution network modeling
The voltage level is the 380/220V part, and the boundary between this part and the medium voltage distribution network is the distribution transformer. The main application requirements for low-voltage distribution networks are low-voltage monitoring, fault research and judgment, etc., and accurate low-voltage analysis and calculations are not required. Therefore, low-voltage modeling can usually ignore low-voltage lines, and the key considerations are the distribution transformer and low-voltage switch, the low-voltage switch and the low-voltage branch phase, the hierarchical relationship between the low-voltage branch box and the meter box, and the meter box and the user.

5. Applications
Following the above design, the data two-way interactive power distribution automation system DMS-3220 was developed. The system has been demonstrated and applied in many domestic engineering sites. The pilot project deployed 2 database servers and disk arrays to support the storage of all data. Deployed a DSCADA cluster composed of 4 DSCADA servers for parallel computing and processing. Network real-time data, 4 front-end servers are deployed to receive measurement data sent by all power distribution terminals. Four distributed collection servers were deployed, located in Zone I and Zone III, to collect real-time data from the power distribution terminal.

The main functions include data collection, distribution network operation control and distribution network analysis and statistics, including network analysis, operation mode management, risk early warning and auxiliary decision-making, and safety and error prevention.

6. Conclusions
Based on the analysis of the current situation of domestic and foreign distribution automation system construction and application, this paper proposes the data two-way interactive distribution automation system technical architecture, and focuses on the rapid access of power distribution terminals and grid integration based on IEC 61850 and IEC61968 standards. A number of key technologies such as modeling and global decision analysis, and based on the design, developed the DMS-3220 distribution automation system. The system has been piloted in domestic engineering sites, which fully proves that the technical ideas are feasible and correct.

Acknowledgments
NARI Group Technology Project (524608210002)"Application development of distribution operation and power supply service based on SaaS layer of Internet of things cloud platform"

References
[1] GUOJiancheng,QIAN Jing,CHEN Guang,ZHANG Wei,DU Peng,CUILizhong,SHANG Xuewei. TechnicalSchemeofSmartDistributionGridDispatchingandControlSystems [J]. Automation of Electric Power Systems, 2015, 39(1): 226-231.
[2] Yin Zili, Qian Jing， Chen Yuxing, Huang Wenying, Guan Shenliang, Ren Xiaohui. (2016)Dispatching/Distribution Integration Technology Scheme Based on D5000 Platform[J]. Automation of Electric Power Systems. 40(18):162-168
[3] WANG Haoming, TANG Chong,et al. Dynamic Reliability Assessment of Distribution Network Based on Equipment State Evaluation Model[J]. Proceedings of the CSU-EPSA, 2017,27(7):68-74.

[4] WANG Liang. Discussion on application practice of distribution automation. Power system protection and control. 2016, 44(20):12-16.
[5] Lü Yan, DENG Chunjian, ZOU Kun. Design of Portable Antenna Device Control Equipment[J]. Laboratory Research and Exploration. 2017, 36(4):119-122.
[6] LIU Junhong, DENG Zhaoyun, LI Zeke, LI Huanming. Automatic information verification technology of smart substation based plug and play. Power system protection and control, 2018, 46(2):137-143.
[7] LIANG Wumin, ZHOU Shuibin,PAN Jing, WANG Weijie, ZHENG Hao, MAO Lina. A plug and play sensor design method of high voltage equipment. Electrical Measurement & Instrumentation. 2018, 55(6):133-135.
[8] Han Yinfeng, Wang Dong, Kang Xiaoping, Zhang Qihua. (2018)Research on Engineering Issues of Large-scale Application in Automatic Feeder Automation[J]. Distribution & Utilization. 35(3):58-63
[9] Shen Bingbing, Zhang Zizhong, Zhang Weiwei. (2015)Framework Construction Strategy of Power System Interoperability Based on IEC 61968 Standards[J]. Southern Power System Technology. 9(11):13-17
[10] ZHANG Tao, TIAN Bin, ZONG Zhigang, YU Yuan,et al. Planning and Typical Engineering Application of Distribution Automation. Electric Power. 2017, 50(9):44-51.

Research and analysis of steering-by-wire system stability

You Fu[1], Lizhu Zhang[1*], Chuanbin Wei[1] and Peng Tang[1]

[1]School of Automobile and Transportation, Tianjin University of Technology and Education, Tianjin, 300222, China

[*]Corresponding author's email: qcxy303@tute.edu.cn

Abstract—In this paper, the basic structure of steer-by-wire system is firstly introduced, and the steering stability of steer-by-wire system is summarized by means of stability control method and variable angle transmission ratio method. Finally, the unstudied direction of steer-by-wire system at this stage is summarized, and the intelligent steer-by-wire system is prospected, which provides basic information and ideas for the study of steering stability of steer-by-wire system.

1. Introduction

At present, China is in a high-speed development stage. Due to the rapid development, to a certain extent, the demand for energy will be large. In particular, oil resources, as non-renewable resources, have been in a state of tension, and China has a high degree of dependence on foreign oil. The rapid development of new energy vehicles eased the demand for oil in China. The emergence of new energy vehicles is also crucial for environmental protection. In the 5G era, in line with the concept of "intelligence, safety, energy saving and environmental protection" large domestic and foreign automobile enterprises and major scientific research institutions have invested cutting-edge technologies such as artificial intelligence in the new energy automobile industry, which has greatly promoted the related technology industries such as intelligent driving vehicles. In terms of the current automobile steering development situation, the traditional steering structure has been unable to meet the needs of intelligent vehicle electronic control technology.

In the process of continuous development, linear steering technology will be on the main stage of the times. Steering-By-Wire System realizes the decoupling between manual steering and vehicle response by replacing the traditional mechanical connection with electrical signal transmission. Due to the cancellation of the mechanical connection between the steering actuator and the steering wheel, the design freedom of each transmission ratio of the steering system is strengthened, and it is convenient to combine the vehicle state parameters with the optimized control algorithm, which can provide more steering maneuverability for drivers. Operating stability can be seen as two interrelated parts, and it is also a key indicator to measure vehicle driving performance.

2. Modeling of Steering Structure and Dynamics

Front wheel steer-by-wire system and distributed steer-by-wire system are the two parts of steer-by-wire research. According to the traditional front wheel mechanical steering, mechanical hydraulic power steering system, electronic hydraulic power steering system and electronic power steering system, the linear steering system is developed slowly, which is mainly composed of steering wheel and steering column integration part, steering actuator and front wheel integration part and main controller. [1]In addition, fault tolerant system, power system segment and other auxiliary systems are also indispensable to the steer-by-wire system, as shown in Figure 1.

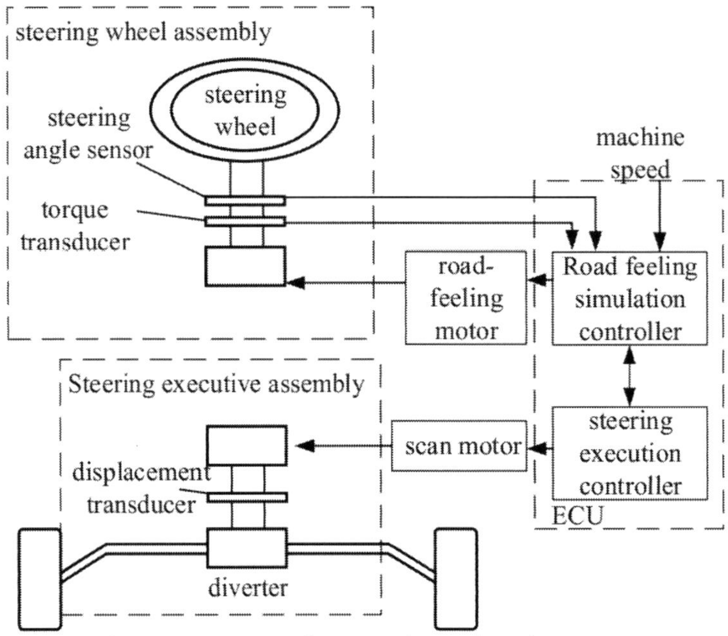

Figure 1. Structure diagram of steer-by-wire system

The distributed steer-by-wire system has the characteristics of high mobility, which can meet the needs of complex urban road conditions, such as moving in a narrow space and parking on the side, which greatly improves the maneuverability of drivers. In the research field, the actuator of distributed steer-by-wire system can be divided into two different technical forms. One is the independent steering of front and rear wheels. Two steering actuators are placed on the front and rear axles of the steer-by-wire chassis to drive the wheels on both sides to rotate at the same time. The other is a four-wheel independent steering, each wheel has a steering actuator, which can independently drive each wheel of the wire-controlled chassis to rotate. Four-wheel distributed steer-by-wire can make four-wheel independent steering, independent positive and negative driving, and the steering angle of the wheel can be ±90°. It can realize the omni-directional displacement function, and has eight driving modes. As shown in Figure 2, it has high controllable degrees of freedom, strong trajectory tracking ability, and can improve the comprehensive performance of the human-vehicle system by high-precision control, combined with the driver's driving intention and characteristic identification algorithm, which greatly improves the handling stability and active safety of the vehicle. In the four-wheel distributed steer-by-wire configuration, the dynamic response of planar motion can be improved, especially in the lateral direction. It is not difficult to see that the four-wheel distributed steer-by-wire has stronger stability.

2.1.Steering wheel assembly modeling

The steering wheel assembly can receive the driver ' s steering instructions and transfer the road feel. Its structure mainly includes steering wheel, road feel motor and reducer, angle and torque sensor. The torque input by the driver through the steering wheel is the driving torque, which is used to overcome the various resistance torques and friction torques existing in the steering wheel assembly. The dynamic model is established as follows :

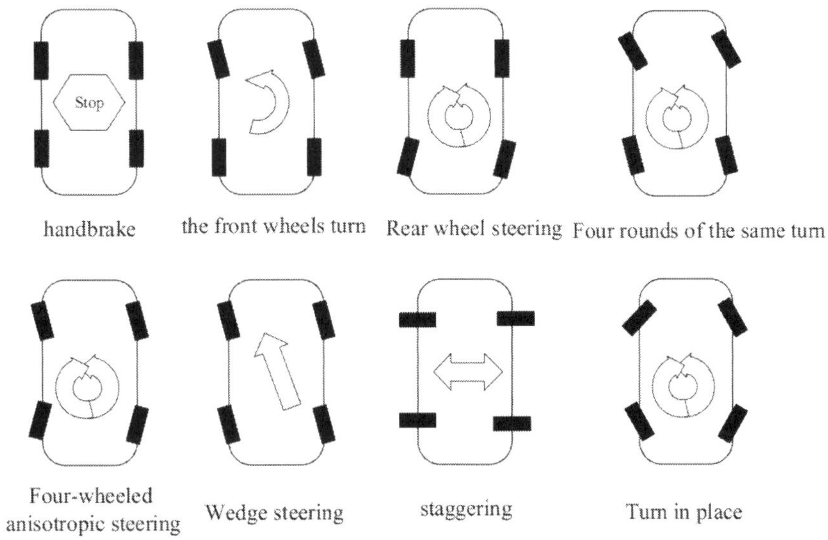

Figure 2. Four-wheel steer-by-wire driving mode diagram

The steering wheel to torque sensor model:

$$T_{sw} = J_{sw}\ddot{\theta}_{sw} + B_{sw}\dot{\theta}_{sw} + T_{mw} + T_f \tag{1}$$

$$T_{mw} = K_c\left(\theta_{sw} - \theta_m / g_m\right) \tag{2}$$

In the formula, T_{SW} - the driver ' s torque θ_{SW} - steering wheel angle. θ_m -Road induction motor angle. g_m – Reduction ratio of road induction motor reducer. T_{mw} – torque value of torque sensor. T_f - Equivalent friction torque of steering wheel assembly.

Torque sensor to road induction motor part model :

$$T_m = J_m\ddot{\theta}_m + B_m\dot{\theta}_m + \frac{T_{mw}}{g_m} \tag{3}$$

2.2.Modeling of steering execution assembly

The steering execution assembly is mainly composed of gear rack mechanism, steering component and steering execution motor. The torque provided by the steering motor has two directions. As shown in Figure 3. One is consumed by the inherent rotational inertia and damping of the motor, and the other is transmitted to the gear and rack mechanism through the steering gear to drive the wheel steering through the deceleration and torque increase of the deceleration mechanism.

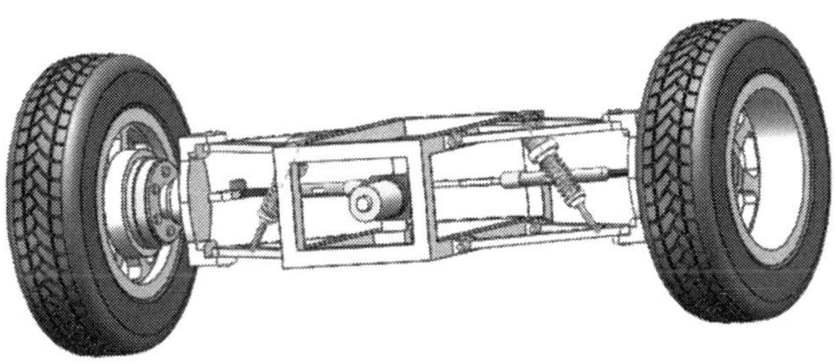

Figure 3. Diagram of steering actuator for steer by wire

Steering motor model, According to the torque balance, the dynamic equation can be obtained as follows :

$$T_{fm} = J_{fm}\ddot{\theta}_{fm} + B_{fm}\dot{\theta}_{fm} + \frac{T_p}{g_m} \tag{4}$$

$$T_p = K_{fc}\left(\theta_{fm} / g_{fm} - x_r / r_p\right) \tag{5}$$

T_{fm} - Steering motor electromagnetic torque. θ_{fm} - Steering motor angle. T_p - Equivalent to the steering resistance distance on the steering gear.

3. Stability control of steer-by-wire system

There are two research methods for the steer-by-wire system. One is the stability control method, that is, based on the reference model, through tracking the driver input parameters and vehicle running state, combined with the control algorithm, the ideal yaw rate, centroid sideslip angle and other technical parameters are calculated. The stability controller is designed with the ideal front wheel angle as the main parameter, and then the angle is allocated. The required torque is transmitted to the actuator through the steering motor to complete the tracking of the front wheel angle, as shown in Figure 4. The second is the variable transmission ratio control method, that is, the ideal front wheel angle is obtained by calculating the driver 's front wheel angle and the system transmission ratio, and then the steering motor is controlled by designing the angle tracker to track the ideal front wheel angle.

3.1.Linear steering stability control technology

Literature [2] the integral terminal sliding mode control theory is used to optimize the upper controller. The actual yaw rate of the vehicle is infinitely close to the expected yaw rate, and the corrected angle is output. Robust adaptive sliding mode control algorithm is used for the lower controller to ensure that the desired front wheel angle is almost the same as the normal front wheel angle. This paper studies the phenomenon of sideslip and tail flick of steer-by-wire.

Literature [3] the fuzzy control method is used to control the angle of front wheel and rear wheel, and the error rate of yaw rate is used as an important parameter to reduce the overshoot of yaw rate and sideslip angle of centroid, so that the vehicle has strong stability.

3.2.Linear steering variable transmission ratio control technology

Most of the studies on the control of steer-by-wire variable-angle transmission ratio in China and abroad have realized the handling stability and safety adaptability of steer-by-wire vehicles the structure and control algorithm.

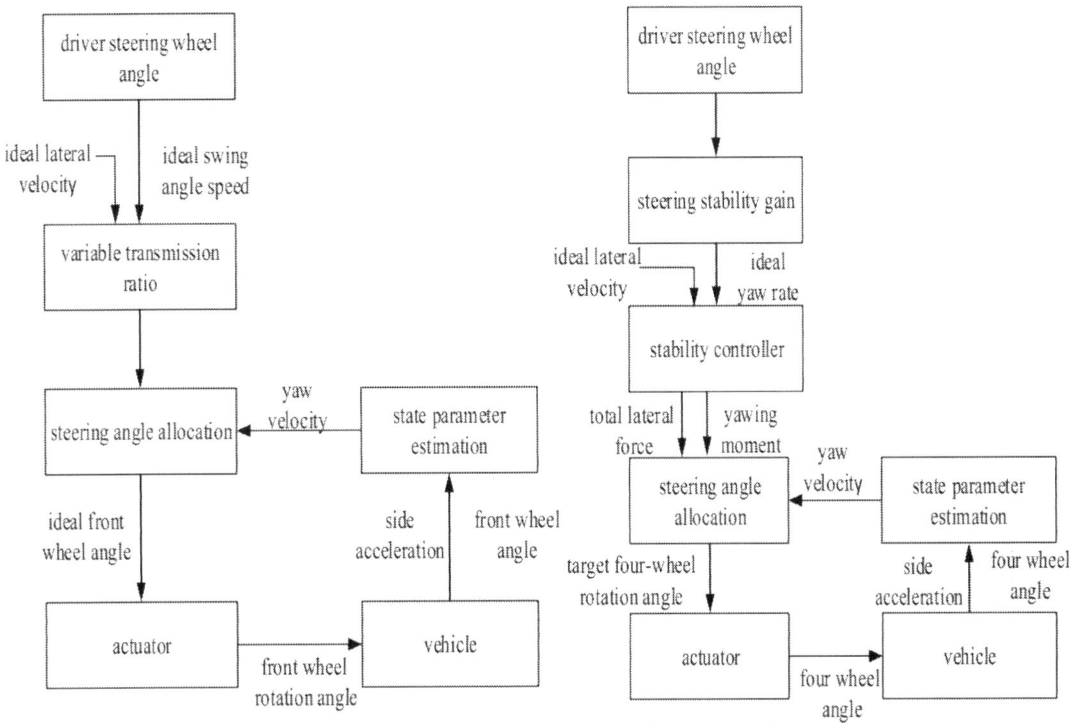

Figure4. Stability control method of linear steering

3.2.1. Variable Angle Transmission Ratio Control with Constant Steering Gain

Literature [4] adopted different design schemes in the middle-speed section and the high-speed section. In the middle-low speed section, the transmission ratio was designed under the design condition of constant steady-state yaw rate gain. Then the fuzzy control optimization transmission ratio scheme is used in high speed section. The experimental results show that the transmission ratio designed in this paper can improve the tracking effect and stability of the vehicle in the trajectory. Literature [5] Because it is a four-wheel independent steering, the variable transmission ratio control of front and rear wheel steering is studied respectively. In the study of variable transmission ratio control of front wheel steering, the constant transmission ratio strategy is used according to different speed ranges. In the middle speed range, the variable transmission ratio is determined by the constant yaw angular velocity gain method. In the high speed range, the variable transmission ratio is determined by limiting the lateral acceleration gain. In the study of variable transmission ratio control of rear wheel additional steering, the control method is low speed additional rear wheel reverse angle and high speed additional rear wheel co-direction angle. Literature [6] The Car Sim real vehicle model is used to simulate the vehicle stability, and the constant gain method is used to study the middle and low speed sections. Taking the vehicle handling stability evaluation index as the research factor, the genetic algorithm is combined with it to design the ideal transmission ratio. In the middle and high speed range, the ideal steering ratio is designed by combining the yaw rate gain with the lateral acceleration gain.

3.2.2. Variable Angle Transmission Ratio Control of Other Technologies

Literature [7] Based on the analysis of the influence of vehicle speed and road adhesion coefficient on the transient steering characteristics of the vehicle, the variable angle transmission ratio of the SBW system under different road adhesion coefficients is designed based on the steady-state yaw angular velocity gain. The simulation experiments are carried out under the conditions of double shift line test and steady-state rotation test. Considering the nonlinearity and time-delay time-varying of vehicle system, an active steering control method based on linear time-varying model predictive control (LTV-

MPC) is proposed. On the basis of considering the vehicle yaw, lateral and roll stability in three directions, the yaw angular velocity ωr, the sideslip angle β and the rollover index RI are selected as the control objectives. A model predictive switching controller including four control modes is designed to greatly reduce the risk index of vehicle rollover.

4. Conclusion

With the combination of line control chassis and artificial intelligence, as an important component of line control chassis, line control steering will also consider intelligence, networking and personalization. On the premise of meeting the above objectives, ensuring the safety and stability of steer-by-wire must be the primary goal of scientific researchers. As far as current research is concerned, most literatures only improve vehicle handling stability by optimizing the stability control of steer-by-wire system and changing the control strategy of steer-by-wire variable angle transmission ratio. It is not difficult to find that most researchers only consider the control factor of vehicle parameters, and do not take into account the driver parameters and road parameters, temperature and humidity, manufacturing costs and other aspects of the stability factors, such as the driver ' s driving state and driving habits and other human factors and road conditions, weather conditions. Drivers are the direct participants in the process of vehicle driving and steering. Therefore, in future research, complex unstable factors such as driver characteristics should be taken into account, and more stable and accurate control algorithms for safety and stability should be designed to enhance the human-vehicle trust bond.

Under the background of the development of vehicle line control, the fault-tolerant ability and actuator fault-tolerant ability of vehicle control system are still facing great challenges. The application of line control steering technology in mass vehicles is still unable to popularize. However, with the rapid development of automotive electronics technology and the strong support of national policies, the advantages of line control steering in active steering, variable angle transmission ratio and lateral dynamic control will be expanded, and the problems of handling stability, safety and economic cost will be solved. It will follow the trend of the times and combine with intelligent connected vehicles. So steer-by-wire technology will take a place in future vehicle development.

Reference

[1] Zhao W.Z., Zhang H., Zou S.C., etc. (2021) Overview of steer-by-wire system control technology. Journal of Automotive Safety and Energy Conservation, 12(1):17.

[2] Shi L.H., (2019) Research on yaw stability control system of steer-by-wire vehicle based on sliding mode control theory. Hefei University of Technology, Hefei.

[3] Qu W.T., Li W.R., Gao R., etc. (2021) Research on the handling stability of four-wheel steer-by-wire vehicle based on fuzzy algorithm. Electronic mechanical engineering, 37(2):5.

[4] Yin F.Q., Jiang L.C., (2019) The ideal transmission ratio design of linear steering system. Automobile practical technology, No.288 (09):103-105.

[5] Wang C.S., (2019) Research on variable transmission ratio control of steering lever linear control steering of four-wheel independent drive / independent steering electric vehicle. Liaoning University of Technology, Anshan.

[6] Feng X.Z., Shi P.l., Cao F., etc. (2018) Design of ideal transmission ratio for steer-by-wire vehicles based on CarSim. Journal of Guangxi University: Natural Science Edition, 43(06):2122-2132.

[7] Zhang J., (2018) Study on active steering control of steer-by-wire system. Jiangsu University, Zhenjiang.

The Study on Void Fraction Prediction of Gas-liquid Two Phase Flow Based on Convolutional Neural Network

Zhe Kan *, Xinyang Liu

School of Information and Control Engineering

Liaoning Petrochemical University,Fushun,China

*Corresponding author email: kanzhe @lnpu.edu.cn

Abstract. For the gas-liquid two phase flow in the horizontal pipeline, at the center angle the void fraction of the different liquid phases is calculated with the finite element simulation software, and then a soft measurement model of the void fraction is established. By comparing with traditional recursive augmented least squares (RELS), particle swarm optimization (PSO), and simulated annealing-based PSO, the void fraction soft measurement model is identified and calculated separately.The segmentation optimization results of PSO based on simulated annealing have higher accuracy and stability than RELS and PSO, but as the number of center angles increases, the relative accuracy and stability of the system will deteriorate.And the characteristic is not conducive to the calculation and analysis of data results.By combining the actual model, the convolutional neural network weight update algorithm is added to the LSTM, and the RNN-LSTM convolutional neural network is used to predict the void fraction of the second half the region.It improves the effect of RNN gradient problem on learning ability and improves learning ability. Through comparison, it is found that the convolutional neural network based on RNN-LSTM has a better prediction effect, improves the accuracy and stability of the system, and provides a new method for the measured void fraction of twophase flow.

1. Introduction

Void fraction is an important parameter in the measurement of two phase flow. It is widely present in the chemical and petroleum fields, and is particularly important for process optimization in production processes, flow measurement and control, and personal safety of operators [1]. In actual industrial production, the traditional soft sensor model method based on a single model has the problems of high model training complexity, poor matching of process characteristics and low prediction accuracy. With the continuous development of soft sensor theory, more and more attention has been paid to the soft sensor model method based on multiple models. By adding the prediction results of multiple models, the prediction accuracy and generalization ability of the model are improved. Literature [2] has achieved a good prediction of the water content of oil-water two phase flow, but the samples taken are too few and not convincing enough; the number of clusters output by the radiation propagation cluster depends on the input bias parameter p[3], and is not mentioned in the text.

In this paper, the finite element simulation of the adopted model is carried out, and the relationship between the model parameter (void fraction) to be identified and the known physical parameter (center angle) is established through the theoretical model to obtain the soft measurement model of the void fraction. Through simulation experiments, it is found that using traditional RELS [4] to identify the soft sensor model of void fraction will greatly reduce the accuracy and stability of the system, because

 2121 (2021) 012029 doi:10.1088/1742-6596/2121/1/012029

the change of the centre angle does not apply to the entire model. It is improved accuracy and stability of the system by analysis of the distributional data samples properties and segmental optimization, and the soft measurement model of void fraction is used the particle swarm algorithm (PSO)[5]. But the relative accuracy and stability of the system deteriorated need continue to increase as the circularity angle. This is because the basic PSO is easy to fall into the influence caused by local optimization; PSO based on simulated annealing [6] solves the problem that PSO easily falls into local optimization during the calculation process, reduces the influence of the change of the center angle on the measurement result, and improves the accuracy and stability of the system. However, when the center angle is large, there will be a large system error and fluctuation. In order to further improve the accuracy and stability of the system and solve the problems of small model applicability, insufficient global search capability and unsatisfactory prediction results in the prediction process, a convolutional neural network based on RNN-LSTM is used to predict the void fraction. The weight update algorithm in the convolutional neural network is added to the LSTM, which improves the influence of the RNN gradient problem on learning ability and improves learning ability. Simulation experiments show that this method has higher system accuracy and stability than the first three methods.

Research done in this article:

Part1: establish the simulation model of the capacitive sensor through the finite element simulation software;

Part2: use LSTM convolutional neural network to improve RNN;

Part 3: Compare the prediction effects of the four methods through simulation research;

Part 4: draw conclusions.

2. Gas-liquid two phase flow measurement model

2.1 Schematic diagram of gas-liquid two phase stratified flow

When the horizontal pipeline is filled with gas and liquid two kinds of media, stratification will occur, with the lower density above and the higher density below.Figure 1 shows the stratification of two different media, gas and liquid. Assuming that the density of the gas is less than the density of the liquid, the center angle β corresponds to the liquid part, and the rest is the gas part.

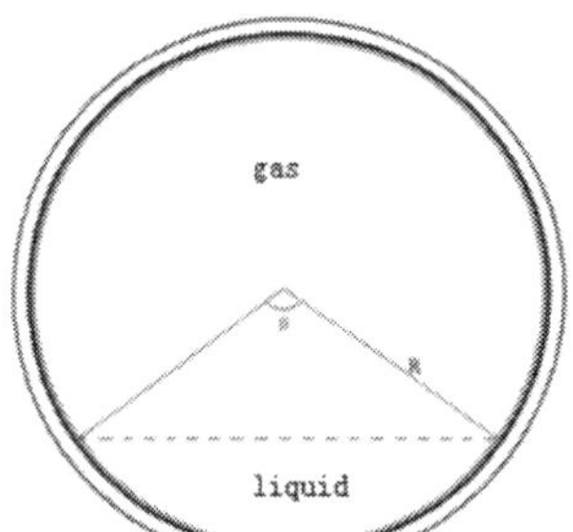

Fg.1 Schematic diagram of gas-liquid stratified flow

2.2 Finite element simulation model

According to the model of the toroidal capacitance sensor, we use the finite element simulation software [7] to establish a similar model for the experiment, as shown in figure 2. In the picture, two ring electrodes are sheathed on the outside of the pipe, one is the excitation electrode and the other is the receiving electrode. The outer side of the electrode is a shielding cover, the purpose is to shield the magnetic field and other signal interference. The medium in the lower part of the pipe is a liquid, here we set its relative dielectric constant to 2.5, and the upper part is gas, we set its relative dielectric constant to 1, and it is assumed that the relative dielectric constant will remain unchanged during the measurement process. Solve and record the centre angle and void fraction using finite element simulation software, and establish the relationship between the model parameters (void fraction) to be identified and the known physical parameters (centre angle) through the theoretical model to obtain the soft sensor model of the void fraction.

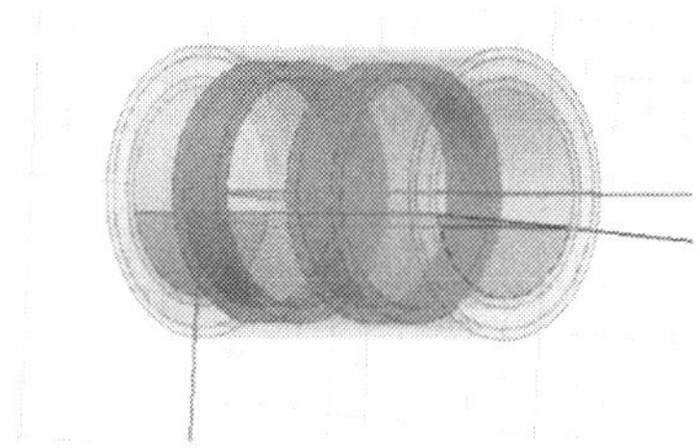
Fig.2 capacitive sensor model

3. Based on RNN-LSTM convolutional neural network

3.1 Recurrent neural network (RNN) model

Recurrent neural network (RNN) is a type of recurrent neural network that takes sequence data as input, performs recursion in the evolution direction of the sequence, and all nodes (recurrent units) are connected in a chain.

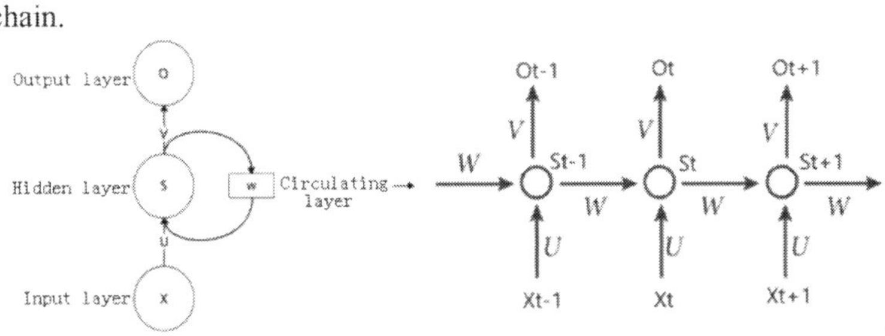

Fig.3 RNN working principle

X is a vector, which is the feature vector of a certain data, as the input layer. U is the parameter matrix from the input layer to the hidden layer. S is a vector of hidden layers. V is the parameter matrix from the hidden layer to the output layer. O is the vector of the output layer, and W is expanded according to the timeline, which is the weight matrix between each time point. The reason why RNN can solve the sequence problem is that it can remember the information at each moment. The hidden layer at each moment is determined not only by the input layer at that moment, but also by the hidden layer at the previous moment.

3.2 Improvement of RNN by LSTM convolutional neural network

Although the gradient problem of RNN has been solved[8] to some extent by LSTM [9], it is still not enough. The reason is that it can handle fewer orders of magnitude, but it will still be difficult for longer sequences. Therefore, we still need to find a new method to solve this problem. Here we use the LSTM convolution [10] neural network to improve the RNN, mainly adding the weight update algorithm in the convolutional neural network to the LSTM. The influence of the RNN gradient problem on learning ability is improved, and the learning ability and the accuracy of prediction data are improved.

3.2.1 LSTM

All RNNs have a chained form of repeating neural network modules. In a standard RNN, this repeated module has only a very simple structure, such as a tanh layer. LSTM has the same structure, but the repeated modules have a different structure, as shown in figure 4. Unlike a single neural network layer, there are four, including three σ and a tanh, interacting in a very special way.

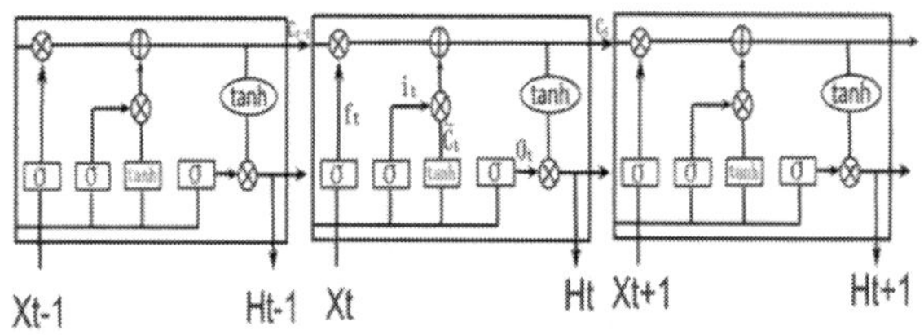

Fig.4 Repeating module in LSTM

In figure 4, we can simply understand the working principle of the repeating module in LSTM: the vector is transmitted from one point operation to another point operation via the neural network layer, and finally distributed to different locations through replication.

The calculation process of the repeat module is as follows:

(1) Determine discard information $f_t = \sigma\left(W_f \cdot [H_{t-1}, x_t] + b_f\right)$

(2) Confirm the updated information $\begin{cases} i_t = \sigma\left(W_i \cdot [H_{t-1}, x_t] + b_i\right) \\ \tilde{C}_t = \tanh\left(W_C \cdot [H_{t-1}, x_t] + b_C\right) \end{cases}$

(3) Update cell status $C_t = f_t * C_{t-1} + i_t * \tilde{C}_t$

(4) Output information $\begin{cases} O_t = \sigma\left(W_o [H_{t-1}, x_t] + b_0\right) \\ H_t = O_t * \tanh(C_t) \end{cases}$

3.2.2 Convolutional neural network weight update

In the CNN, the convolution kernel can be regarded as the weight in the artificial neural network, and the sampling layer is essentially a convolution operation, which updates the weight [11] of the convolution layer in CNN. Then combined with LSTM is applied in the traditional RNN algorithm.

The process of solving the weight update of the convolutional layer in CNN is as follows:

(1) For each feature map of the convolutional layer output: $x_i^l = F\left(\sum_{i \in M_j} x_i^{l-1} * k_{ij}^l + b_j\right)$

(2) Calculated sensitivity: $\delta_j^l = \delta_j^{l+1} W_j^{l+1} \circ F'(u^l) = \beta_j^{l+1} up(\delta_j^{l+1}) \circ F'(u^l)$

here ◦ means multiplying each element, 'up' means up sampling operation.

(3) Calculate the partial derivative of the error cost function for the bias b: $\frac{\partial E}{\partial b_j} = \sum_{u,v} (\delta_j^l)_{u,v}$

(4) Calculate the partial derivative of the error cost function for the convolution kernel k: $\frac{\partial E}{\partial k_{ij}^l} = \sum_{u,v} (\delta_j^l)_{u,v} (p_i^{l-1})_{uv}$

$(p_i^{l-1})_{uv}$ is the convolution of x_j^{l-1} when doing convolution with kij, (u, v) is the centre of the patch, and the value of (u, v) position in the feature map is output.

The process of updating the weight of the down-sampling layer in CNN:

(1) For each feature map of the convolutional layer output: $x_j^l = F\left(\theta_j^l down(x_j^{l-1}) + b_j^l\right)$

'down' means down sampling.

(2) Calculated sensitivity:

$\delta_j^l = \delta_j^{l+1} W_j^{l+1} \circ F'\left(u_j^l\right) = F'\left(u_j^l\right) \circ conv2\left(\delta_j^{l+1}, rot180(k_j^{l+1}), 'full'\right)$

(3) Calculate the partial derivative of the error cost function for the bias b: $\frac{\partial E}{\partial b_j} = \sum_{u,v} (\delta_j^l)_{u,v}$

4. Simulation experiment and comparison

This experiment is to learn and simulate the data by using Matlab. Since this article discusses the relationship between the centre angle and the void fraction, the time series is replaced by the centre angle. Through the simulation of the previous three methods, we found that when the centre angle is greater than 180°, the soft measurement model of void fraction is no longer applicable, therefore, LSTM convolutional neural network is used to improve the RNN, and the algorithm is used to learn the void fraction data with a centre angle of less than 180°, and the predicted data of the void fraction with a centre angle of more than 180° is obtained.

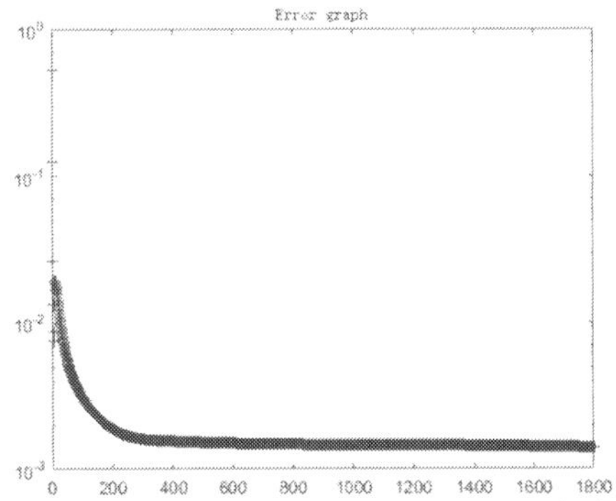

Fig.5 Error analysis chart based on RNN-LSTM convolutional neural network

Predictive_value =	Actual_value =
0.4906	0.4822
0.4876	0.4815
0.4102	0.3934
0.3766	0.3594
0.3056	0.3108
0.2560	0.2421
0.1925	0.2073
0.1509	0.1582
0.1389	0.1225
0.0453	0.0668
0.0818	0.0812
0.0592	0.0502

Fig.6 Data inspection results

It can be seen from figure 5 that after 1,800 network trainings, the RNN-LSTM convolutional neural network operation error is close to 1.4x10-3. Figure 6 is the result obtained by learning and predicting 12 sets of data. It can be seen that the error between the predicted value (Predictive_value) and the actual value (Actual_value) is small, and this method improves the accuracy of the predicted result.

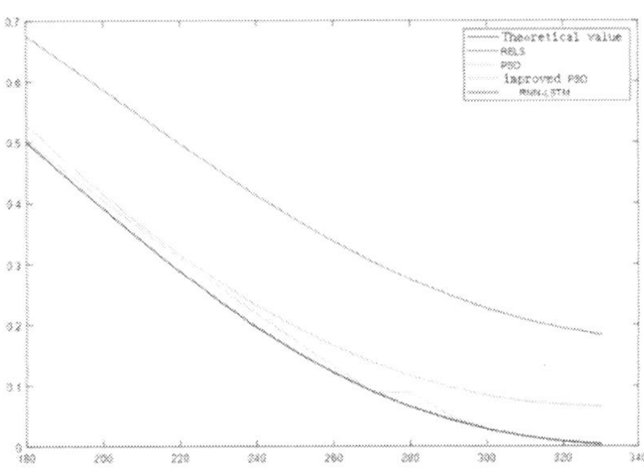

Fig.7 Identification result of void fraction

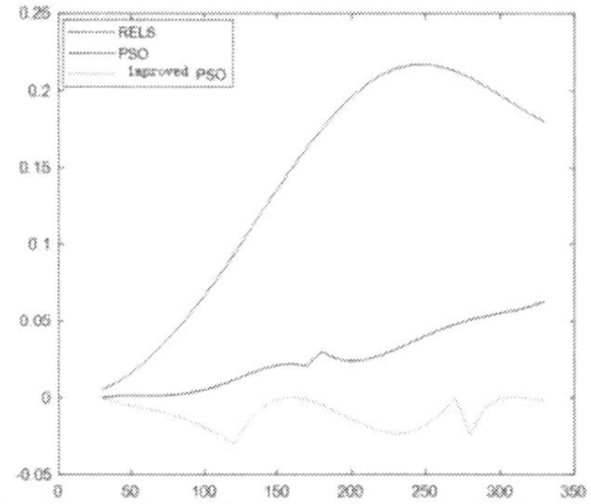

Fig.8 Error analysis graph of the first three optimization algorithms

It can be seen from figure 7 that the data obtained by using soft measurements of RELS, PSO and PSO based on simulated annealing, the change trend of the data is almost the same as the theoretical value, but through figure 8 we can clearly see that the error is relatively large, especially when the centre angle is greater than 180°, the error of the measurement results of RELS and PSO will increase significantly. Although the error of PSO based on simulated annealing is smaller than that of RELS and PSO, when the centre angle is within a certain angle range, the stability of the system is not good, resulting in a certain fluctuation in the predicted results. These phenomena are caused by the shortcomings of their algorithms. Based on the RNN-LSTM convolutional neural network, the predicted value of the void fraction (black curve) and the theoretical value (blue curve) almost overlap, and the forecast data error and error fluctuation are smaller, which shows that the method has a higher accuracy and stability.

5. Conclusion

The weight update algorithm in the convolutional neural network is added to the LSTM, and the void fraction is predicted by using the convolutional neural network based on RNN-LSTM. Simulation experiments show that compared with RELS, basic PSO and PSO based on simulated annealing, the prediction data obtained by the algorithm is closer to the theoretical value, andsolves the problems of small model applicability, insufficient global search ability and unsatisfactory prediction results in the identification process of the first three optimization algorithms.At the same time, it also improves the impact of RNN gradient problems on learning ability, and improves learning ability and accuracy of prediction data. It can better realize the prediction of the void fraction of gas-liquid two phase flow, which greatly improves the accuracy and stability of the prediction system. It is of great significance for process optimization, flow measurement and control, and personal safety during the production process, and provides a new method for the measurement of two phase flow void fraction.

Acknowledgments

This work was supported by the General Project of theLiaoning Province Education Department (L2020019), China.

References

[1] Hewitt G F.Measurement for Two- phase Flow Parameters. London: Academic Press,1978
[2] Xin Zhao, Ningde Jin,Weibo Li.Soft measurement method of phase content of oil-water two phase flow[J].Journal of Chemical Industry and Engineering,2005(10):1875-1879.
[3] Kaijun Wang, Junying Zhang, DanLi, et al. Adaptive affine propagation clustering[J]. Journal of Automation,2007, 33(12): 1242-1246.
[4] Dongqing Wang. Recursive augmented least squares identification method based on auxiliary model [J]. Control Theory and Applications, 2009, 26(01): 51-56.

[5] Xiaoliu Zhu, Weili Xiong, Baoguo Xu. QDPSO algorithm based on simulated annealing technology[J].Computer Engineering,2007(15):209-210.
[6] Changtao Wang, Xiaotong Sun, Zhonghua Han, Yi Zhu. Research on building pipeline layout based on adaptive simulated annealing PSO algorithm[J]. Journal of System Simulation, 2018, 30(05): 1941-1949.
[7] Yanli Zhang. Simulation study of a new type of electric compensation electrode based on Ansoft Maxwell's calculated capacitance[D]. Qingdao University, 2015.
[8] Zelong Li, Chunjie Yang,Wenhui Liu, Heng Zhou, Yuxuan Li.Prediction of silicon content in molten iron based on LSTM-RNN model[J].Journal of Chemical Engineering,2018,69(03):992-997.
[9] Ping Yang, Dan Wang, Zijian Kang, Tong Li, Lihua Fu, Yueren Yu. Prediction Model of Paroxysmal Atrial Fibrillation Based on Pattern Recognition and Integrated CNN-LSTM[J].Journal of Zhejiang University (Engineering Science Edition),2020, 54(05): 1039-1048.
[10] Chuanjun Pang, Jianming Yu, Changyou Feng, Yan Liu, Yefeng Jiang. Power load clustering modeling and characteristic analysis based on LSTM automatic encoder [J/OL]. Power system automation: 1-11[2020-06-20]
[11] Kun Xiao. Multi-layer convolutional neural network deep learning algorithm portability analysis[J/OL]. Journal of Harbin Engineering University: 1-6[2020-06-21].

Orderly grid-connected cooperative scheduling control strategy based on distributed energy storage for electric vehicles

Xiaomei Li[1], Rong Cao[2], Wenbo Hao[2], Mingyu Xu[2], Heng Hu[2], Peng Jiang[2], and Feng Zhou[1*]

[1] College of Electrical Engineering and Automation, Shandong University of Science and Technology, Qingdao, Shangdong Province, 266590, P.R. China

[2] State Grid Heilongjiang Electric Power Company Limited, Heilongjiang Province, 150090, P.R. China

[*]Corresponding author's e-mail: zhoufeng@sdust.edu.cn

Abstract. Aiming at the problem that large-scale disorderly grid connection of electric vehicles negatively affects grid operation and causes a large amount of abandoned wind and abandoned light, an orderly grid connection cooperative scheduling control strategy based on distributed energy storage of electric vehicles is proposed. The strategy takes the charging and discharging price as the lever to guide the users to charge and discharge in an orderly manner, takes the optimal economics on the user side and the optimal cost of power generation on the grid side as the objective function, and uses linear weighting normalization to convert the multi-objective function into a single objective function for simulation solution. The simulation results show that the effect of peak shaving and valley filling can be achieved on the basis of satisfying users' demand, and renewable energy can be effectively consumed.

1. Introduction

With the increasing proportion of renewable energy in the power grid, the disadvantages of renewable energy such as discontinuous power generation and large randomness[1], present new challenges to the stable operation of the power system. With the rapid development of the electric vehicle industry, the large-scale disorderly grid connection of a large number of electric vehicles will lead to increased load fluctuations in the grid and intensify the negative impact on the stable operation of the grid. Studies have shown that 90% of electric vehicles are offline during the day[2]. Therefore, it is important to study the orderly grid connection of electric vehicles as distributed energy storage units to participate in grid regulation for the stable operation of the grid.

Many scholars have carried out research on the orderly grid-connection control of electric vehicles. Yang Bing[3] constructed a distribution grid system containing electric vehicles, wind power and photovoltaic from the perspective of collaborative planning of electric vehicles and each distributed power source, but did not fully consider the economics of the user side. Qian Kejun[4] studied the charging and discharging problem of electric buses in charging stations participating in the vehicle network. Hu Biao[5] studied the impact of time-sharing tariff on users' electricity consumption behavior, and constructed a multi-objective system model with the optimization objectives of maximizing the lifetime of energy storage system and maximizing users' benefits in microgrid. Many scholars have proposed various tariff mechanisms based on economic principles, such as graded linkage tariffs on both sides of supply and consumption[6] and electricity consumption elasticity matrix tariffs[7]. Zeng Zheng[8] gave the use of tariff regulation to guide electric vehicle users to reasonably arrange charging time, but did not consider the actual load profile in a variety of load contexts, including those involving electric vehicle load. Sun Xiaoming[9] differed from the conventional peak, valley, and flat time-sharing tariff time division, carried out research on charging time-sharing tariff time division for regional load fluctuations, but did not propose a specific division method in conjunction with the actual situation.

It can be seen that the existing researches do not sufficiently consider the user's willingness to participate in grid dispatch. For this reason, this paper proposes a cooperative control strategy for grid-connected EVs, which fully considers users' demands, takes charging and discharging prices as levers to guide users to charge and discharge in an orderly manner, takes grid-side and user-side economics as the objective function, selects the optimal weighting coefficients, and the multiple objectives are converted into single objectives and solved by genetic algorithm. Under the premise of satisfying the user-side demand, the renewable energy is effectively consumed and the grid peak reduction and valley filling is realized.

2. Electric vehicle disorderly charging analysis

Under the guidance of strategies such as time-sharing tariff without consideration, EVs are connected to the grid in a disorderly manner, and the charging behavior is completely determined by the respective travel requirements and demands of vehicle owners, and the charging behavior of EV owners is stochastic. A Monte Carlo method is used to simulate the disorderly charging behavior of users, and probabilistic models of charging start time t_0, charging power SOC and charging power P are established, and corresponding random data are generated according to the established probabilistic models.

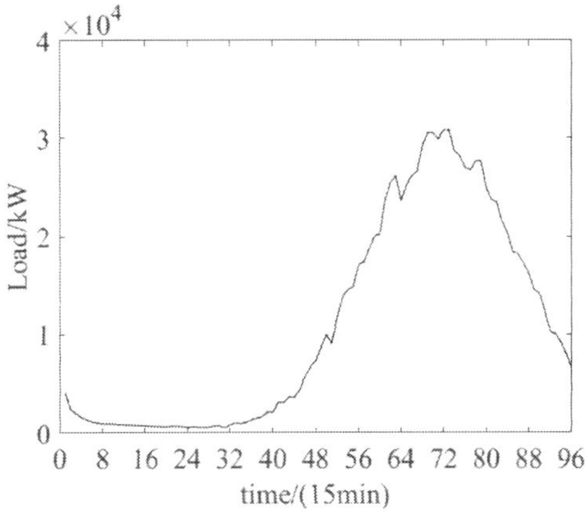

Figure 1. Daily load curve of disorderly charging

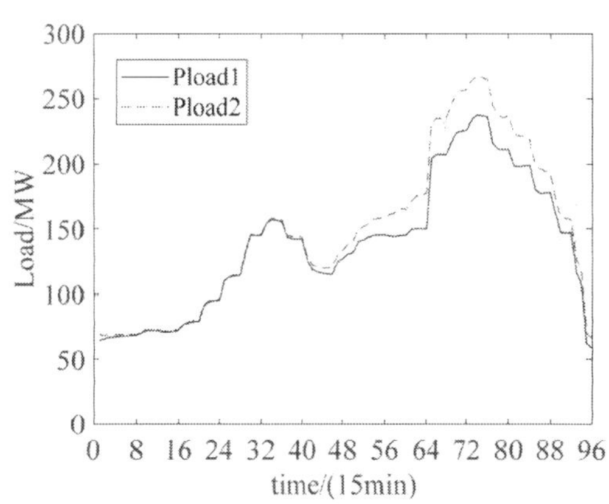

Figure 2. Daily load curve of a region

For verification purposes, the following settings are made.

- Using one day as a scheduling cycle and 15min as a time interval, we divide a cycle into 96 time slots.
- The starting time t_0 of electric vehicle charging uses data from the National Highway Traffic Safety Administration (NHTS), and the starting time $t_0$of electric vehicle charging satisfies a normal distribution $N(70.4,13.6)$.
- 1,000 electric vehicles were selected and their charging power data were statistically analyzed, and their charging power SOC was found to satisfy a normal distribution $N(12.94,11.8)$.
- Electric vehicle charging power is usually taken in three ways: AC level 1, AC level 2 and DC charging, and the ratio of the three charging powers is set at 10%, 40% and 50%.

Monte Carlo simulation was carried out with a sample of 10,000 EVs to obtain the load curve of EV disorderly charging and discharging. The simulation results in Figure 1 shows that the charging behavior in the case of EV disorderly charging is not regulated and depends entirely on the behavior habits of vehicle owners, and the charging behavior of EVs generally starts immediately at the end of the owner's trip, and reaches 18:00 in the evening peak.

Figure 2 shows the daily load curve of a region in summer, Pload1 is the daily load curve of the region, and Pload2 is the daily load curve after superimposing the disorderly charging behavior of electric vehicles. It can be seen that there are two peak periods of electricity consumption in the area, among which 18:00-23:00 is the maximum peak period of the day, and it is the peak period of disorderly charging of electric vehicles.

3. Constraints for grid-connected cooperative control of electric vehicles

It is necessary to consider the system power conservation constraint, renewable energy output constraint, in-transit energy storage system constraint, electric vehicle capacity, charging and discharging power constraint and other conditions to make the model construction more realistic and reasonable.

3.1. Equation constraints for the conservation of supply and demand power

The electric vehicle grid-connected cooperative control model needs to comply with the equation constraint of conservation of supply and demand power, satisfying the power provided by energy storage devices, renewable energy sources and the grid to be equal to the power consumed by the load.

$$P_{load}(t) = P_{WT}(t) + P_{PV}(t) + P_x(t) + P_{bat}(t) + P_{EV}(t) \quad (1)$$

$P_x(t)$To exchange power between the grid and the microgrid, the grid can both supply power to the microgrid system and recover power from the microgrid system at a lower price than the current time-of-use tariff, meet:

$$P_x(t) = \lambda P_1(t) - \mu P_2(t) \quad (2)$$

λ, μ is the grid exchange power factor, when $\lambda = 1$ the grid provides power to the microgrid system $P_1(t)$, when $\mu = 1$ power is purchased from the microgrid system $P_2(t)$.

3.2. Energy storage system capacity and power constraints

In order to avoid overcharge and overdischarge, the upper and lower limits of the charge state and charge/discharge power of the energy storage system should be set, the SOC and P_{bat} should meet:

$$SOC_{min} \leq SOC(t) \leq SOC_{max} \quad (3)$$

$$P_{bat,min} \leq P_{bat}(t) \leq P_{bat,max} \quad (4)$$

3.3. Renewable Energy Capacity Constraints

The actual output of wind power and photovoltaic should be within a certain range if abandoned wind and light are allowed.

$$P_{WT,min} \leq P_{WT}(t) \leq P_{WT,max} \quad (5)$$

$$P_{PV,min} \leq P_{PV}(t) \leq P_{PV,max} \quad (6)$$

3.4. Electric vehicle constraints

As a distributed energy storage device, the electric vehicle has two working states, charging and discharging, which can be used as a load to obtain power from the grid for charging, and as a power source to discharge to the grid during peak load periods.

$$P_{EV}(t) = \alpha P_{dis}(t) - \beta P_{cha}(t) \quad (7)$$

α, β is the electric vehicle operating state parameter. At any moment, the energy storage system works in only three ways: charging, discharging, and idle, and the operating state parameters and power corresponding to various operating states are shown in Table 1.

Table 1. Parameters in different operating conditions

	Charging	Discharge	Free
α	0	1	0
β	1	0	0
P_{cha}	$P_{cha} > 0$	$P_{cha} = 0$	$P_{cha} = 0$
P_{dis}	$P_{dis} = 0$	$P_{dis} > 0$	$P_{dis} = 0$

Thus the working state constraint is satisfied:

$$P_{cha} \cdot P_{dis} = 0 \tag{8}$$

Without considering the self-discharge of electric vehicle battery, the electric vehicle power and its power meet:

$$SOC_{EV}(t) = SOC_{EV}(t-1) + \frac{P_{cha}\Delta t\eta - P_{dis}\Delta t\eta}{E_N} \tag{9}$$

η is the charging and discharging efficiency of the battery, and E_N is the rated capacity of the battery.

In order to avoid overcharging and overdischarging of electric vehicles that affect their service life and cause safety hazards, it is necessary to meet the inequality limits of battery capacity and charging and discharging power as follows:

$$SOC_{EV,min} \leq SOC_{EV}(t) \leq SOC_{EV,max} \tag{10}$$

$$P_{EV,min} \leq P_{EV}(t) \leq P_{EV,max} \tag{11}$$

4. Consider the objective function of the optimal economy on the grid side and the customer side

In order to achieve the optimal overall benefit, it is necessary to take into account both the interests of the user side and the economics of the grid side. A day is used as a dispatching cycle, 15min is a time interval and the cycle is divided into 96 time periods.

4.1. Optimal objective function of user-side economy

Considering at the customer level, it is desired that the total cost of its electricity consumption is the lowest. The sources of power obtained by the load are mainly the grid and renewable energy sources. The load obtains power from the main grid with a time-of-use tariff billing, and the power exchange cost of the load from the renewable energy side depends on the generation cost of wind PV.

$$f_1 = min\, C_{sum} = \sum_{t=1}^{96}(P_{WT}(t)C_{WT} + P_{PV}(t)C_{PV} + C_{bat}(t) + C_x(t) + C_{EV}(t)) \tag{12}$$

$$C_x(t) = \lambda P_1(t)C_1(t) - \mu P_2(t)C_2(t) \tag{13}$$

$$C_{EV}(t) = \alpha P_{dis}(t)C_1(t) - \beta P_{cha}(t)C_2(t) \tag{14}$$

C_1 are the time-of-use tariffs issued by the grid. C_2 is the tariff for electricity sold to the grid.

4.2. Grid-side generation cost optimal objective function

Considering the generation efficiency and cost of the new energy side, the system's ability to consume new energy should be improved to reduce the abandoned wind and light as much as possible. The power of wind and light abandonment at a certain moment can be expressed as:

$$X(t) = P_{WT,max} - P_{WT}(t) \tag{15}$$

$$Y(t) = P_{PV,max} - P_{PV}(t) \tag{16}$$

Measured by grid-side economic indicators, it can be seen as the least loss of wind and light abandonment.

$$min\ C_{re} = C_{WT} \sum_{t=1}^{96} \left(P_{WT,max} - P_{WT}(t)\right) + C_{PV} \sum_{t=1}^{96} \left(P_{PV,max} - P_{PV}(t)\right) \tag{17}$$

5. Simulation analysis of electric vehicle orderly grid-connected cooperative control strategy

The linear weighting method is used to process the objective function and convert the multi-objective function into a single objective function.

$$F = \varphi f_1 + \psi f_2 \tag{18}$$

Weighting factor φ, ψ meet: $\varphi + \psi = 1$, the weights of the two objective functions are different and the weighting coefficients are set differently according to the different demands of optimal scheduling. If in the promotion stage of EV on-grid operation mode, in order to motivate users to participate in EV on-grid optimal dispatching, it is necessary to increase the revenue of users' participation in EV charging and discharging and increase f_1. In the long term, grid-side economy is the main objective, so this paper focuses on grid-side economy with the weighting coefficients set to $\varphi = 0.3$ and $\psi = 0.7$. The results of the daily load curve comparison before and after optimization are shown in Figure 3.

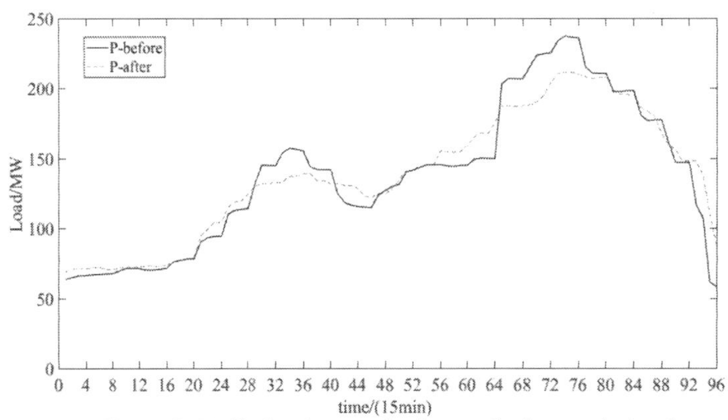

Figure 3. Daily load curve before and after optimization

The cooperative control strategy of electric vehicle orderly grid connection studied in this paper can realize load leveling, which has a certain effect of peak-shaving and valley-filling for the grid.

Figure 4 and 5 show the wind power before and after optimization. Before optimization, the renewable energy cannot be consumed and the wind and light abandonment rate is high, especially during the 0:00-7:00 hours, when the load of the grid is low, the cost of power purchased from the grid is low, and the main source of power consumed by the load is the grid exchange, so it will cause the wind power cannot be consumed. When the grid-side economics are taken into account to optimize the regulation of the system, the abandoned wind and light rate is significantly reduced, effectively improving the utilization rate of wind power and photovoltaic.

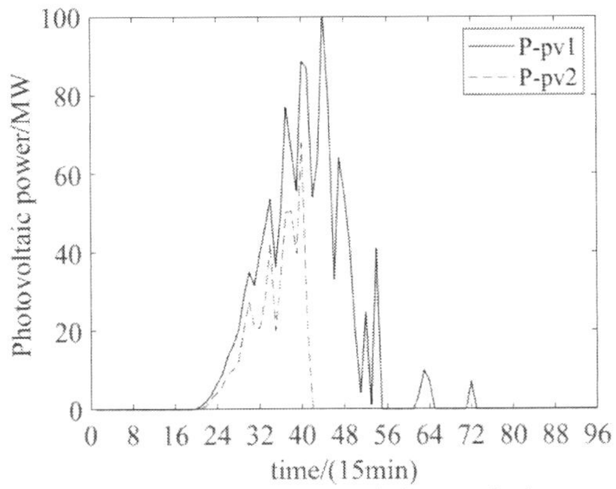

Figure 4. Abandonment of photovoltaic power

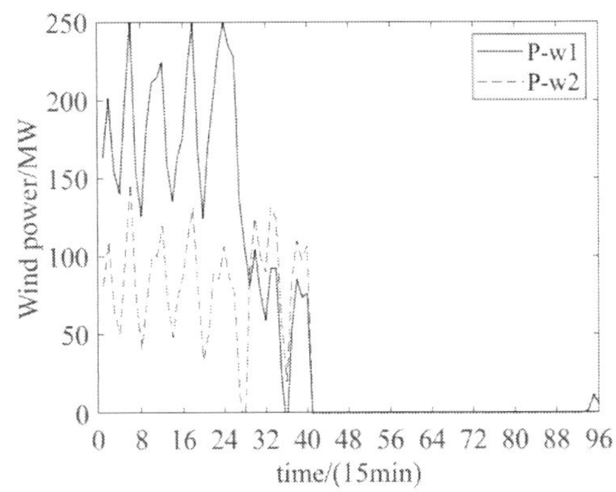

Figure 5. Abandonment of wind power

6. Conclusion

In this paper, we construct a model with the optimal economics on the user side and the optimal cost of power generation on the grid side as the objective function, select the optimal weighting coefficients, and use linear weighting normalization to convert the multi-objective function into a single objective function for simulation solution.

- An orderly grid-connected cooperative scheduling control strategy based on distributed energy storage for electric vehicles is proposed, which takes into account the willingness of users to participate in grid scheduling and satisfies the needs of users to the greatest extent.
- The charging and discharging price is used as a lever to guide users to orderly charging and discharging, so that users can stagger their electricity consumption and achieve load leveling.
- The optimal weighting coefficients are selected according to the demand of optimal scheduling, and the linear weighting method is used to transform the complex multi-objective function into a single objective function.

This strategy can significantly reduce the rate of abandoned wind and light, which provides an idea for the research of orderly grid integration of electric vehicles as distributed energy storage units.

Acknowledgments

Funded by the 2021 R&D Science and Technology Project of State Grid Heilongjiang Electric Power Company Limited: Research on the Influence of Waveform Distortion and Harmonics on Line Loss and Evaluation Method under Nonlinear Load (52243720003B).

References

[1] Chengshan, W., Zhen, W., Peng, L. (2014) Application prospects and challenges of distributed electric energy storage technology. J. Power System Automation, 38(16): 1-8+73.

[2] Kempton, W., Letendre, S.E. (1997) Electric vehicles as a new power source for electric utilities – ScienceDirect. J. Transportation Research Part D: Transport and Environment. 2(3): 157-175.

[3] Yang, B., Wang, L., Liao L., Chenglin, C., Ji, L., Xu D. (2015) Dong F. Research on collaborative and orderly control of electric vehicles and distributed power sources. J. Journal of Electrotechnology. 30(14): 419-426.

[4] Kejun, Q., Meng, Q., Yuanjun, S., Zhenkai, Z., Yi, L., Ying, X., Lijuan, C. (2020) Multi-objective optimal scheduling of charging stations based on fuzzy affiliation. J. Power Construction, 41(02): 118-124.

[5] Biao, H., Xujun, J. (2020) Research on the voltage regulation method of distribution network with distributed power supply and electric vehicle. J. Automation and Instrumentation, 35(10): 5-9+19.
[6] Zhongfu, T., Mianbin, W., Rong, Z., Jianxun, Q., Chengwen W. (2017) A hierarchical optimization model for peak-valley tariff linkage between generation side and supply side .J. Power System Automation,(21): 26-29+34.
[7] Shirong, D., Xia, L., Daowei, C., Tao, Y., Yi Y. (2013) Research on peak-to-valley charging and discharging tariff for electric vehicles[J]. Power Grid and Clean Energy, 29(07): 77-82+91.
[8] Zheng, Z., Rongxiang Z., Huan, Y., Lei J. (2012) Coping strategies for high penetration of electric vehicles and optimal planning of switching stations. J. Power Automation Equipment, 32(09): 7-13.
[9] Xiaoming, S., Wei, W., Su, S., Juchun, J., Lijie, X., Xuanhu, H. (2013) Design of an orderly charging control strategy for electric vehicles based on time-sharing tariff. J. Power System Automation, 37(01): 191-195.

Analysis on the research method of road sense feedback in steer-by-wire system

Chuanbin Wei[1], Lizhu Zhang[1*], You Fu[1] and Faying Xia[1]

[1]School of Automobile and Transportation, Tianjin University of Technology and Education, Tianjin, 300222, China

[*]Corresponding author's email: qcxy303@tute.edu.cn

Abstract. Nowadays, the automobile industry is gradually developing towards the trend of electrification and intelligence. Compared with the traditional steering system, the steer-by-wire system cancels the mechanical transmission structure, reduces the space utilization, reduces the probability of damage to the driver caused by the steering system in the collision accident, and improves the driving portability and enhances the driver 's handling experience. The road feeling feedback of steer-by-wire system has the greatest impact on the driver ' s driving experience. This paper discusses the research methods of road feeling feedback of steer-by-wire system, introduces the basic structure of road feeling feedback of steer-by-wire system, the basic idea of dynamic modeling, the establishment of simulation model of road feeling feedback, and the establishment method of control strategy and simulation platform of road feeling feedback. Finally, it summarizes and prospects in order to provide basic information and perspectives for the development and research of steer-by-wire system.

1. INTRODUCTION

At present, the rapid development of China 's economy and the rising voice of environmental protection have greatly promoted the development of new energy vehicles. With the development of new energy vehicles, the traditional mechanical structure can no longer meet the intelligent control requirements of new energy vehicles.

The steering-by-wire system has become a new generation of steering system, which cancels the mechanical connection between the steering wheel and the steering actuator and adopts the wire beam connection control. It reduces the use of the cab space, improves the angular transfer characteristics, improves the force transfer characteristics, and is conducive to communication and integration with other active control systems, and is the typical foundation for the realization of automobile autonomous steering[1]. The existing steering system is that the return torque generated by the tire is transmitted to the driver through the mechanical steering system, and the information transmitted to the driver also includes the road adhesion coefficient, road roughness, steering degree and tire pressure. The steer-by-wire system cancels the mechanical connection and cannot transmit the feedback torque through the mechanical structure. It is necessary to collect the data of the vehicle through the sensor, calculate the torque that can truly reflect the road feel, and transmit the torque to the driver through the simulation of the road feel motor, so that the driver can control the vehicle more accurately.

2. Brief introduction of road feeling feedback of steer-by-wire system

The steer-by-wire system is composed of steering wheel assembly, steering execution assembly and controller. In addition, there are fault diagnosis system, power supply system and other auxiliary systems.

The schematic diagram of steer-by-wire system structure is shown in Figure 1.

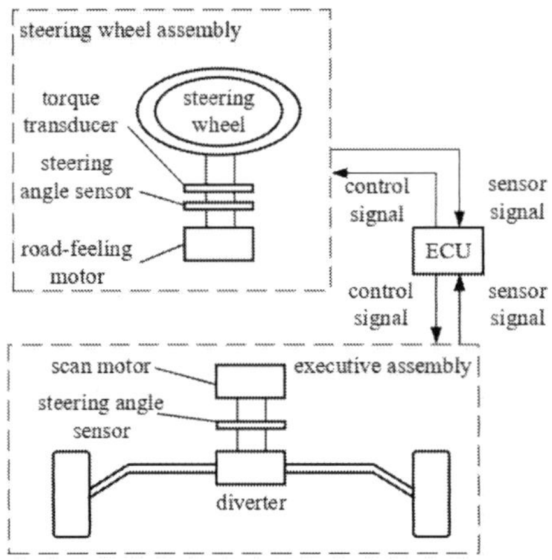

Figure 1. Structure diagram of steer-by-wire system

The steering wheel assembly collects steering wheel torque signal through torque sensor and angle sensor and converts it into electrical signal to transmit to ECU, so as to realize the control function of the lower steering executive assembly. At the same time, the torque signal transmitted by the ground is transmitted to the road sensor motor through the ECU. The road sensor motor simulates the actual road feeling, so that the driver can understand the real-time state of the vehicle and perceive the road information.

The steering executive assembly can receive the control signal from the ECU, and drive the steering mechanism to steer through the steering motor. At the same time, the information of the vehicle can be collected through the sensor and transmitted to the ECU for the generation of road feel signal to control the road feel motor.

ECU is the core part of the steer-by-wire system. The steering wheel transfer information, steering feedback motor and front wheel steering motor control are all completed. Moreover, with the development of science and technology, the scope of the role of ECU is also increasing. When the steering execution or steering feedback command is obtained, the main controller makes reasonable adjustments according to the vehicle driving state and road condition information to ensure the safety and stability of the vehicle in the driving process to the maximum extent[2].

For the control mode of steering feel feedback, it can be divided into three categories:

- Measurement method: The sensor is used to measure the return torque of the gear rack in the steering structure, and then the road induction motor is used to carry out the feedback torque of the simulator. However, this method requires high accuracy and high cost of the sensor.
- Fitting method: By establishing the functional relationship between the feedback torque and the relevant parameters, the size of the feedback torque is estimated.
- Dynamic model method: by combining a corner information, vehicle driving state, and according to the dynamic model of the vehicle to estimate the way out feel torque and compensation system inherent torque to obtain the desired feedback torque.

The dynamic model method is designed based on the dynamic model of the vehicle, and the final feedback torque combines the results of vehicle driving state and driver operation. Compared with other control methods, the estimation results of the dynamic model method are more accurate, so the most important control method is based on the vehicle dynamic model. The follow-up content of this paper mainly introduces the research method of road feeling feedback based on dynamic model.

3. Vehicle Dynamics Modeling

Before the simulation experiment, it is very important to establish the dynamic model of the steer-by-wire system to analyze the steering characteristics of the vehicle, which is related to the effectiveness of the subsequent simulation experiments. The steering wheel dynamic model and vehicle lateral dynamic model need to be established for the linear control chassis system.

The parameters related to the steering wheel dynamic model include input torque, system inertia, road induction motor torque, damping and friction. The dynamic equation is as follows [3,4].

$$\tau_{hw} = J_{hw}\ddot{\theta}_{hw} + B_{hw}\dot{\theta}_{hw} + K\left(\theta_{hw} - \frac{\theta_m}{i_s}\right) + f_{hw} \quad (1)$$

$$\tau_m = J_m\ddot{\theta}_m + B_m\dot{\theta}_m + \frac{K\ (\theta_{hw} - \frac{\theta_m}{i_s})}{i_s} \quad (2)$$

In the formula, τ_{hw} is the steering wheel torque, θ_{hw} is the steering wheel angle, J_{hw} / J_m steering wheel / motor inertia, B_{hw} / B_m steering wheel / motor damping coefficient, f_{hw} is the direction friction torque, τ_m is the road induction motor torque, K torsional stiffness, θ_m is the road induction motor angle, i_s is the deceleration ratio.

For vehicle lateral dynamics model, the following assumptions should be established[3]:

- It is considered that the characteristics of the left and right sides of the car are consistent, and the change of the return torque caused by the change of the left and right wheel load is ignored.
- Ignore the road and tire contact normal force on the tire during driving, and ignore the air friction resistance.
- It is considered that the left and right wheel angles of the vehicle are consistent when steering.

Based on the dynamic analysis of the vehicle, the longitudinal and lateral dynamic equations of the vehicle and the dynamic equations of the yaw direction obtained by the torque balance of the vehicle around the Z axis are established.

4. Road sense feedback control model

For the road inductance torque of the linear control chassis steering system, it can be divided into the upper return torque and the lower road inductance torque. The upper return torque controls the road-induction motor to return to the steering wheel quickly without external force through appropriate control algorithm. The lower layer can estimate the road feel moment by the estimation method of design parameters.

The torque dynamic equation of road sense feedback of steer by wire vehicle can be expressed by the following formula [5]:

$$\tau = \tau_1 + \tau_2 \quad (3)$$

$$\tau_1 = J\ddot{\theta} + B\dot{\theta} + \tau_d \quad (4)$$

$$\tau_2 = \tau_{da} + \tau_{in} + K\ \ (\tau_{al} + \tau_{ja}) \quad (5)$$

where, τ_1 represents the upper return torque of the steering wheel, τ_2 represents the lower road inductance torque, θ represents the rotation angle of the steering wheel, B represents the viscous friction coefficient, J represents the rotational inertia of the steering wheel, τ_d denotes motor torque disturbance, τ_{da} and τ_{in} represent the equivalent damping and inertia torque of the system, K represents the road inductance torque gain, τ_{al} represents the sub return torque, and τ_{ja} represents the lifting torque.

Through the above dynamic equation, the parameters in the formula can be further decomposed and analyzed in detail to obtain a model that can more accurately reflect the road feeling.

5. Road sense feedback control strategy

The control strategy of road sense feedback is shown in Figure 2.

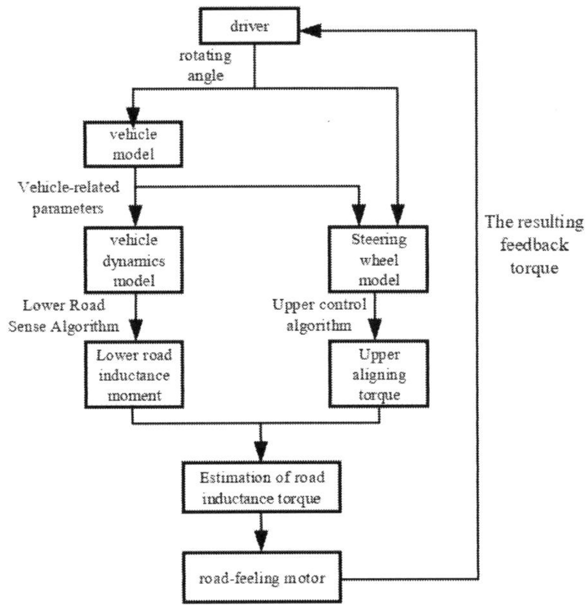

Figure 2. Schematic diagram of inductive feedback control strategy

Firstly, the rotation angle is input to the system, and the relevant information is transmitted to the steering wheel model and the vehicle dynamics model by combining the relevant parameters of the vehicle. The upper return torque and the lower road inductance torque are calculated by the corresponding algorithm, and then the road inductance torque is calculated. The corresponding information is transmitted to the road inductance motor, and the torque generated by the road inductance motor is fed back to the driver through the steering wheel.

For the calculation of the upper aligning torque, the common control strategy is PID control, which is close to the real algorithm by controlling the coefficients of proportional, integral and differential. The principle is simple, stable and reliable, but it is difficult to ensure the feedback transmission of the torque in the calculated response time domain. In the literature [6], the fuzzy algorithm is used to adjust the parameters, and the simulation results are better improved in response time and overshoot. In addition, it can be controlled by the traditional sliding mode control method, which has strong robustness to the dynamic change of model parameters and interference problems, and the algorithm design is simple. On this basis, [2] in the literature adopts the terminal sliding mode control algorithm, so that the calculation time can be controlled within a certain period of time, and the uncertainty of modeling parameters is considered. The influence of system uncertainty is eliminated, and the road information can be transmitted to the driver more accurately. In the follow-up study, the neural network algorithm can be considered to further optimize the upper torque algorithm. For the estimation of the lower road inductance torque, it is more difficult than the calculation of the upper aligning torque. It is necessary to reflect the real information of the road that the vehicle is driving. The dynamic characteristics of the tire sideslip, the tire lateral force and the change of the aerodynamic trajectory should be considered. The control strategy in Reference assumes that the sideslip angle is small, and some unknown parameters are estimated by the least square method. The parameters that are easy to obtain are used as far as possible to estimate, and the lower road inductance torque is obtained and fed back to the road inductance motor for output according to a certain gain.

6. Road Feedback Simulation Platform

The above contents include the establishment of dynamic model, the establishment of road feedback control model and the idea of road feedback control strategy. In the actual research process, the simulation research can be carried out after the above operations are completed.

The above contents include the establishment of dynamic model, the establishment of road feedback control model and the idea of road feedback control strategy. In the actual research process, the simulation research can be carried out after the above operations are completed. For the simulation of road feedback research, most of the literature used the method of Matlab / Simulink and CarSim co-simulation for research. The overall simulation platform structure diagram as shown in Figure 3 :

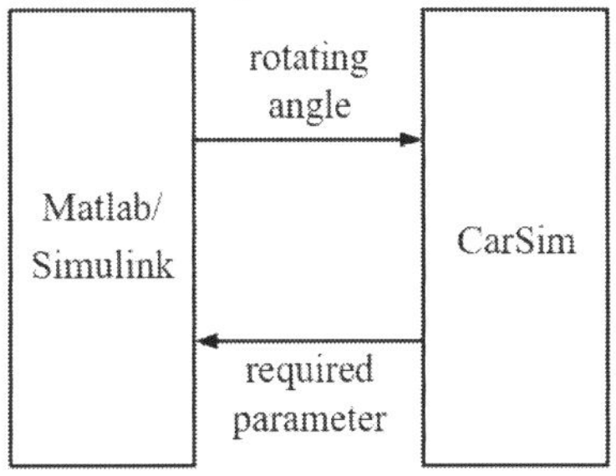

Figure 3. Structure of the overall simulation platform

In the simulation, the Matlab / Simulink simulation is used to generate the corner signal and the signal is transmitted to CarSim, and then CarSim obtains the corner information. The simulation is carried out according to the set automobile model and environmental parameters, and the required parameters are transmitted to Matlab / Simulink in real time. Matlab / Simulink processes the data and controls the operation of the vehicle according to the parameters and the designed control algorithm.

After completing the above operation, the simulation results are evaluated and the optimization control algorithm is continuously analyzed to obtain better road feel feedback.

This method can make the simulation conform to the actual situation to the greatest extent, and can carry out real-time control of the vehicle model, obtain more accurate results, ensure safety and save costs.

7. Conclusion

With the continuous development of the automobile industry, the requirements for automobile chassis are also increasing. Intelligent lightweight chassis has become one of the directions of future development. The wire control chassis cancels the mechanical structure and uses the wire harness to control. While reducing the weight of the chassis, it can be more intelligent and lightweight, which meets the requirements of future development. At present, the development of steer-by-wire technology is still in the stage of continuous improvement, and some technologies need to be broken through. This paper introduces the basic structure of road feedback of steer-by-wire, the basic idea of dynamic modeling, the establishment of road feedback simulation model, and the establishment method of road feedback control strategy and simulation platform in order to provide basic research ideas for more researchers in the future. For the study of road sense feedback, it is still difficult to make more accurate dynamic modeling, and the ideal assumption of the model leads to the real state of the system cannot be obtained, which reduces the accuracy of feedback. At the same time, the research on the upper control algorithm can also be more in-depth. For example, the neural network method is used to design the control algorithm in order to obtain more realistic design results.

Some technical problems of the wire control system still need to be solved, and it is difficult to promote the application in large quantities on vehicles at this stage. However, it is believed that with the rapid development of new energy vehicles and automobile electronic technology, the problem of line control system will be gradually solved, and it will conform to the trend of the times and combine with intelligent Internet vehicles to become a new generation of chassis system.

References

[1] Feng, Z.X., (2019) Research on Control Strategy of Vehicle Steer-by-Wire System Based on CarSim, Shandong University of Technology, Zibo.

[2] Wang, J.W., (2020) Design and Research of Steering Feel for Steer-By-Wire Vehicle Control System, Hefei University of Technology, Hefei.
[3] Mitschke, M., Henning, W., (2010) Dynamik der Kraftfahrzeuge(4. Auflage), Tsinghua University Publishing House, Beijing.
[4] Tian, D.Q., (2006) Research on Control Method for Electric Power Steering System Based on Robust Control Theory, Sichuan University, Chengdu.
[5] Wang, H., Zhi, H.M., Chen, W.X., Zheng, J.C., (2013) Robust sliding mode control for Steer-by-Wire systems with AC motors in road vehicles, In: 2013 IEEE 8th Conference on Industrial Electronics and Applications (ICIEA), Melbourne, pp. 674-679.
[6] Tian, j., Wu, J., Chen, N., Luo, S., (2011) Research on road feeling control of Steer-by-wire System, In: 2011 Second International Conference on Mechanic Automation and Control Engineering, Inner Mongolia, pp. 4349-4353.

Single-ended fault location and early warning method of transmission line based on back propagation neural network

Han Zhou*, **Minghui Liu, Xin Yu, Weiyang Wang and Jingyao Gao**

State Grid Dandong Electric Power Supply Company, Dandong, 118000, China

*Corresponding author' s e-mail: zhengy3_dl@ln.sgcc.com.cn

Abstract. For power transmission systems, accurate and reliable fault location methods can ensure rapid recovery of faulty lines and improve power supply reliability. In order to solve the problems of the structural complexity of the transmission system and the difficulty of line fault location, a single-ended fault location and early warning method of transmission line based on back propagation neural network is proposed. First, the fault line selection is performed quickly when the fault occurs. Then, the voltage fault components collected at the measuring point when the fault occurs are decomposed and reconstructed by wavelet packet to obtain the wavelet packet energy, which is used as the input sample to train through the nonlinear fitting ability of back propagation. With the help of backpropagation neural network, arbitrary complex functions can be processed, and the learning results can be accurately used for new knowledge, and circuit faults can be diagnosed conveniently and quickly. Finally, the corresponding fault distance can be output by substituting the wavelet packet energy reflecting the fault location. The simulation results show that the method has strong resistance to transition resistance and high positioning accuracy.

1. Introduction

With the gradual increase in the coverage of the grid, compared to the traditional power grid, a significant increase in the probability of failure of the transmission line and the difficulty of locating the fault[1]. Most of the faults that occur on the transmission line are transient faults, that is, the faults that can be reclosed by the automatic reclosing device[2]. Although the damage traces caused by the transient short-circuit fault are not obvious, it seriously affects the insulation strength of the line and is likely to induce more serious faults again. The rapid and accurate location of faults in long-distance transmission lines is of great significance for reducing the burden of line inspection, quickly restoring faults, and ensuring the safe operation of power systems[3]. The traveling wave method calculates the distance to the fault according to the traveling wave propagation time between the measuring observation point and the fault point [4]. Although it has a high ranging accuracy, it requires additional investment in a special primary sensor, which has poor reliability. Transition resistance is a key factor affecting fault location. It has shortcomings such as complicated operations, pseudo-roots, and high requirements for synchronization. Improving the robustness of the ranging algorithm is a key issue in fault location research.

Literature [5] is based on multi-level back-propagation neural network, data fusion of single-ended multi-dimensional electrical quantities and calculation of fault location. However, due to insufficient research on the variation characteristics of electrical quantities, there are shortcomings of large input electrical quantities, complex network training, and only suitable for small resistance faults. Literature [6] studied the structure of neural network and proposed an optimal network structure suitable for fault location problems. However, the training sample set in this article does not cover the complete fault mode space, and only considers the grounding of small transition resistance. Therefore, this paper presents a fault location method for transmission line based on BP neural network. First, analyze the change of electrical quantity with transition resistance. Then, a distance measurement neural network based on the classification of transition resistance is constructed. Finally, the training method is improved, and the repeated sample pairs composed of error-free and error-free samples are used as the training input samples of the BP neural network, so that the trained ranging network has a certain ability to adapt to random errors.

2. Fault traveling wave analysis and location

2.1. Fault traveling wave analysis

This paper provides a system simulation diagram of a 525 kV transmission line, as shown in Figure 1. The high-voltage transmission line is a three-phase transmission line, and there is phase coupling, so decoupling transformation is required first. This paper uses Karen Bell transform. Taking the current traveling wave as an example, if the transformation matrix of the current is Q, then Karen Bell principle of phase-mode transformation:

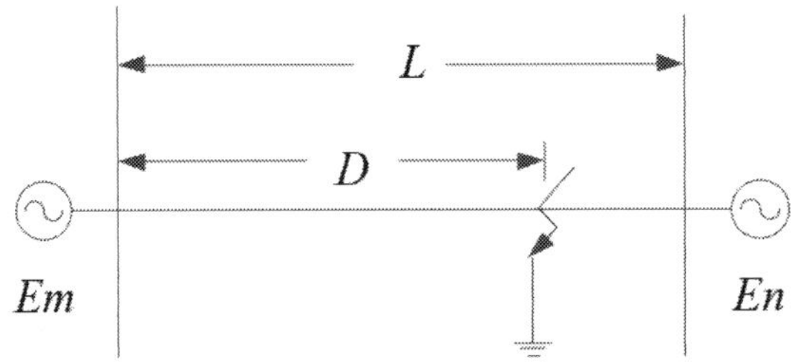

Figure 1 Simulation diagram of fault system

$$i = Qi_m \quad (1) \qquad i_m = \begin{bmatrix} i_0 \\ i_\alpha \\ i_\beta \end{bmatrix} \quad (2) \qquad Q = \begin{bmatrix} 1 & 1 & 1 \\ 1 & -2 & 1 \\ 1 & 1 & -2 \end{bmatrix} \quad (3) \qquad Q^{-1} = \frac{1}{3}\begin{bmatrix} 1 & 1 & 1 \\ 1 & -1 & 0 \\ 1 & 0 & -1 \end{bmatrix} \quad (4)$$

After transformation, we get:

$$\begin{bmatrix} i_0 \\ i_\alpha \\ i_\beta \end{bmatrix} = \frac{1}{3}\begin{bmatrix} i_a + i_b + i_c \\ i_a - i_b \\ i_a - i_c \end{bmatrix} \quad (5)$$

α , β sequence impedance:

$$Z_\alpha = Z_\beta = \sqrt{\frac{L_1}{C_1}} \quad (6)$$

α , β sequence wave velocity:

$$v_\alpha = v_\beta = \frac{1}{\sqrt{C_1 L_1}} \quad (7)$$

In formula (7), L_1 and C_1 are the positive sequence inductance and capacitance of the unit length line respectively. Through the change of wave speed, it can be judged whether there is a fault, etc.

3. Fault location and early warning based on BP neural network

3.1. Single-ended traveling wave fault location algorithm

According to the law of traveling wave refraction and reflection, the fault area is divided into (0, L/3), (L/3, L/2) in the time range of 3~0τ (τ refers to the time from the initial traveling wave to the measurement end through the fault point) , (L/2, L/3), (2L/3, L) four sub-sections (L refers to the distance from the initial traveling wave to the measurement end through the fault point). Assuming that the fault occurs from the initial time t_0, the moments when the initial transient current traveling wave, the first or second type traveling wave, and the first, second or third type traveling wave arrive at the M terminal are t_1, t_2,..., t_p, etc. It can be further subdivided into this article, using about 4 moments at the M end of the measurement bus to calculate the average wave velocity v of the transient traveling wave and the fault distance x, and the ranging algorithm for each section is derived as follows:

Section 1 (0，L/3):

$$\begin{cases} v(t_1 - t_0) = x \\ v(t_2 - t_0) = 3x \\ v(t_3 - t_0) = 5x \end{cases} \quad (8)$$

In the formula, the average propagation speed of transient traveling wave v, the unknown distance to fault x, the first to third moments when each traveling wave head arrives at the M terminal in turn are t_1, t_2, t_3, and the maximum value of the wavelet transform is used for calculation :

$$\begin{cases} v(t_1 - t_0) = x \\ v(t_2 - t_0) = 3x \\ v(t_4 - t_0) = 2L - x \end{cases} \quad (9)$$

Solve the equation and get the result: $x = \dfrac{(t_2 - t_1)L}{-2t_1 + t_2 + t_4}$ (10)

Section 2 (L/3，L/2):

$$\begin{cases} v(t_1 - t_0) = x \\ v(t_2 - t_0) = 3x \quad t \le 3\tau \\ v(t_3 - t_0) = 2L - x \end{cases} \quad (11)$$

Solve the equation and get the result: $x = \dfrac{(t_2 - t_1)L}{-2t_1 + t_2 + t_3}$ (12)

Section 3 (L/2，L/3):

$$\begin{cases} v(t_1 - t_0) = x \\ v(t_2 - t_0) = 2L - x \quad t \le 3\tau \\ v(t_3 - t_0) = 3x \end{cases} \quad (13)$$

Solve the equation and get the result: $x = \dfrac{(t_3 - t_1)L}{-2t_1 + t_2 + t_3}$ (14)

Section 4 (2L/3，L):

$$\begin{cases} v(t_1 - t_0) = x \\ v(t_2 - t_0) = 2L - x \\ v(t_p - t_0) = 3x \end{cases} \quad (15)$$

Solve the equation and get the result: $x = \dfrac{(t_p - t_1)L}{-2t_1 + t_2 + t_p}$ (16)

According to the analysis results of the above interval, it is found that the fault distance is only related to the line length L and time t_p, and has nothing to do with the traveling wave velocity v. Therefore, the distance measurement method is not affected by the transient traveling wave speed, and can be directly applied to the fault location and distance measurement of high-voltage transmission lines.

Journal of Physics: Conference Series **2121** (2021) 012032 doi:10.1088/1742-6596/2121/1/012032

Refraction and reflection occur when the faulty traveling wave reaches the bus. According to the literature, the expression of the voltage reflection coefficient ρ_w is:

$$\rho_w = \frac{Z_2 - Z_1}{Z_2 + Z_1} \quad (17)$$

In the case of a large transition resistance, the reflected wave at the fault point will be small, so that the direct ranging method will be affected, so it is also necessary to use the reflected wave head of the opposite bus to realize the ranging. The formula of the indirect method is as follows:

$$D_{mF} = L - \frac{1}{2} v \Delta t_2 \quad (18)$$

In formula (18), Δt_2 expresses the time difference between the reflected wave head of the opposite end bus and the first wave head. The traveling wave line mode voltage with small dispersion is selected for fault location. Therefore, the application of fault traveling wave single-ended electrical measurement can theoretically achieve the effect of double-ended electrical measurement.

3.2. The establishment of BP neural network model

In order to make the training sample set as comprehensive as possible to include the frequency characteristics of different fault distances, the first fault point is set 5 km away from the protection installation of the distance measuring end of the rectifier station of the Yun-Guang UHV DC transmission line. After that, a fault point is set every 10 km, and the setting of the fault point continues until the 132 sets of data are trained on the BP neural network by the inverter side ranging end protection, and the remaining 10 sets of data are used to verify the BP neural network. In order to make each input component play an equally important position in the neural network operation, the sample vector is normalized, and the input amplitude of each component is adjusted reasonably, so that the variation range is generally distributed in the (0,1) interval. Normalization processing helps to improve the convergence speed of training. After the above steps, the constructed BP neural network ranging model is shown in Figure 2:

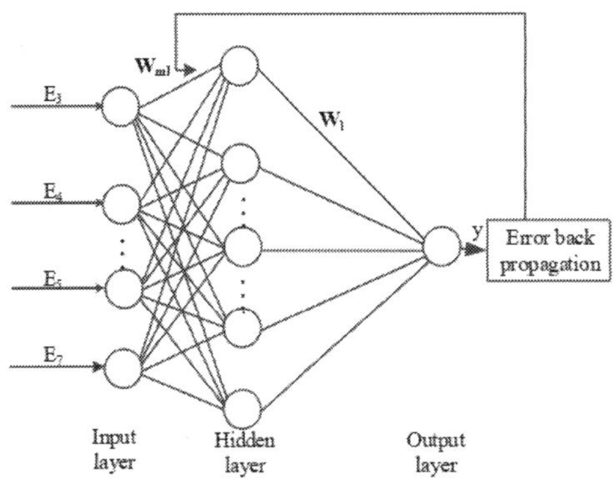

Figure 2 BP neural network ranging model

The built BP neural network ranging model has 1 input layer, the number of neurons in the input layer is 5; 1 hidden layer, the number of neurons in the hidden layer is determined to be 15 by trial and error; 1 output layer, the number of neurons in the output layer is 1, and the output is the fault distance predicted by the neural network model, so the topology structure of the BP neural network ranging model built is 5*15*1. The connection strength between the *m* th neuron of the input layer and the first neuron of the output layer is represented by the connection weight w_{ml}. The connection strength between the first neuron in the hidden layer and the output layer neuron is represented by the connection weight ω_l.

3.3. Early warning method

The correlation model between current amplitude and echo intensity, echo peak height, and connection weight is established through data, as shown in equation (19), and then the predicted value of current amplitude is obtained according to the predicted values of these parameters.

$$I = f\left(x_1, x_2, \omega_l\right) \qquad (19)$$

In formula (19): x_1 is the intensity of the echo; x_2 is the length of the echo.

In order to obtain the specific expression of formula (19), the historical M transmission line fault events are selected as samples for analysis and research. The input variables are three parameter values such as the echo intensity recorded when the transmission line trips. The output value is the maximum amplitude of the current at the time of tripping.

4. Simulation analysis during failure

When the model runs for 4 ms, a ground fault is added, and the fault point is set at a distance of 10 km from the head end of the rectifier side, with a step length of every 10 km. Set up single-phase ground faults in the two overhead lines until the end of the line, so that the training sample set contains the frequency characteristics of different fault distances as much as possible, and the fault grounding resistance is 0.001 Ω, 10 Ω, and 100 Ω, respectively. The fault traveling wave voltage component can obtain wavelet packet energy at different scales through wavelet packet transformation. Taking a fault 200 km away from the line measurement point and a transition resistance of 0.001 Ω as an example, through simulation, the transient voltage component of the fault is decomposed into 8 frequency bands after wavelet packet decomposition, and each frequency band represents a different frequency range, see table 1.

Table 1 The frequency band corresponding to each node in the third layer of wavelet packet decomposition

Node	Frequency band/Hz	Node	Frequency band/Hz
1	0-1250	5	5000-6250
2	1250-2500	6	6250-7500
3	2500-3750	7	7500-8750
4	3750-5000	8	8750-10000

In the Matlab sequence, set up and run 10 networks. After the training is completed, save the most accurate data of the results. The remaining 6 sets of data are used to verify the effect of neural network training. The final fault location results are shown in Table 2. In the same way, if the fault is on the right side of the T-connected busbar, 6 groups of data can be selected for verification after neural network training. The distance measurement results are shown in Table 3.

Table 2 Fault location results on the left side of the T-connected busbar

Fault distance /km	Measured value/km	Transition resistance/Ω	Absolute error/km	Relative error%
31	30.9567	0	0.0433	0.0014
89	89.0234	10	0.0234	0.0003
135	134.8416	0	0.1584	0.0012
239	237.2394	10	1.7606	0.0074
364	362.0287	100	1.9713	0.0054
569	571.1286	100	2.1286	0.0037

Table 3 Fault location results on the right side of the T-connected busbar

Distance from busbar /km	Fault distance/km	Measured value/km	Transition resistance/Ω	Absolute error/km	Relative error%
85	1025	1026.2542	100	1.2542	0.0012
289	1235	1233.1279	0	1.8721	0.0015
335	1289	1286.7524	100	2.2476	0.0017
439	1377	1374.9854	100	2.0146	0.0015
464	1435	1438.0157	10	3.0157	0.0021
569	1538	1540.2567	10	2.2567	0.0015

When the fault is on the right line of the T-connected busbar, it is verified in combination with Table 2 and Table 3. The results show that this method can also be accurately located with high accuracy. Through experiments, it can be seen that as the traveling wave head travels farther and farther, the traveling wave head is smoother; the greater the transition resistance, the smaller the initial traveling wave amplitude, which makes the wavelet packet affect the wave head recognition and the ranging accuracy. Through a large number of simulation analysis, it is found that line length changes and error factors do not reduce the effectiveness of wavelet packet energy spectrum analysis, and have a small impact on the positioning accuracy, and the ranging accuracy meets the requirements.

5. Conclusion

This paper studies a single-ended fault location and early warning method for transmission lines based on back-propagation neural network. The wavelet packet energy spectrum is used to extract the fault frequency band. Because the frequency band is rich in fault information, the spectrum energy is easier to extract than the natural frequency, and the operation is simpler, which is more effective when used for fault location. BP neural network has non-linear fitting ability. The energy of each layer obtained by wavelet packet decomposition of fault voltage is used as the input sample of BP. After training and fitting, the output is the fault distance. In the transmission line fault location, the sample set of this method is small and clear, and the convergence is good, which greatly improves the accuracy of fault location. In this paper, the simulation data obtained by establishing the simulation model is used for method verification, and the simulation model and the influencing factors considered are relatively simple. The number, type, noise and other simulation examples of bus outlets should be increased to make it closer to the actual situation.

References

[1] Kanagasabapathy, O. (2021). Fault Location in Transmission Line Through Deep Learning—A Systematic Review. Inventive Systems and Control, 223-238.

[2] Gonzalez-Sanchez, V. H., Torres-García, V., & Guillen, D. (2021). Fault location on transmission lines based on travelling waves using correlation and MODWT. Electric Power Systems Research, 197, 107308.

[3] Makmur, S., Ahmad, R. S., Muhammad, R. D., Rahimuddin, R., & Ikhlas, K. (2020). Faults detection and classification on parallel transmission lines using modified Clarke's transformation-ANN approach. Przegląd Elektrotechniczny, 96(NR 4), 23-27.

[4] Shi, Y., Zheng, T., & Yang, C. (2020). Reflected Traveling Wave Based Single-Ended Fault Location in Distribution Networks. Energies, 13(15), 3917.

[5] Mukherjee, A., Chatterjee, K., Kundu, P. K., & Das, A. (2021). Probabilistic Neural Network-Aided Fast Classification of Transmission Line Faults Using Differencing of Current Signal. Journal of The Institution of Engineers (India): Series B, 1-14.

[6] Kumar, A. N., Sridhar, P., Kumar, T. A., Babu, T. R., & Mohan, V. C. J. (2020). Adaptive neuro-fuzzy inference system based evolving fault locator for double circuit transmission lines. IAES International Journal of Artificial Intelligence, 9(3), 448.

Data-driven predictive control for the swing process of a cutter suction dredger

Chengxi Yu[1], Menghong Yu[2] and Wei Yuan[1*]

[1] School of Computer, Jiangsu University of Science and Technology, Zhenjiang, Jiangsu, 212100, China

[2] Ocean Equipment Research Institute, Jiangsu University of Science and Technology, Zhenjiang, Jiangsu, 212003, China

[*] Corresponding author's e-mail:192070001@stu.just.edu.cn

Abstract. The working process of cutter suction dredger is complex, so it is difficult to establish accurate model by traditional physical modeling method. The final yield of cutter suction dredger is directly related to the density of the slurry in the pipeline, while the density is controlled by the swing process. Therefore, by analyzing the measured data of the dredger, it establishes the swing process mathematical model based on subspace theory . The algorithm is based on input–output data, simulation results show the algorithm is effective for the control application.

1. Introduction

The cutter suction dredger is a kind of ship widely used in the dredging project at present. The operation first cuts the sediment on the bottom of the river by the cutter, then the slurry is sucked into the pipeline by the submerged dredge pump, and finally the slurry is pumped by the cabin pump and the relay pump for long-distance transportation. The operation process includes the swing process, the cutting process and the pipeline transportation process. The traverse cutting process is the most frequent operation, and the swing speed is directly related to slurry density. However, the current swing speed is often adjusted by the operator based on the dredging experience, which makes the dredging inefficient and can easily lead to the blockage of the pipe. And because in the actual construction operation, faced with the complex construction environment, varied soil quality, uncertain working conditions, making the control of swing speed becomes particularly difficult.

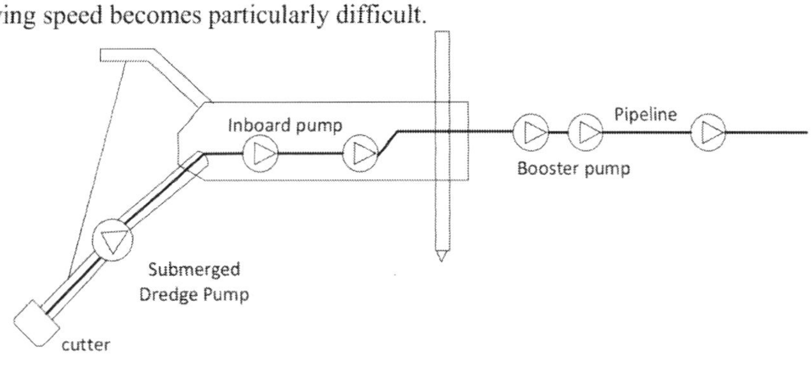

Figure 1. Brief description of cutter suction dredger.

Therefore, it is very important to control the process of lateral movement with high efficiency, high precision and high automation. Some research has been done on swing process in China, WEI Changyun[1] applied predictive control to the lateral movement process based on the RBF-ARX theory, Zhu Wenliang[2] proposed a modeling method of lateral movement process based on BP neural network, Song Dongpeng[3] proposed lateral movement control system based on BP neural network. The above control system requires a large amount of historical data to model the swing system, and due to the uncertainty of the construction environment, soil quality and working conditions, it can't achieve good control effect in the actual construction.

Based on the theory of subspace identification, this paper constructs the swing process model from dredging data, which can accurately predict the slurry density. After the prediction model is obtained, rolling optimization and feedback regulation are carried out. The swing process model established in this paper can predict the slurry density through the swing speed, which meets the requirements of the prediction model. Rolling optimization and feedback regulation can make the swing system continuously optimized during operation, thus solving the problem that the dredger construction cannot be kept optimal due to external factors.

2. Subspace identification method

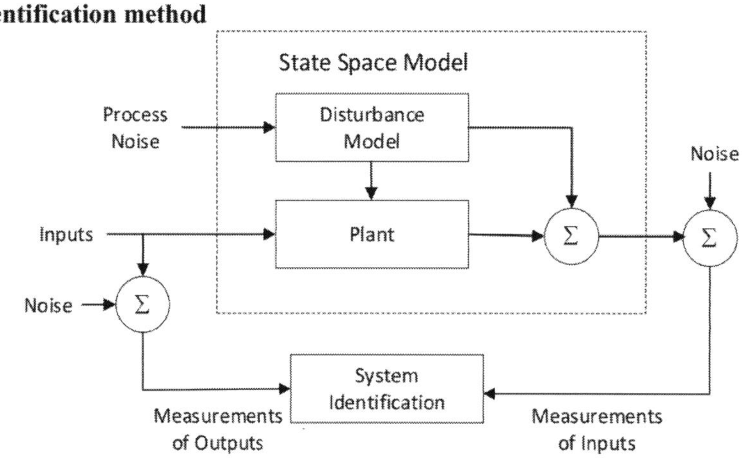

Figure 2. The frame of subspace state-space identification.

In the past decades, subspace identification method has made great progress in both theory and practice [4-5], provides an attractive alternative for the parameterization of the Multiple input multiple output (MIMO) system.

For a system with l input and m output, the prediction time domain is set as i, and the subspace prediction identification problem can be described as follows: in the case of known past input and output data W_p and future input data U_f, looking for the best prediction of future output Y_f.

$$\hat{Y}_f = L_1 W_p + L_2 U_f \tag{1}$$

Where, $L = [L_1 \quad L_2]$ is the corresponding subspace matrix, for $L_1 \in R^{(l \times N)\times((l+m)\times N)}$, $L_2 \in R^{(l \times N)\times(l \times N)}$, obtained by solving the following least squares problem.

$$\min_{L_1,L_2} \left\| Y_f - (L_1 \quad L_2)\begin{pmatrix} W_p \\ U_f \end{pmatrix} \right\|_F^2 \tag{2}$$

Then use QR decomposition to solve the least squares problem. The best prediction of future output $\hat{Y}_f$ could be described as an orthogonal mapping from the column space Y_f to the column spaces W_p and U_f:

$$\hat{Y}_f = Y_f / \begin{pmatrix} W_p \\ U_f \end{pmatrix} = L_1 W_p + L_2 U_f \tag{3}$$

The solution for equation (3) could be found by performing a QR-decomposition:

$$\begin{pmatrix} W_p \\ U_f \\ Y_f \end{pmatrix} = \begin{pmatrix} R_{11} & 0 & 0 \\ R_{21} & R_{22} & 0 \\ R_{31} & R_{32} & R_{33} \end{pmatrix} \begin{pmatrix} Q_1^T \\ Q_2^T \\ Q_3^T \end{pmatrix} \tag{4}$$

Then:

$$L = \begin{pmatrix} R_{31} & R_{32} \end{pmatrix} \begin{pmatrix} R_{11} & 0 \\ R_{21} & R_{22} \end{pmatrix}^{\dagger} \tag{5}$$

Where $L = [L_1 \quad L_2]$, L_1 , L_2could be denoted as follow:

$$\begin{gathered} L_1 = L(:,1:i(l+m)) \\ L_2 = L(:,i(l+m)+1:i(2l+m)) \end{gathered} \tag{6}$$

Set the current time to k, then the past input/output and future input data as shown below:

$$\left(\begin{bmatrix} u(k-i) \\ u(k-i+1) \\ \vdots \\ u(k-1) \end{bmatrix}, \begin{bmatrix} y(k-i) \\ y(k-i+1) \\ \vdots \\ y(k-1) \end{bmatrix} \right) \qquad \begin{bmatrix} u(k) \\ u(k+1) \\ \vdots \\ u(k+i-1) \end{bmatrix} \tag{7}$$

Then the future output is shown as follows:

$$\begin{bmatrix} \hat{y}(k) \\ \hat{y}(k+1) \\ \vdots \\ \hat{y}(k+i-1) \end{bmatrix} = L_1 \begin{bmatrix} u(k-i) \\ u(k-i+1) \\ \vdots \\ u(k-1) \\ y(k-i) \\ y(k-i+1) \\ \vdots \\ y(k-1) \end{bmatrix} + L_2 \begin{bmatrix} u(k) \\ u(k+1) \\ \vdots \\ u(k+i-1) \end{bmatrix} \tag{8}$$

3. Model predictive control

In fact, MPC has been widely used in the dredging field, and the control effect is satisfactory[6]. Model predictive control can be described as a process of minimizing the following cost functions[7]:

$$J = \sum_{k_p=1}^{N_p} \left(y_{t+k} - r_{t+k} \right)^T W_Q \left(y_{t+k} - r_{t+k} \right) + \sum_{k_c=1}^{N_c} \Delta u_{t+k}^T W_R \Delta u_{t+k} \tag{9}$$

Where N_p and N_c are the prediction and control horizon, y_t is the output at time t, r_t is the setpoint at time t, W_Q and W_R are weighting matrices, Δu_t is the incremental input.

The system output is predicted by the predictive model obtained from the subspace identification:

$$\hat{y}_f = l_1 w_p + l_2 u_f \tag{10}$$

Where $\hat{y}_f = [\hat{y}_{t+1} \quad \hat{y}_{t+2} \quad \cdots \quad \hat{y}_{t+N_p}]$, $u_f = [u_{t+1} \quad u_{t+1} \quad \cdots \quad u_{t+N_c}]$, then the cost function can be redefined as below:

$$J = \left(r_f - \hat{y}_f\right)^T W_Q \left(r_f - \hat{y}_f\right) + u_f^T W_R u_f \tag{11}$$

By using the prediction equation (10) to minimize the cost function (11), the future input u_f can be obtained:

$$u_f = \arg\min_{u_f} \left\{ \left(r_f - \hat{y}_f\right)^T W_Q \left(r_f - \hat{y}_f\right) + u_f^T W_R u_f \right\} \tag{12}$$

Data-driven predictive control can be achieved by taking the first set of calculated data as input to the actual system and then cycling.

4. Experimental results and analysis

4.1 Results of subspace identification

The first step of data-driven predictive control is subspace identification. Only when the model identified in the subspace achieves good prediction results, the following predictive control can obtain accurate control results.

600 sets of data in the actual dredging process were selected, and the subspace identification was carried out with the swing speed as the input and the density as the output. After the identification, the output of the swing system model and the prediction errors as follows:

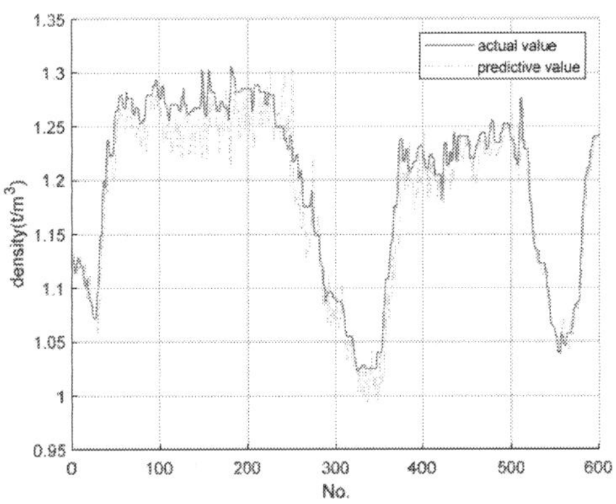

Figure 3. Comparison between actual density and predicted density.

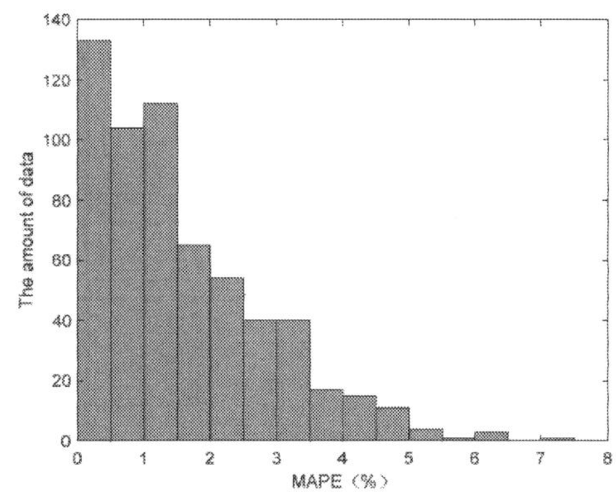

Figure 4. Mean Absolute Percentage Error.

The prediction error rate is concentrated around 1%, and the MAPE(Mean Absolute Percentage Error) is 1.8771%. The prediction results are relatively accurate.

4.2 Predictive control results

There are 50 groups of data in the cycle control, and the data collection is 1 group in 2 seconds, and the total operation time is 100 seconds.

When the setpoint is a fixed value of 1.27(t/m3), the control results are as follows:

2121 (2021) 012033 doi:10.1088/1742-6596/2121/1/012033

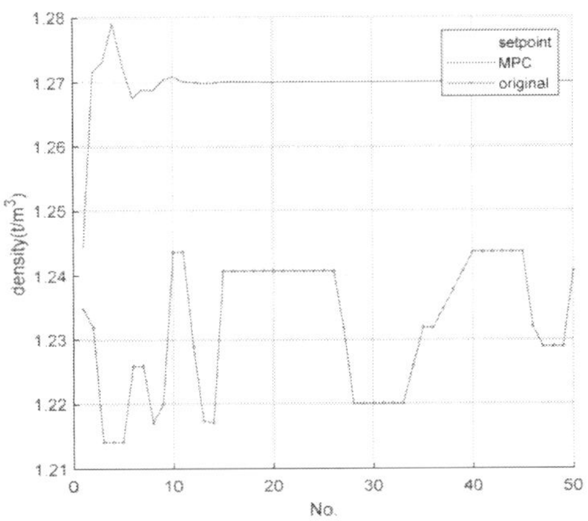

Figure 5. Comparison of setpoint, original value and MPC control value(when the setpoint is a fixed value).

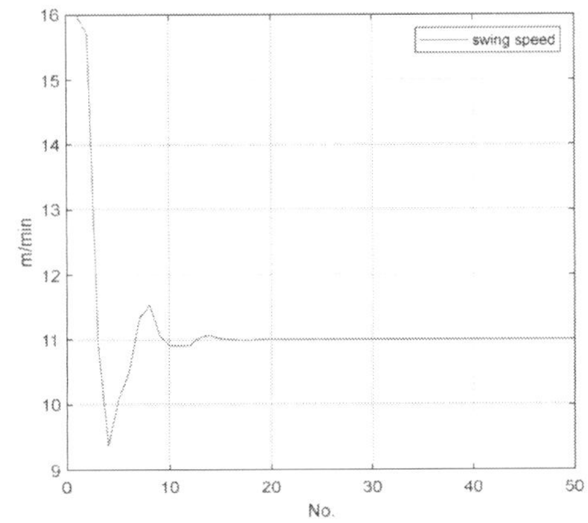

Figure 6. The change of the swing speed(when the setpoint is a fixed value).

It can be seen from the control results that when the target value is fixed, the density after control quickly reaches the target value, and the swing speed also tends to be stable quickly.

When the target value is set as the optimal density calculated by the dredging data, that is, when the target value is dynamic fluctuation, the control results are as follows:

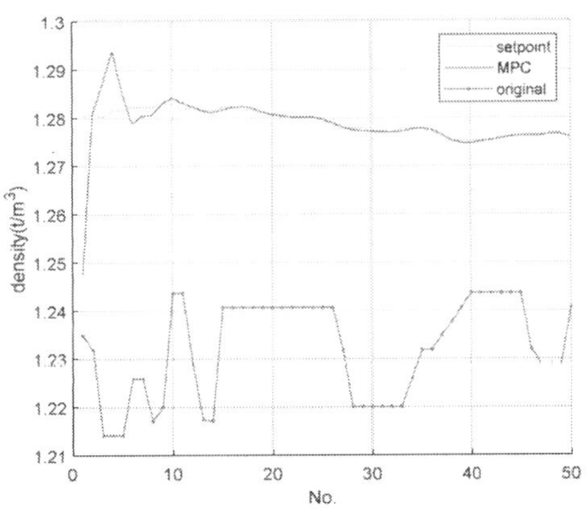

Figure 7. Comparison of setpoint, original value and MPC control value(when the setpoint is changing).

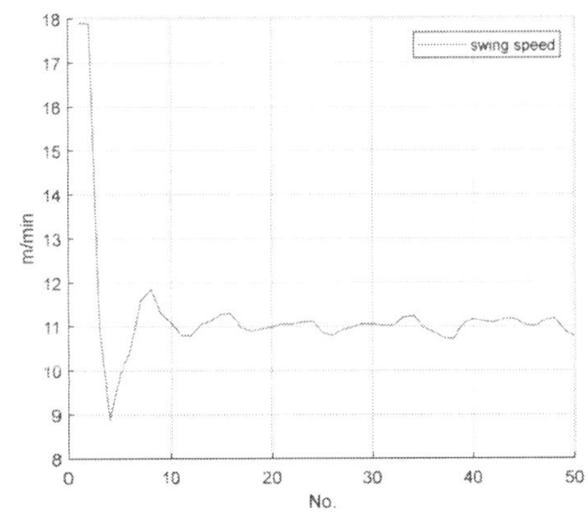

Figure 8. The change of the swing speed(when the setpoint is changing).

It can be seen from the control results that when the set value is changed, the value after predictive control can also reach the set value more accurately, and the change of swing speed is also stable.

5. Conclusion

This paper uses the method of subspace identification to build the prediction model of the swing system, and uses data-driven predictive control to skip the complicated traditional modeling and control the swing speed of the cutter suction dredger to make the slurry density reach the predetermined value. The experimental results show that the prediction model obtained by using the subspace identification has high accuracy and can achieve good control effect, which provides a new method for the control of the swing system of cutter suction dredger.

References

[1] Wei, C., Ni, F., DAI, W., Jiang, S., Chen, X. (2018) The predictive control for the swing process of a cutter suction dredger. Machine Design and Manufacturing Engineering, 47(01): 70-74.

[2] Zhu, W., Ni, F., Wei, C., Tao, C. (2016) Linear Quadratic Optimal Control of Cutter Suction Dredger during Swing Process. SHIP ENGINEERING, 38(S2): 230-235.

[3] Song, D., Zhang, L. (2020) Lateral movement control of cutter suction dredger based on neural network and maximum output estimation. Port & Waterway Engineering, 2020(S1): 123-127.

[4] Luo, X. (2012) State space model predictive control based on subspace identification. Computer Engineering and Applications, 48(19): 234-237.

[5] Zeng, J., Gao, C., Su, H. (2010) Data-driven predictive control for blast furnace ironmaking process. Computers and Chemical Engineering, 34(11): 1854-1862.

[6] Wei, C., Wei, Y., Ji, Z. (2021) Model predictive control for slurry pipeline transportation of a cutter suction dredger. Ocean Engineering, 227:108893.

[7] Camacho, E.F., Bordons, C. (2007) Model Predictive Control. Springer London, London.

Study on dimensions of void in GRP rod of composite insulator

Hongchuan Dong[1], Shan Jiang[1], Xuejun Tang[1], Fangchao Ke[1], Sida Zhang[2] and Yunfan Liu[2*]

[1] Economic and Technological Research Institute, State Grid Hubei Electric Power Co., Wuhan, Hubei Province, 430000, China

[2] State Key Laboratory of Power Transmission Equipment & System Security and New Technology, Chongqing University, Chongqing, 410044, China

[*] Corresponding author's e-mail: cqulyf@cqu.edu.cn

Abstract. As an important part of composite insulator, glass fiber reinforced polymer (GRP) rod needs to bear the main mechanical and keep certain insulation performance. The void in the material is key factor to the performance of GRP rod. The void with larger size could affect the electric field distribution significantly and lead to the "Lantern-like" heating. In this paper, the structure of void in GRP rod is studied by 3D X-ray microscope. It is found that the void dimensions in the GRP rod is generally less than 300μm, but the interconnection of linear hole makes the length of the void increase significantly. For the GRP rod with high porosity, the length of connection hole is hard to be detected directly by the existing technology, but it could be calculated based on the diameter of the hole. Based on the calculation method, the dimensions of void in GRP rod could be inspected by the observation of section using optical microscope.

1. Introduction

Compared with traditional insulation equipment such as ceramic and glass insulators, composite insulator composed of high-temperature vulcanised silicone rubber (HTVSR) and glass fibre reinforced polymers (GRP) have the advantages of anti-pollution flashover, high strength, and low density [1-3]. The rod in the composite insulator is the main carrier of axial tension load, and GRP has always been considered the most suitable rod material since the birth of the composite insulator [4]. However, void defects are inevitable for the production process of GRP material are inevitable [5-6]. The axial channels composed by void defects in GRP rod could affect the electric field distribution significantly and lead to the "Lantern-like" composite insulator heating defects at the early stage of operation [7]. It is necessary to propose an effective methods to inspect the void dimensions in the GRP rod.

The existing detection tests for dimensions of void in composite insulator GRP rods include the dye penetration test [8-9], transparency test [10], ultrasonic test [11], and X-ray transmission test [12]. Although each evaluation method has its own basis, their effectiveness is limited to the analysis on the void structure. For the material of GRP rod, the diameter of glass fiber is 14~20 μm and the distance between the fiber usually does not exceed the glass fiber diameter, so that the dimensions of void should be less than 15 μm [13]. In order to accurately describe the void structure and calculate the void size, the 3D scanning is necessary and the pixel unit of the scanning device need to be less than 2 μm.

In this paper, an analysis method of void structure was proposed based on X-ray microscope and the interconnection of void has been proved to be the main reason for the formation of large-size holes.

Furthermore, the calculation method of the length of connection hole was proposed based on the relationship between diameter and length. Finally, it is demonstrated that the dimensions of connection hole in GRP rod could be inspected by the observation of section using optical microscope.

2. Analysis method of void structure

X-ray computed tomography is the basis of three dimensional imaging. For this method, the material properties of each pixel unit could be distinguish by the difference of X-ray absorptivity. For a single material with small volume, its X-ray absorption coefficient expression is shown in equation (1). a is a parameter with weak energy dependence and b is a constant. E is the energy of X-ray, which is depended on the wavelength of the electromagnetic wave. ρ and z are the density and atomic number of the material respectively.

$$\mu(E) = \rho\left(a + b\frac{z^{3.8}}{E^{3.2}}\right) \tag{1}$$

As X-ray with determined radiation intensity I_0 penetrates a substance through a path L, the X-ray intensity detected by the detector without scattering is shown in equation (2). By multi-dimensional scanning of the object and solving the multivariate equations, the three-dimensional model of the sample could be constructed based on absorption coefficient of each pixel unit. However, the imaging resolution of Micro-CT system with geometric magnification only reach 2~3 μm, which could not meet the analysis requirements of the void structure in GRP rod.

$$I_{\mathrm{d}}(L) = I_0 \exp\left(-\int \mu(x,E)\mathrm{d}L\right) \tag{2}$$

In order to get higher magnification, the optical principle is proposed for secondary amplification, known as X-ray microscope. The image formed on the X-ray detector is observed with an optical microscope and recorded with a high-resolution camera (as shown in figure 1). Under 20-fold optical magnification, the theoretical resolution can reach up to 500 nm and the spatial resolution can reach up to 2-3μm when a CCD camera with 10 μm pixels is used for recording [14].

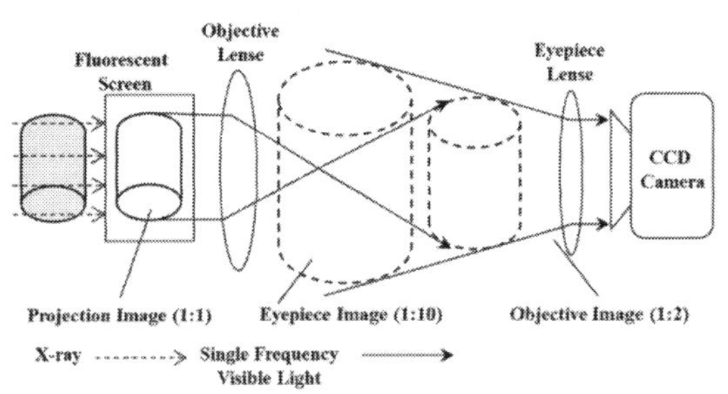

Figure 1. Schematic diagram of X-ray microscope.

Figure 2. Grayscale image of X-ray microscope.

Although the principle of optical microscope improves the resolution significantly, the three-dimensional model of the sample could not be constructed based on absorption coefficient of each pixel unit anymore. Moreover, the imaging from X-ray detector is different from that of scanning electron microscope (SEM). When X-ray penetrates the sample, it scatters at the boundary of different materials, thus forming a transition region in the projection (as shown in figure 2). Due to there are large number of interfaces in the GRP rod, it is difficult to form characteristic peaks based on the gray value of pixels. In order to get the actual volume ratio of fiber to epoxy resin, we made thermos-gravimetric analysis (TGA) of GRP rod material and obtained corresponding gray value range of glass fiber and epoxy resin. In equation (3), m_f is the gray value range of the glass fiber pixel, M_f is threshold of glass fiber and epoxy pixel unit, $M_{\max}$ and $M_{\min}$ are the maximum and minimum gray value of the solid material .

$$m_f \in \left[M_f, M_{max}\right] \qquad M_f = \left(M_{max} - M_{min}\right) \times 23.7\% + M_{min} \qquad (3)$$

The void in the GRP rod is composed of gas molecules, so that its X-ray absorption coefficient as well as gray value is close to 0. Although the X-ray will scatter at the boundary between the void and the solid material, making the gray value of some void pixels increase, the gray value is still far less than the solid material itself. Because of this, the pixel unit with gray value of (0, M_{min}+3) are used as the model building of the void.

3. Void structure and dimensions

According to void structure, the hole in GRP rod can be divided into linear hole and spherical hole. As shown in figure 3, the number of linear hole far exceeds that of spherical hole. The length and section diameter of spherical hole are about 70 μm and 115 μm, while the length and section diameter of linear hole are about 10 μm and 260 μm.

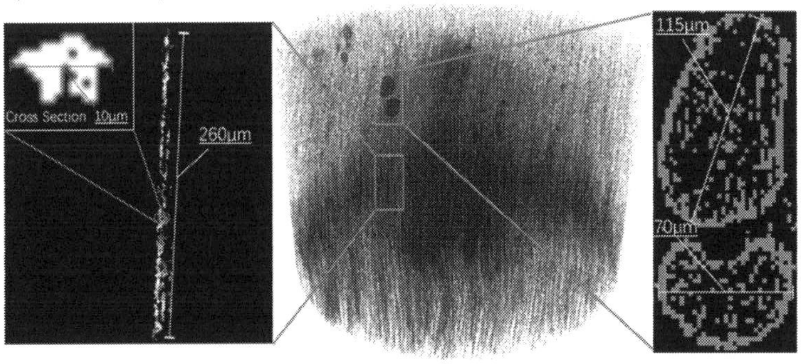

Figure 3. Linear hole and spherical hole in GRP rod.

The interaction of these two type of void is totally different. For spherical hole (as shown in figure 4), even if some voids are very close to each other, the voids are not interconnected. For linear hole (as shown in figure 5), its extension direction is consistent with that of glass fiber, but there is a certain deviation from the rod axial direction. Therefore, as the porosity of GRP rod increase, there will be a large number of voids connected with each other.

Figure 4. The interaction of spherical hole in GRP rod.

Figure 5. The interaction of linear hole in GRP rod.

The number of linear holes (include independent and connection hole) in samples with different porosity is shown in figure 6 and figure 7. As the porosity lower than 0.1%, the number of independent holes with 10 μm will less than 50 and the connection hole will not form. As the porosity over 0.5%, the number of independent holes with 2 μm and 10 μm exceeds 500000 and 25000 respectively.

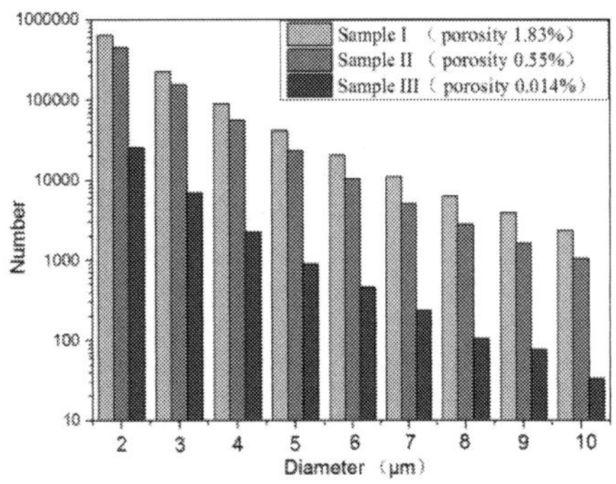

Figure 6. The number of independent holes in samples with different porosity.

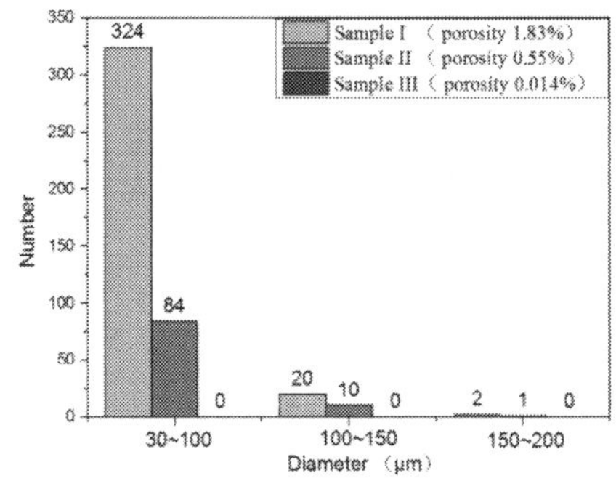

Figure 7. The number of connection holes in samples with different porosity.

For the axial positions of the linear holes are generally different, the axial length of the whole structure will increase significantly as the linear holes are connected with each other. The dimensions of connection holes are shown in figure 8.

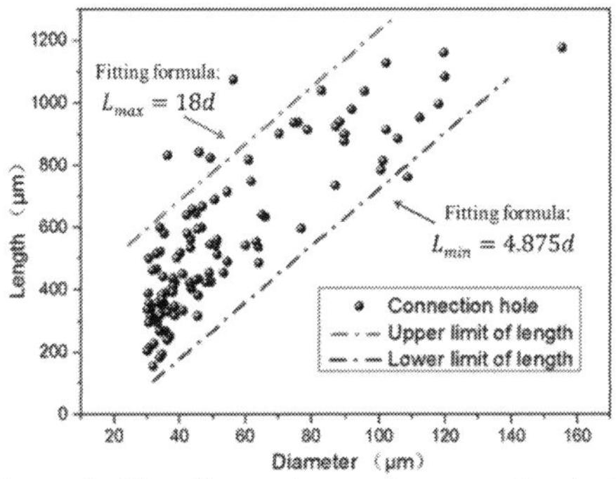

Figure 8. The dimensions of connection hole.

The section diameter and length of connection hole is generally more than 20 μm and 200 μm. For the largest connection hole intersects the entire scanning area, its length could not be inspected directly. However, due to the connection hole is formed by the random connection of adjacent linear holes, the relationship between the diameter and the length satisfies the statistical law. It can be seen from the Figure 8, the upper limit value $L_{\max}$ and the lower limit value $L_{\min}$ of the length of connection hole meet the equation (4) and equation (5) respectively. The proportion of voids whose length exceeds the upper and lower limits shall not exceed 5.3%. For the diameter of the largest connection hole in Sample I is 351 μm, the length of the void shall be 1702~6318 μm.

$$L_{\max} = 18d \tag{4}$$

$$L_{\min} = 4.85d \tag{5}$$

4. Inspection method of connection hole

Due to the length of connection hole satisfies the statistical law and always over the scanning area of X-ray microscope, the observation of section by optical microscope should be a better method for GRP rod inspection. In this paper, the Sample I was processed to 1 mm and the position of the connection hole was determined by dye penetration test. The section of connection hole taken by optical microscope is shown in Figure 9. Since the diameter of the void is about 200 μm, the length of the void should be 970~3600 μm. This calculation result was demonstrated by dye penetration test of adjacent slice sample.

5. Conclusion

- The void in GRP rod can be divided into linear hole and spherical hole. The void dimensions in the GRP rod is generally less than 300μm, but the interconnection of linear hole makes the length of the void increase significantly.
- Due to the connection hole is formed by the random connection of adjacent linear holes, the relationship between the diameter and the length satisfies the statistical law.
- The largest connection hole will intersect the entire scanning area as the porosity increase. Based on the relationship between the diameter and the length, the dimensions of connection hole in GRP rod could be inspected by the observation of section using optical microscope.

Acknowledgments
The authors thank Economic and Technological Research Institute of State Grid Hubei Electric Power Co.

References

[1] Liang, X., Wang, S., Fan, J., Guan, Z., (2015) Development of composite insulators in China, IEEE Trans. Power Del, 6: 586–594.
[2] Wang, J., Zhang, S., Zuo, Z., (2016) Rapid Development of Silicone Rubber Composite Insulator in China, High Voltage Engineering, 42:2888-2896.
[3] Yang, L., Bi, J., Zhang, F., (2019) Effects of structure and material of polluted insulators on the wetting characteristics, IET Science, Measurement & Technology, 13: 131-138.
[4] Cherney, E.A., (2013) 50 years in the development of polymer suspension-type insulators, IEEE Electr. Insul. Mag., 29: 18-26.
[5] Centea, T., Hubert, P., (2012) Modelling the effect of material properties and process parameters on tow impregnation in out-of-autoclave prepregs, Composites: Part A, 43:1505-1513.
[6] Grunenfelder, L.K., Nutt, S.R., (2012) Void formation in composite prepregs – Effect of dissolved moisture, Composites Science and Technology, 70:2304-2309.
[7] Huang, H., Wang, T., Lv, Q., Ma, X., Zhang, F., Zeng, H., Wu, J., (2021) Failure Analysis of "Lantern-Like" Composite Insulator Heating Defects and the Improved Dye Penetration Test Method, IEEE ACCESS, pp.50646-50651.
[8] Standard, (2012) Polymeric HV Insulatiors for Indoor and Outdoor Use-General Definitions," Test Methods and Acceptance Criteria, Test Methods and Acceptance Criteria, IEC 62217.
[9] Standard, (2008) Insulators for Overhead Lines Composite Suspension and Tension Insulators for A.C. System with a Nominal Voltage Greater than 1000V Definitions, Test Methods and Acceptance Criteria, IEC 61109.
[10] Standard, (2016) Technical specificaton for core of long rod composite insulators for A.C or D.C. systems, Test Methods and Acceptance Criteria, DL/T 1580.
[11] Liang, X., Dai, J., Zhou, Y., Chen, Y., (2005) Ultrasonic Detecion on Crack of FRP Rod in Brittle Fracture of Composite Insulator, Proceedings of the CSEE, 25:110-114.
[12] Gao, Y., Liang, X., Bao, W., (2019) Failure analysis of a field brittle fracture composite insulator: characterisation by X-ray photoelectron spectroscopy analysiss, High Voltage, 4:89-96.
[13] Armentrout, D., Kumosa, L.K., (2004) Water Diffusion into and Electrical Testing of Composite Insulator GRP Rods, IEEE Trans. Dielectr. Electr. Insul., 11:506-522.
[14] An, K., Yin, Y., Li, F., Shi, L., Hu, X., Tang, J., Zhou, R., (2020) A Novel Micro-Multifocus X-Ray Source Based on Electron Beam Scanning for Multi-View Stationary Micro Computed Tomography, IEEE Electron Device Letters, 41:167-170.

Fault location method of active distribution network based on graph theory and matrix algorithm

Yalei Li[1], Xiaohong Zhang[1*]

Shanghai Dianji University,Shanghai 201306,China

*Corresponding author's e-mail: zhangxiaohong@sdju.edu.cn

Abstract. In view of the large scale distributed power distribution network, distribution network from the traditional single main transformer power supply system becomes more complex broken power supply network, the trend of the distribution network flow and network frame produced change, failure fault feature information of great change, the traditional distribution network fault location method can not accurately obtain the location of the fault zone, the low accuracy of fault location, the error bigger problem a new fault location method of active distribution network combining graph theory and matrix was proposed. According to the knowledge of graph theory, the distribution network is simplified to topology diagram and described in the form of matrix, and then the fault judgment matrix is obtained through a series of matrix addition operation, and the type of fault interval can be identified accurately and quickly by combining the fault criterion table. The simulation results show that this method is simple in principle, fast and effective in criterion, suitable for the flexible operation mode of active distribution network structure, can accurately and quickly determine the fault segment, and can meet the requirements of complex distribution network fault location.

1. Introduction

In recent years, with the development and utilization of new energy, the scale of distributed power supply, mainly photovoltaic power generation and wind power, connected to the distribution network is increasing, which also brings many new problems. Distribution network from the traditional single main transformer power supply system becomes more complex broken power supply network, the trend of the distribution network flow and network frame produced change, dramatic change failure fault characteristic information, such as the failure of short-circuit current source, direction and size all can change, causing the original distribution network fault location scheme and relay protection technology are no longer applicable[1].And with the rapid development of economy and the improvement of residents' pursuit of life, users have higher and higher requirements for power supply reliability of distribution network. In order to solve the problem caused by the massive distributed power supply connected to the distribution network, CIGRE C6.11 project team proposed the concept of Active Distribution Network (ADN) in 2008.It aims to solve the compatibility of power grid and the application of large-scale intermittent renewable energy, improve the utilization rate of green energy and optimize the primary energy structure[2].Therefore, it is of great significance to study fault location technology of active distribution network under the new situation.

At present, distribution network fault location based on feeder terminal (FTU) is widely used in active distribution network fault location methods at home and abroad. This method is mainly divided into two research directions: intelligent algorithm and matrix algorithm: One is the fault location algorithm based on artificial intelligence, such as artificial neural network, expert system, genetic algorithm, fuzzy theory, etc. Due to the long calculation time, this kind of algorithm is difficult to meet the requirements of real-time online processing, and is generally used less[3].The other is the matrix algorithm based on graph theory, which is widely used because of its small amount of calculation and simplicity. The topology of ADN with distributed energy is flexible and changeable, so the traditional fault segment determination algorithm can not be applied completely. Therefore, the improved matrix algorithm has more advantages in computing speed and efficiency. Based on the above, this paper proposes a new matrix algorithm based on graph theory. The algorithm is simple in calculation, avoids complex matrix multiplication operation, and does not need to redefine the direction when the network structure changes. It can quickly and accurately locate faults and terminal faults between multiple power nodes and achieve fault isolation.

2. Fault location method of active distribution network

2.1 Graph theory and network structure simplification

Distribution network is a connected network composed of switches, distribution feeders and distribution transformers. The structure of active distribution network with distributed energy can be simplified and further studied by graph theory in discrete mathematics.

A figure consisting of several different vertices with edges connecting some of them is called a graph. Among the elements of the graph, the nodes of the active distribution network are taken as the vertices in the graph, and the branches of the active distribution network are taken as the edges in the graph. If two vertices can be connected by an edge, the two vertices are said to have adjacent relations. Usually figure can be expressed as $G = (V, E)$, V is the set of all its vertices $V = [v_1, v_2, \cdots, v_n]$, E is a collection of all its edge to $E = [e_1, e_2, \cdots, e_n]$.

Active distribution network is usually divided into radial network and ring network according to the structure characteristics. Based on the above graph theory knowledge, distribution network composition can be simplified, load nodes are regarded as vertices, distribution feeders and distribution transformers between two adjacent vertices are regarded as edges, and the distribution network topology can be described by undirected graph, and then the topology description of distribution network in accordance with graph theory knowledge can be obtained. On the basis of the undirected graph, the power direction of the main power supply of the network is selected as the reference forward direction, and the network is described as a directed graph. Then, the directed dynamic half-angle adjacency matrix is obtained, which can reflect the structure state of the network, the direction relation between nodes and the electrical connection relation.

2.2 Fault information coding method and matrix principle

$\beta_i (i = 1,2,3, \dots D)$ represents the fault state information of the section in article I, and the diagonal element represents the state of this section, and its encoding mode is shown in Formula (1), where D is the total number of sections contained in the system.

$$\beta_i = \begin{cases} 1 \quad \text{Sector } i \text{ is normal} \\ 0 \quad \text{Sector } i \text{ out of service} \end{cases} \tag{1}$$

According to graph theory, A connected graph with n nodes, b branches and no self-loop can be represented by an association matrix A of order n×b.The rows and columns of matrix A correspond to nodes and branches in the network respectively, and the elements of matrix A satisfyA_{ij}.

$$A_{ij} = \begin{cases} 0, \quad \text{Node } i \text{ is not connected to branch } j \\ 1, \text{ Node } i \text{ is connected to branch } j \end{cases} \tag{2}$$

The operation of row vector A_i and vector β of matrix A gives the electrical correlation matrix E, which describes the electrical connection relation between nodes.

$$A = [\alpha_1^T, \alpha_2^T, \cdots, \alpha_m^T]^T \tag{3}$$

$$E = [(\alpha_1 \cap \beta)^T, (\alpha_2 \cap \beta)^T, \cdots, (\alpha_m \cap \beta)^T]^T \tag{4}$$

Journal of Physics: Conference Series **2121** (2021) 012035 doi:10.1088/1742-6596/2121/1/012035

Adjacency matrix R describes the unicom relation between network nodes, and its row and column numbers correspond to corresponding node numbers, satisfying:

$$R_{ij} = \begin{cases} 0, \text{ Node } i \text{ is not connected to node } j \\ 1, \text{ Node } i \text{ is connected to node } j \end{cases} \quad (5)$$

Matrix R changes with the status of network branches, and electrical correlation matrix E describes the electrical connection relationship between nodes and network branches.The adjacency matrix R can be obtained by the operation of Equation (6) on matrix E. See reference [4] for details.

$$R = (EE^T) \cap \bar{I} \quad (6)$$

When the fault occurs in the active distribution network, the data acquisition and Monitoring control system (SCADA) obtains the switch fault current information uploaded by the feeder terminal unit (FTU) installed on the node switch, and the master station collects and generates the n-dimensional diagonal matrix F, which is defined as follows:

$$f_{ii} = \begin{cases} 1, \text{The } I_f \text{ direction is consistent with the specified positive direction} \\ 0, I_f = 0 \\ -1, \text{The direction of } I_f \text{ is opposite to the specified direction} \end{cases} \quad (7)$$

Finally, the fault judgment matrix J is obtained by the operation $J = R + F$ between the directed dynamic adjacency matrix R and the fault information matrix F.

2.3 Common Branch Node types

The following are five common branch node types in active distribution networks.As shown in Figure 1, they are power outlet node, terminal load node, 2 branch node, 3 branch node failure, and 4 branch node respectively.

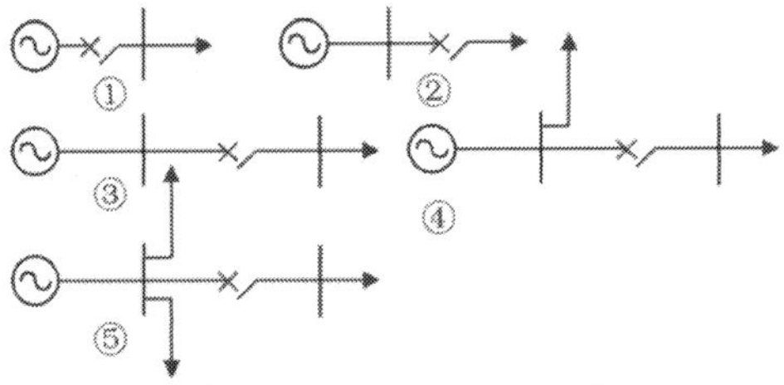

Figure 1. Common branch node types in distribution network

2.4 Fault criteria

Due to the different types of branch nodes in the network, the corresponding fault characteristics are different, so the fault criterion corresponding to one and one is proposed [5].After the fault occurs, the fault judgment matrix J is obtained by calculating the directed dynamic adjacency matrix R and the fault information matrix F in the protected host, that is, $J = R + F$.After the fault judgment matrix J is generated, fault location can be realized according to the six criteria in Table 1 below.

Table 1. Fault location criteria

Fault Type	Fault Feature	Fault criterion
The power export	The system only detected fault current supplied by DG which was opposite to the reference forward direction	The matrix J has only elements -1 and 0 on the diagonal
Two node range	The ith row of the matrix J contains one and only one of the other elements except J_{ii}	$J_{ii} = 1$, $J_{ij} = 1 (i \neq j)$ and $J_{jj} = 0$ or J_{jj}=-1, make $J_{ii}\&J_{jj} = 0$

Three node range	The ith row of the matrix J has two ones except for the J_{ii} element	$J_{ii} = J_{ij} = J_{ig} = 1$, $[J_{ii} \odot J_{gg}]! = 0$($\odot$ is the same or, ! is not) , make $J_{ii} \& [J_{jj} \odot J_{gg}]! = 0$(& is with)
Four node range	The ith row of the matrix J has three ones in all the elements except J_{ii}	$J_{ii} = J_{ij} = J_{ik} = J_{ig} = 1$, and $[J_{jj} \odot J_{kk} \odot J_{gg}] = 0$, make $J_{ii} \& [J_{jj} \odot J_{kk} \odot J_{gg}] = 0$
Line end interval	In the ith row of the matrix J corresponding to the end, all the others are 0 except the element J_{ii} is 1	$J_{ii} = 1$ and for $j (j \neq i)$ all have $J_{ij} = 0$

3. Example analysis and simulation

3.1 Directed dynamic matrix acquisition

The following takes a 14 node active distribution network as an example to explain the principle of fault location algorithm adopted in this paper. The distribution network structure is shown in Figure 2. In Figure 2, there are 14 nodes and 13 branches. DG1 and DG2 are access nodes 5 and 14 respectively, FTU is installed on nodes 1 to 13, S_1 is the main power supply. Based on the above explanation, Figure 2 can be further simplified into an undirected graph, as shown in Figure 3.

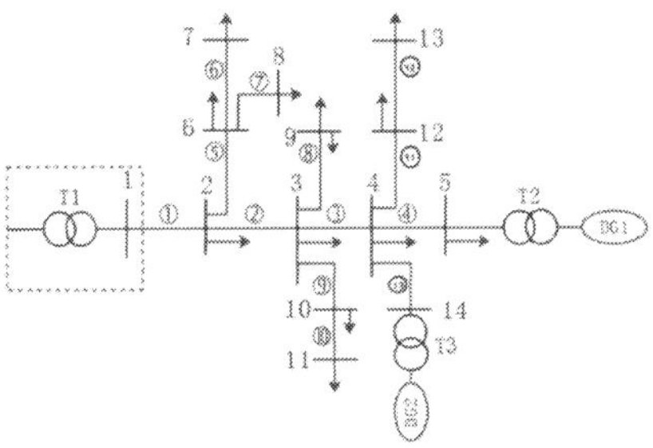

Figure 2. active distribution network of 14 nodes

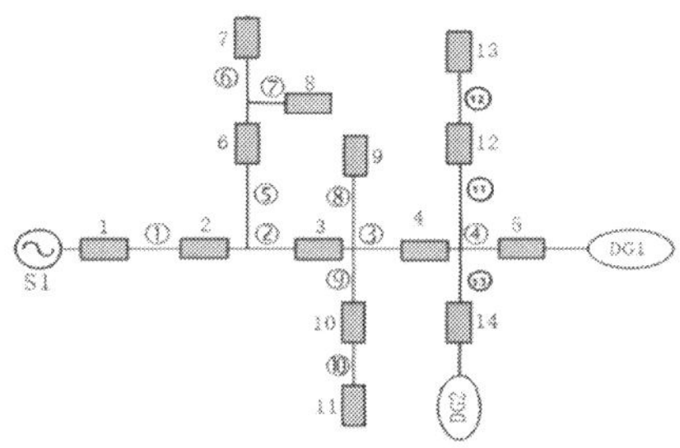

Figure 3. Undirected diagram of 14-node active distribution network

The branch vector representing the state of this section is as follows

$$F = [\ 1\ 1\ 1\ 1\ 1\ 1\ 1\ 1\ 1\ 1\ 1\ 1\ 1\ ] \tag{8}$$

Figure. 3 shows the connected graph without isolated nodes and self-loops, corresponding to which A 14-row and 13-column association matrix A can be generated. Substitute the matrix A and the matrix F into Equation (4), the electrical correlation matrix $E_{14\times 13}$ is obtained, whose form is the same as that of matrix A.

$$A = \begin{bmatrix} 1&0&0&0&0&0&0&0&0&0&0&0&0 \\ 1&1&0&0&1&0&0&0&0&0&0&0&0 \\ 0&1&1&0&0&0&0&1&1&0&0&0&0 \\ 0&0&1&1&0&0&0&0&0&0&1&0&1 \\ 0&0&0&1&0&0&0&0&0&0&0&0&0 \\ 0&0&0&0&1&1&1&0&0&0&0&0&0 \\ 0&0&0&0&0&1&0&0&0&0&0&0&0 \\ 0&0&0&0&0&0&1&0&0&0&0&0&0 \\ 0&0&0&0&0&0&0&1&0&0&0&0&0 \\ 0&0&0&0&0&0&0&0&1&1&0&0&0 \\ 0&0&0&0&0&0&0&0&0&1&0&0&0 \\ 0&0&0&0&0&0&0&0&0&0&1&1&0 \\ 0&0&0&0&0&0&0&0&0&0&0&1&0 \\ 0&0&0&0&0&0&0&0&0&0&0&0&1 \end{bmatrix} \tag{9}$$

$$E = \begin{bmatrix} 1&0&0&0&0&0&0&0&0&0&0&0&0 \\ 1&1&0&0&1&0&0&0&0&0&0&0&0 \\ 0&1&1&0&0&0&0&1&1&0&0&0&0 \\ 0&0&1&1&0&0&0&0&0&0&1&0&1 \\ 0&0&0&1&0&0&0&0&0&0&0&0&0 \\ 0&0&0&0&1&1&1&0&0&0&0&0&0 \\ 0&0&0&0&0&1&0&0&0&0&0&0&0 \\ 0&0&0&0&0&0&1&0&0&0&0&0&0 \\ 0&0&0&0&0&0&0&1&0&0&0&0&0 \\ 0&0&0&0&0&0&0&0&1&1&0&0&0 \\ 0&0&0&0&0&0&0&0&0&1&0&0&0 \\ 0&0&0&0&0&0&0&0&0&0&1&1&0 \\ 0&0&0&0&0&0&0&0&0&0&0&1&0 \\ 0&0&0&0&0&0&0&0&0&0&0&0&1 \end{bmatrix} \tag{10}$$

The adjacency matrix $R_{13\times13}$ obtained from Equation (6) is as follows. In addition, The adjacency matrix R is an undirected symmetric Boolean matrix. By selecting the power direction of main power supply S1 as the reference positive direction, the node relation expressed in adjacency matrix R is simplified to the forward connection relation, and the element value of the negative connection relation is set to 0 to obtain the directed dynamic half-angle matrix. Therefore, matrix R is a simplified adjacency matrix R after digitizing to omit the lower half Angle zero element for convenience.

$$R=\begin{bmatrix}
0&1&0&0&0&0&0&0&0&0&0&0&0&0\\
1&0&1&0&0&1&0&0&0&0&0&0&0&0\\
0&1&0&1&0&0&0&0&1&1&0&0&0&0\\
0&0&1&0&1&0&0&0&0&0&0&1&0&1\\
0&0&0&1&0&0&0&0&0&0&0&0&0&0\\
0&1&0&0&0&0&1&1&0&0&0&0&0&0\\
0&0&0&0&0&1&0&0&0&0&0&0&0&0\\
0&0&0&0&0&1&0&0&0&0&0&0&0&0\\
0&0&1&0&0&0&0&0&0&0&0&0&0&0\\
0&0&1&0&0&0&0&0&0&0&1&0&0&0\\
0&0&0&0&0&0&0&0&0&1&0&0&0&0\\
0&0&0&1&0&0&0&0&0&0&0&0&1&0\\
0&0&0&0&0&0&0&0&0&0&0&1&0&0\\
0&0&0&1&0&0&0&0&0&0&0&0&0&0
\end{bmatrix} \tag{11}$$

$$R=\begin{bmatrix}
0&1&0&0&0&0&0&0&0&0&0&0&0&0\\
&0&1&0&0&1&0&0&0&0&0&0&0&0\\
&&0&1&0&0&0&0&1&1&0&0&0&0\\
&&&0&1&0&0&0&0&0&0&1&0&1\\
&&&&0&0&0&0&0&0&0&0&0&0\\
&&&&&0&1&1&0&0&0&0&0&0\\
&&&&&&0&0&0&0&0&0&0&0\\
&&&&&&&0&0&0&0&0&0&0\\
&&&&&&&&0&0&0&0&0&0\\
&&&&&&&&&0&1&0&0&0\\
&&&&&&&&&&0&0&0&0\\
&&&&&&&&&&&0&1&0\\
&&&&&&&&&&&&0&0\\
&&&&&&&&&&&&&0
\end{bmatrix} \tag{12}$$

3.2 Fault section determination

The fault diagram of active distribution network is shown in Figure 4, K1, K2, K3, K4 and K5, respectively, indicating that the fault occurs at the exit, two-node, three-node, four-node interval and terminal.

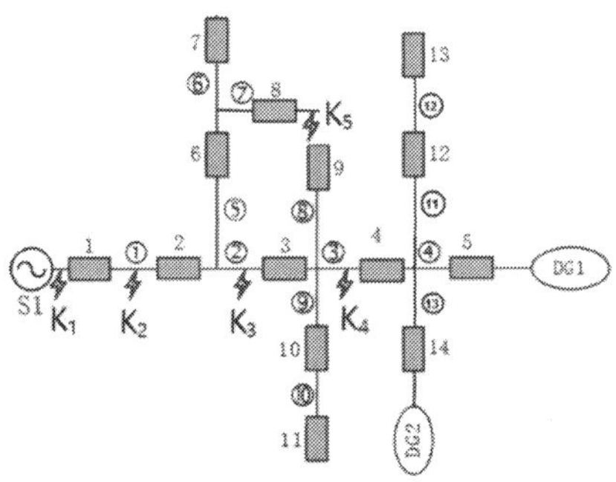

Figure 4. Schematic diagram of fault

According to the network fault diagram of 14-node active distribution network, the corresponding fault information matrix of K1, K2, K3, K4 and K5 faults.

Table 2. Fault information matrix

Fault location	Fault information matrix
K1	$F_1 = [\ -1\ \ -1\ \ -1\ \ -1\ \ -1\ \ 0\ \ 0\ \ 0\ \ 0\ \ 0\ \ 0\ \ 0\ \ 0\ \ -1\ ]$
K2	$F_2 = [\ \ 1\ \ -1\ \ -1\ \ -1\ \ -1\ \ 0\ \ 0\ \ 0\ \ 0\ \ 0\ \ 0\ \ 0\ \ 0\ \ -1\ ]$
K3	$F_3 = [\ \ 1\ \ \ 1\ \ -1\ \ -1\ \ -1\ \ 0\ \ 0\ \ 0\ \ 0\ \ 0\ \ 0\ \ 0\ \ 0\ \ -1\ ]$
K4	$F_4 = [\ \ 1\ \ \ 1\ \ \ 1\ \ -1\ \ -1\ \ 0\ \ 0\ \ 0\ \ 0\ \ 0\ \ 0\ \ 0\ \ 0\ \ -1\ ]$
K5	$F_5 = [\ \ 1\ \ \ 1\ \ -1\ \ -1\ \ -1\ \ 1\ \ 0\ \ 1\ \ 0\ \ 0\ \ 0\ \ 0\ \ 0\ \ -1\ ]$

After the fault occurs, the fault judgment matrix J is obtained by calculating the directed dynamic adjacency matrix R and the fault information matrix F in the protected host, that is, $J = R + F$. According to the above information, the fault determination matrices J_1、J_2、J_3、J_4、J_5 are obtained by the operation of fault information matrices F_1、F_2、F_3、F_4、F_5 with the reduced adjacency matrix R. The fault location results of the active distribution network of the 14 nodes can be obtained according to the fault determination matrices and fault criteria.

Table 3. Fault location results

Fault Type	Fault criterion	Locating result
The power export	The fault information matrix J_1 contains only elements -1 and 0 on the diagonal	The main power outlet is faulty.
Two node range	$J_{11} = 1$, $J_{12} = 1$ and $J_{22} = -1$, there is $J_{11}\&J_{22} = 0$	The fault occurs between nodes 1 and 2
Three node range	$J_{22} = J_{23} = J_{26} = 1$, and $J_{33} = -1$、$J_{66} = 0$, there is $J_{22}\&[J_{33} \odot J_{66}]! = 0$	Nodes 2, 3, and 6 are faulty.
Four node range	$J_{33} = J_{34} = J_{39} = J_{3,10} = 1$, and $J_{44} = -1, J_{99} = 0, J_{10,10} = -1$, there is $J_{33}\&[J_{44} \odot J_{99} \odot J_{10,10}] = 0$	Nodes 3, 4, 9, and 10 are faulty.
Line end interval	In matrix J_5, all elements in row 8 are 0 except J_{88}=1	The cable terminal of node 8 is faulty.

4. Conclusion

This paper introduces a new method for fault location of active distribution network based on graph theory and matrix algorithm, which can accurately identify power outlet, line end and multi-branch interval faults. For the complex distribution network including distributed power supply, due to its complex and changeable architecture, the power direction of main power supply S_1 in the distribution network is defined as the only positive direction, so the adjacency moment R can be automatically updated only by using real-time power flow direction. At the same time, the innovative fault judgment matrix can be generated only through matrix addition operation, which not only greatly reduces the complexity of matrix operation but also shorts the time of fault judgment compared with the tedious multiplication operation. Simulation results show that the algorithm is simple in principle, fast and effective in criterion, and can meet the requirements of accuracy and rapidity in fault location of complex distribution network. In this paper, there is a lack of research on the authenticity of alarm information. In the future, the distorted alarm information will be screened and extracted.

References

[1] Niu, G., Kong, L., Zhou, L. (2017) Review on Power Supply Recovery Technology of Distribution Network with distributed Power. New Technology of Electric Power.36: 51-62.

[2] Fan, M.T., Zhang, Z.P., Su, A.X(2013) Enabling technologies for active distribution systems.Proceedings of the CSEE. 33: 12-18.

[3] Hu, F.N., Sun, S.j. (2016)Application of Matrix Algorithm based on Graph Theory in Distribution Network Fault Location. China Electric Power.9:94-98.

[4] Chu, J.J.(2003)Mathematical Model for Analysing Electrical Connectedness of Main Electrical Scheme in Substation. Automation of Electric Power Systems.27: 31-33.

[5] Hu, F.N., Zhang, K., Bian X.L. (2021) Fault Location Method of Active Distribution Network Based on Graph Theory. Control Engineering.28:833-838.

VSG damping adaptive adjustment based on BP neural network

Mengzhao Zhang* **and Chunlin Guo**

School of Electrical and Electronic Engineering, North China Electric Power University, Changping District, Beijing, 102200, China

*Corresponding author's e-mail: 120202201184@ncepu.edu.cn

Abstract. The moment of inertia and damping of virtual synchronous generator (VSG) can be adjusted flexibly, which also has a significant impact on the transient performance of VSG. Constant damping or moment of inertia can not reduce frequency overshoot and fast response performance, so it is necessary to introduce adaptive damping control. Based on universal approximation theorem, BP neural network can fit continuous nonlinear function well. At the same time, it has the advantages of simple algorithm, powerful learning ability and fast learning speed. Based on the characteristics of the control object, the BP neural network is improved and a new adaptive control strategy is designed. The strategy uses improved BP neural network to adjust VSG virtual damping D online. Python-MATLAB-Simulink was used for co-simulation, BP neural network algorithm was integrated into the control object to establish an adaptive simulation model, and the proposed control strategy was simulated and verified. Simulation results show that the adaptive control strategy can eliminate overshoot and respond quickly when the frequency and active power of virtual synchronous generator change.

1. Introduction

Constant damping has major drawbacks. Too large damping can ensure the transient process without oscillation, but will make the output power of primary FM error. Low damping can reduce such errors, but again there are oscillations. In view of this contradiction, many scholars have proposed the methods of transient damping and inertia in recent years. Literature [1] proposed a solution of introducing a transient damping link that reflects the characteristics of the damping windings of synchronous generators, realizing VSG simulation of transient and steady-state characteristics of synchronous generators at the same time, and establishing a small signal dynamic model of VSG control with transient damping characteristics. Li Zhijun et al. Literature [2] whose author is from Hebei University of Technology proposed a VSG control strategy to improve the damping characteristics of inertia. By adding a differential correction link to the inertia damping control, VSG's inertia damping characteristics in different frequency bands are changed to achieve the purpose of considering the dynamic and static characteristics of the system.

These results solve the contradiction of transient damping and inertia to some extent. However, it relies too much on VSG's small signal model. Due to the linear relationship between the generator power Angle and the output power P_e in a very small range, the accuracy can be guaranteed when the system frequency changes and the power instruction is changed in a small range. However, in the case of wind farm custom change or photovoltaic panel illumination mutation, the output power fluctuates greatly, and the small-signal model loses its accuracy.

2. Virtual Synchronous Generator and BP Neural Network

2.1. Virtual synchronous generator

VSG method is only an improvement on the interface inverter control algorithm, does not need to adjust the grid structure on a large scale, has a strong practical utility in engineering. The structure of the VSG can be roughly divided into two layers, the first layer calculates the active power and reactive power instructions, and the second layer controls the inverter output of the corresponding current or voltage, so that the power output of the inverter meets the requirements of the power command[2]. In this paper, VSG is a current-type virtual synchronous generator, that is, the inverter output current is selected as the control amount, and the control mode is hysteresis control[3].

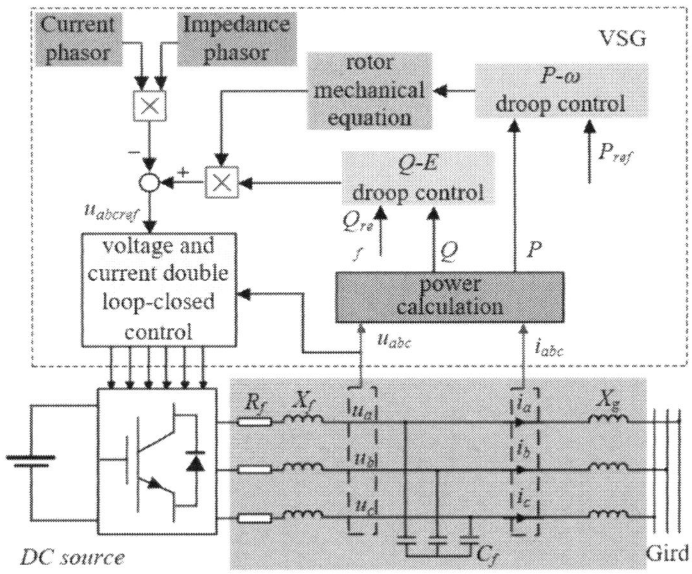

Figure 1. The topology of VSG control circuit

VSG's simulation of a real synchronous generator is reflected in the calculation of power instructions. Active power instruction Pref equivalent to the real synchronous generator's primer input power, the inverter output power is equivalent to the real synchronous generator's output power to the external circuit.

The essence of VSG is still the control of inverter, but the control logic is the algorithm of virtual synchronous generator. Part of the algorithm is the rotor motion equation of the synchronous machine and the electromagnetic relation of the fixed rotor. The other part is frequency and voltage droop control. Combined with the distributed power supply, the rotor inertia and damping characteristics of traditional synchronous machine can be simulated while controlling voltage and frequency.

The rotor mechanical equation of a real synchronous generator is shown in equation 1:

$$\begin{cases} J\dfrac{d\omega}{dt}=\dfrac{P_m}{\omega}-\dfrac{P_e}{\omega}-D(\omega-\omega_0) \\ \dfrac{d\delta}{dt}=\omega-\omega_0 \\ \delta=\int(\omega-\omega_0)dt=\varphi-\int\omega_0 dt \\ \varphi=\int\omega dt \end{cases} \tag{1}$$

In this equation, ω is system angle frequency, and ω_0 is rated angle frequency; J is the moment of inertia; D is the damping coefficient; P_m, P_e and $D(\omega-\omega_0)$ respectively represent the corresponding power of mechanical power, electromagnetic power and damping torque. δ is the internal power angle of the synchronous generator, where φ is the phase of the port line voltage of the synchronous machine when the phase current is the reference phasor.

There is another important relation in actual synchronous generator, that is, the relationship between excitation electromotive force and generator terminal voltage. Synchronous generator armature resistance and synchronous reactance will produce voltage drop, so the excitation electromotive force, generator terminal voltage and stator current meet:

$$\dot{E} = \dot{U} + \dot{I}R_f + j\dot{I}X_f \tag{2}$$

In (2), $\dot{E}$ is congestion electric potential, $\dot{U}$ is generator port voltage, R_f and X_f are a generator armature resistors and synchronous resistors, $\dot{I}$ is the stator current.

2.2. BP Neural Network

BP neural network algorithm was proposed in 1986 by Rumelhart et al. who come from Stanford University. The algorithm effectively solved the connection weight problem of hidden layers in the multi-layer network model, and effectively improved the self-learning and organization ability of neural network.

The topological structure of BP neural network is shown in Figure 2. The input layer of BP neural network acts on the output layer by the hidden layer and outputs. The quantity is obtained by nonlinear transformation. Each sample of neural network training contains input and expected output. By adjusting network weights and thresholds, the error between network output and expected output decreases along the gradient direction until the actual output and expected output are within a predetermined range.

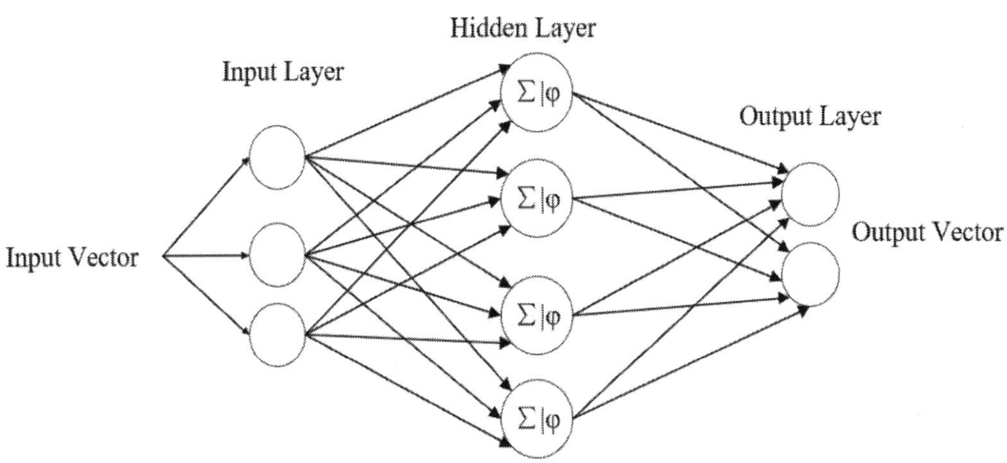

Figure 2. The topology of BP neural network

Training a neural network is the process of getting the output value closer and closer to the expected value. The function that measures how close the output value is to the expected value is called the loss function. The smaller the loss function, the closer the output value is to the expected value. Therefore, the training process of neural network can be summed up as finding the minimum value of loss function. By modifying the connection weight of each node, the loss function will also change in response. When the weight value decreases along the direction of negative gradient, the loss function decreases. The correction process is shown in the equation (2).

$$\omega_{ij}(n+1) = \omega_{ij}(n) - \eta \frac{\partial L}{\partial Y} \frac{\partial Y}{\partial \omega_{ij}(n)} \tag{3}$$

In this equation, $\omega_{ij}(n+1)$ and $\omega_{ij}(n)$ are the weights before and after the update respectively. η is the learning rate, and L is the loss function of neural network. Y is the output of neural network.

3. BP neural network with VSG adaptive model

3.1. Overall framework of adaptive model

Consider the damping as a time-varying function, and the constant damping is a line parallel to the t-axis. A batch of simulation results and corresponding damping are used as the training set. The input of neural network is active power vector and the output is damping vector. After a lot of training, the neural network is fitted into the model of virtual synchronous generator, namely:

$$F(\boldsymbol{P},\boldsymbol{Q}) = \boldsymbol{D} \tag{4}$$

In this equation, P, Q is respectively the input active and reactive power vector, and its specific content is shown in Table 1. When the input X is the ideal waveform, a damping vector can be obtained that causes the virtual synchronous machine to exhibit such a power response curve.

The process of this paper can be summarized as follows: acquire a large number of input data and corresponding output waveform from the perfect and detailed VSG model, and train the neural network to perform relational fitting of its input and output to complete modelling. When the neural network training is completed, the ideal active power transient curve is taken as the input, and the damping time series corresponding to the ideal active power transient curve is finally obtained. The block diagram is shown in Figure 3.

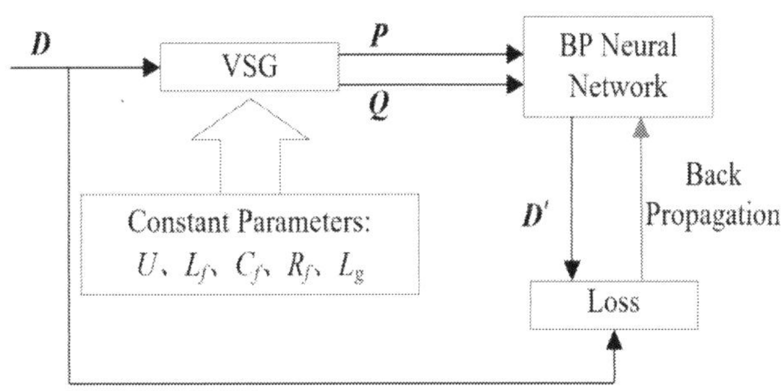

Figure 3. Neural network training process

Any successful training of neural network is inseparable from a comprehensive and rich training set. The training set used in this paper includes power command step, frequency mutation active power change, network voltage drop and reactive power supplement, so as to describe the virtual synchronous generator comprehensively.

3.2. The modeling processes

Based on the simulation data based on VSG mechanism model, the steps of VSG modelling are as follows by using data-driven modelling method:

Step 1): Data acquisition and preparation. The operating data of VSG under different operating conditions were collected, and the input and output variables of neural network were obtained through data integration and preliminary calculation, forming the data of training set and test set.

Step 2): In order to prevent the adverse effects caused by the large difference in the order of magnitude of different physical quantities, all data are normalized.

Step 3): Determine the scale of the neural network, including determining the network structure and training algorithm. This paper adopts a single hidden layer network structure, with 40,000 nodes in the input layer. It is composed of P and Q vectors with length of 20,000. After many trials, when the number of hidden layer neural units is 10, the network modelling effect is better.

Step 4): Neural network training and error calculation. Input data of training set into neural network to get corresponding predictive value. According to the predicted value and the real value of the output data of the training set, the error is calculated. According to the error back propagation algorithm, the parameters of the neural network are adjusted until the error meets the accuracy or reaches the maximum number of iterations.

Step 5): The input data of the test set is input into the trained neural network to obtain the corresponding predicted value, which is compared with the real value of the output data of the test set. The root mean square error is calculated according to Formula (4) to evaluate the modelling effect.

$$E_{RMSE}=\sqrt{\frac{1}{n}\sum_{i=1}^{n}(y_i-y_r)^2} \tag{5}$$

In this equation, y_i is the predicted value of the model; y_r is the true value.

4. The simulation verification

In order to verify the accuracy of the modelling method in this paper, a single VSG grid-connected system as shown in Figure 1 is built on the MATLAB/Simulink simulation platform to obtain the operating data of VSG under different working conditions. The simulation parameters of VSG are shown in Table 1.

The simulation is divided into two conditions. Under the first operating condition, the system frequency drops by 0. 5Hz, and VSG senses the system frequency drop and emits 75KW more power. Under the second working condition, the system frequency increases by 0. 5Hz, and VSG emits 75KW less power. The initial active power of VSG is 100KW and 200KW respectively, and the system frequency changes when t=0. 1s. Firstly, the transient process of primary frequency modulation is observed under constant damping condition. When the damping is 50, 75, 100 and 125 respectively, the transient processes have oscillations and steady-state errors.

Table 1. VSG parameters

Parameter	Value
Rated active power P_n	100KW
Rated reactive power Q_n	50Kvar
Rated frequency f_n	50Hz
Rated line voltage U_n	380V
Filter capacitor reactive Q_c	10Kvar
Line inductance L_f	0.001H
Parasitic resistance R_f	0.004Ω
Virtual Inertia J	0.1s
Damping coefficient D	75
Primary frequency modulation coefficient K_f	150KW/Hz
Reactive power regulation factor k_q	0.00001
DC side rated voltage U_{dc}	700V

If the adaptive damping calculated by BP neural network is used for control, the system frequency change is equal to the rated value without deviation, and the whole transient process has no oscillation and no overshoot. Figure. 4 shows the corresponding waveform when the system frequency drops, and Figure. 5 shows the corresponding waveform when the system frequency rises.

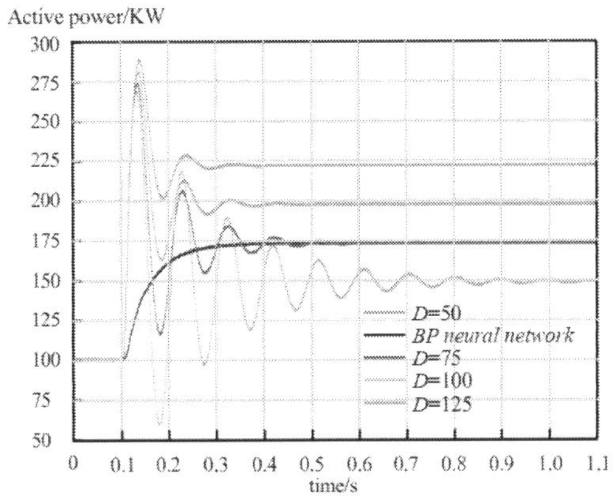

Figure 4. VSG damping using fuzzy control effect (rising step)

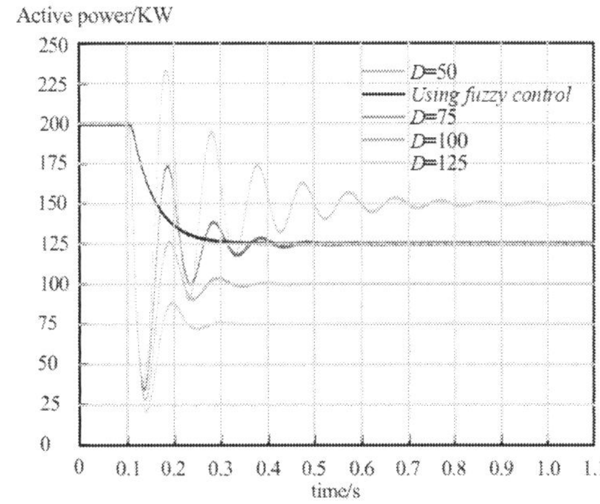

Figure 5. VSG damping using fuzzy control effect (drop step)

5. Conclusion

The virtual synchronous generator with constant damping has many disadvantages. When primary frequency modulation is required, too small damping will not only bring oscillation, but also cause insufficient change of output power. Excessive damping can effectively reduce the oscillation, but correspondingly will make the change in the output power of primary FM deviate from the rated value. After adopting adaptive damping, the output power not only has no overshoot and oscillation in the transient process of primary frequency modulation, but also can accurately reach the rated value. The adaptive damping real-time control strategy based on BP neural network saves complicated calculation and trial, and does not rely on accurate model, showing great advantages.

References

[1] Dongfeng X, Mingxing T. (2020) Relationship between frequency characteristics of virtual synchronous generator and parameters of energy storage equipment[J]. Power System Technology. pp:1-13.

[2] Lin W, Yuntao J, Wenchuan W. (2021) Optimal Design of Inertia and Damping Parameters of Virtual Synchronous Microgrid for Improving Frequency Stability. Proceedings of the CSEE, 41(13): 4479-4490.

[3] Zheng Z, Weihua S, Li R. (2015) Mathematical Model and Strategic Energy Storage Selection of Virtual Synchronous Generators. Automation of Electric Power System, 39(13): 22-31.

[4] Xin L I, Guoliang L, Ranchen Y, (2018) VSG Control Strategy With Transient Damping Term and Seamless Switching Control Method. Power System Technology, 42(7): 2081-2088.

[5] XU Chongfu, LI Wanru, XU Ningyi, WANG Yifei, LIU Yonghui, (2020) Analysis and Improvement of Damping Characteristics of Virtual Synchronous Generator Control Applied to Energy Storage Converter. Power System Technology, 44(5): 1656-1663.

[6] Xufeng L, Limin L, Lexiang C, (2019) Power Decoupling Control Strategy in Virtual Synchronous Generator With Improved Reactive Power Loop Based on Adaptive Virtual Impedance. Power System Technology, 43(10): 3752-3760.

Application of Repetitive Control and ANFIS-PI Compound Control Strategy in APF

Mengtao Zhang*, Tao Sui

College of Electrical Engineering and Automation, Shandong University of Science and Technology, Qingdao, Shandong, 266590, China

*Corresponding author's e-mail: 201982080031@sdust.edu.cn

Abstract. In order to effectively solve a series of problems caused by harmonics in power system, an active power filter based on ANFIS-PI and repetitive control is designed. The traditional single control method cannot balance control accuracy and response speed, so a compound control strategy is adopted. Repetitive control can produce the same output as the harmonic current, so it can achieve zero steady-state error tracking compensation for each harmonic, but its dynamic performance is not good. As a traditional control method, the PI control method has a fast response speed, but in the actual application process, when the load changes suddenly, that is, when the system parameters change, its tracking effect is not ideal. Therefore, a composite control strategy of ANFIS-PI and repeated control is proposed. It has been proved that its compensation accuracy and dynamic response speed have been significantly improved.

1. Introduction

There are a large number of non-linear loads in the power grid, which will cause the distortion of the current waveform and generate harmonic pollution, which will seriously affect the stable operation of the power grid and the power safety of users [1]. In the research of harmonic control, active power filter (APF), as one of the most effective means to suppress power grid harmonics, has broad application prospects [2-3]. The design principle is to obtain the corresponding harmonic current component by collecting the current content of the three-phase load in the system, subtracting the fundamental current value from it, and then sending it back to the power system to complete the suppression of the harmonics in the power grid.

Harmonic current is a high-frequency AC signal, and many scholars have done a lot of work on the control of harmonic current. At present, the compensation current control method of APF has triangular carrier control. This method is easy to implement and the switching frequency is fixed, but the control parameters are difficult to determine and it is easy to produce overshoot. The advantage of hysteresis comparison control is that it responds quickly and does not need a carrier. The disadvantage is that the hysteresis bandwidth is not easy to set. The traditional PI control is a very widely used control method, but the tracking effect is not good for high-frequency AC components such as harmonic currents [4]. The bandwidth of PI control is limited, the attenuation is severe in the middle and high frequency bands, and the phase lag is also obvious. The inability to track the signal in the full frequency band will make it difficult for the pi control to achieve zero error tracking control for the harmonic AC signal.

This paper adopts the compound control strategy of ANFIS-PI and repetitive control to realize the compensation of APF to the harmonic current [5-6]. The algorithm combines the advantages of fast response speed of PI control with the advantages of zero steady-state error tracking of repetitive control to meet the requirements of APF for response speed and accuracy. ANFIS is used to adaptively adjust the parameters of PI controller to solve the problem of load mutation.

2. System model

The system adopts a four-leg circuit model, which is used to compensate the neutral current, and the other three legs track and compensate the corresponding phase current respectively [7]. Compared with the three-arm structure, the circuit is relatively simple. The structure diagram of APF is shown in figure 1.

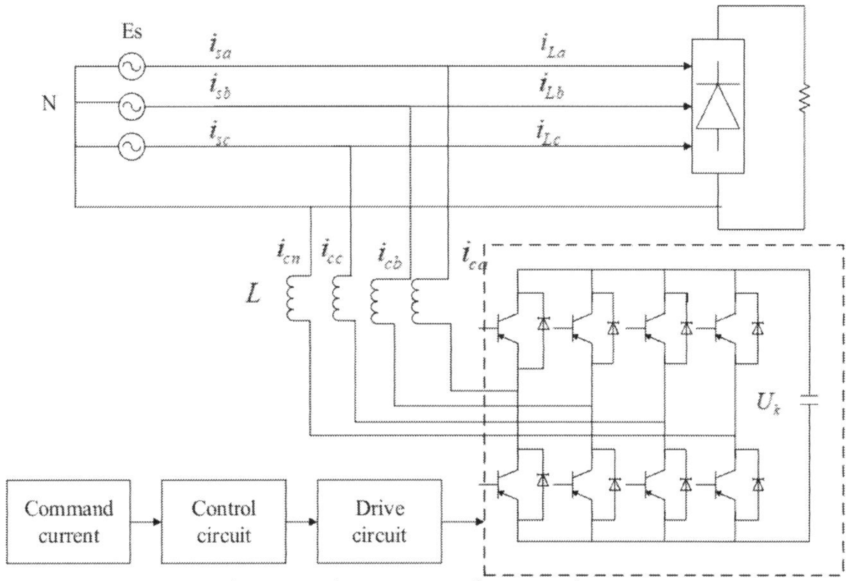

Figure 1. The structure diagram of APF.

According to figure 1, Kirchhoff equation is as follows:

$$\begin{cases} L\dfrac{di_{ca}}{dt} = E_{sa} - (U_a - U_n) \\ L\dfrac{di_{cb}}{dt} = E_{sb} - (U_b - U_n) \\ L\dfrac{di_{cc}}{dt} = E_{sc} - (U_c - U_n) \\ C\dfrac{dU_k}{dt} = S_n i_{cn} - (S_a i_{ca} + S_b i_{cb} + S_c i_{cc}) \end{cases} \tag{1}$$

Where, S_a, S_b, S_c and S_n are state functions. U_a, U_b, U_c, U_n are DC voltage values corresponding to IGBT output respectively. After coordinate transformation, it is transformed into dq0 coordinate system:

$$\begin{cases} L\dfrac{di_{cq}}{dt} = U_{sd} - (S_d U_k + \omega L i_{cq}) \\ L\dfrac{di_{cd}}{dt} = U_{sq} - (S_q U_k + \omega L i_{cd}) \\ L\dfrac{di_{c0}}{dt} = U_{s0} - S_0 U_k \\ 0 = \dfrac{3}{2}(S_d i_{cd} + S_q i_{cq} + 2S_0 i_{c0}) - C\dfrac{dU_k}{dt} \end{cases} \tag{2}$$

3. Compound control algorithm of ANFIS-PI and repetitive control

The composite control structure is shown in figure 2:

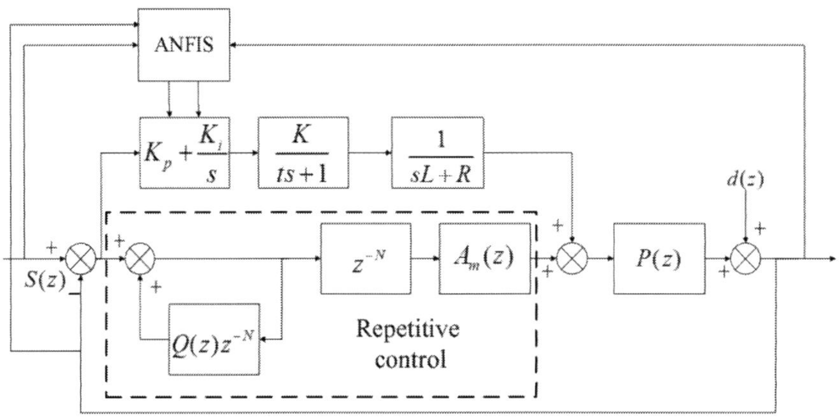

Figure 2. Composite control structure diagram.

The system obtains the compensated command signal through ip-iq method and sends it to the controller. According to the algorithm proposed in this paper, it generates the driving signal to control the on and off of IGBT and realize the harmonic suppression in power system. The traditional PI control algorithm is simple and fast, but the steady-state performance of APF is poor. According to this defect, it is combined with repetitive control. When the system parameters are changed, ANFIS is used to adaptively modify the PI parameters.

3.1. PI control

As can be seen from figure 2, the closed-loop transfer function of PI part is:

$$h(s) = \frac{K_p K / Lt}{s^2 + s/t + K_p K / Lt} \tag{3}$$

Where, $Kp = 0.56$, $k = 20$, $L = 2\text{mH}$, $t = 10^{-4}\text{s}$, so the expression of PI controller in z domain is:

$$H(z) = 2.35 + \frac{0.05}{z-1} \tag{4}$$

3.2. Repetitive control

Repetitive control is a high-precision control method [8-9]. The theoretical basis of this method is the internal model principle. The principle of repetitive control shows that it can track the input signal in any case with high precision and stability, and then realize no static error tracking. In other words, with the change of the input signal, when the change of the input signal tends to reduce the error, the repetitive control can always maintain a reasonable and appropriate output, and then maintain a state without static error. The discrete model is as follows:

$$G(z) = \frac{z^{-N}}{1 - z^{-N}} \tag{5}$$

Where, z^{-N} represents the periodic delay link. The function of z^{-N} is to delay the error value of this period to act on the control quantity of the next period.

Am is the compensator. It sets the parameters according to the specific characteristics of the P(z) of the controlled object, and compensates and corrects the phase and amplitude of the controlled object according to the errors measured in the previous period. The error of each cycle is superimposed until the error is zero. The model is as follows:

$$A_m(z) = K_a z^k T(z) \tag{6}$$

Where, K_a is the gain coefficient, which is used to maintain the stability of the system. z^k is used to compensate the phase. The function of T(z) is to correct medium and low frequency and high frequency to increase the stability of the system. The function of repetitive control in z domain is:

$$R(z) = G(z)A_m(z) = \frac{z^{-N}}{1 - z^{-N}} K_a z^k T(z) \tag{7}$$

Therefore, the function of compound control system in z domain is:

$$Z(z) = z^{-N} \frac{z^2 K_a T(z) G(z)}{z^{-N} - Q(z)} + H(z)G(z) \tag{8}$$

The stability condition of the internal mold structure is $\left|[1 - P(z)]Q(z)z^{-N}\right| < 1$. In this paper, the sampling frequency is 10 kHz and the fundamental frequency is 50 Hz, so $N = 200$.

3.3. ANFIS adaptively adjusts PI parameters

ANFIS has good adaptability to nonlinear and multivariable system [10-11]. It can handle some uncertain data. It combines fuzzy logic and neural network organically, and has the advantages of both. It has three inputs: input value, output value and error. It has two outputs, P parameter and I parameter.

ANFIS has a five-layer structure. After the structure is determined, the initial value is selected and the learning rate is determined. The system continuously modifies the weights, so as to realize the adaptive adjustment of P parameter and I parameter.

- The first layer is the membership function layer.
- The nodes of the second layer are circular nodes labeled Π.
- In the third layer, the ith node calculates the ratio of the ith rule's firing strength to the sum of all rules' firing strength.
- The fourth layer calculates the output of fuzzy rules.
- The fifth layer calculates the total output of all input signals.

4. Results and simulation

According to the above algorithm, the system model is built, and the PI control, PI-repetitive control and PI-repetitive control algorithm with ANFIS are simulated and verified respectively. Figures 3 and 4 show the compensation effect of APF under PI control and PI-repetitive control.

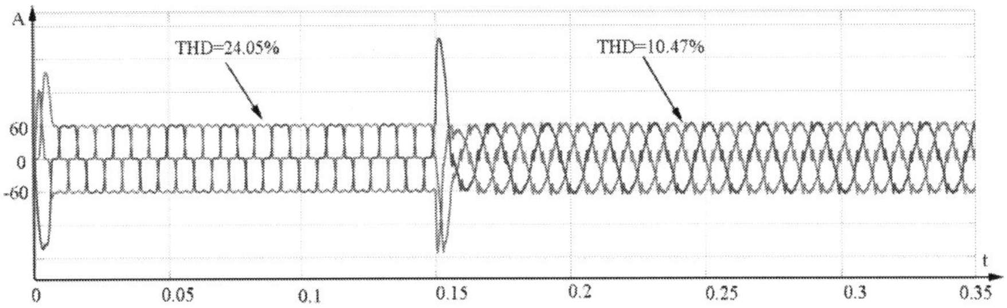

Figure 3. Compensation effect of APF under PI control.

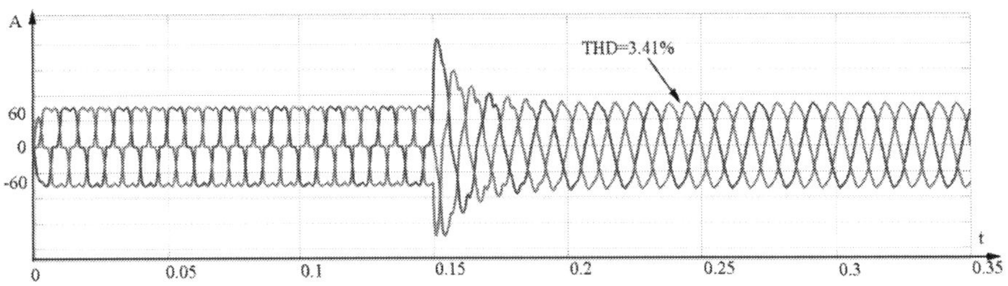

Figure 4. Compensation effect of APF under PI control and repetitive control.

The simulation results show that under PI control and PI-repetitive control, total harmonic distortion (THD) after APF compensation is 10.47% and 3.41% respectively. Green, blue and red waveform curves represent three-phase currents of A, B and C respectively. According to the IEEE standard, for power systems below 6kv, THD should be less than 5%. Therefore, only PI control can not meet the requirements of harmonic current tracking compensation in power system. Considering that the sudden change of load affects the system parameters and the control effect of PI controller, ANFIS is used to adaptively modify the PI parameters.

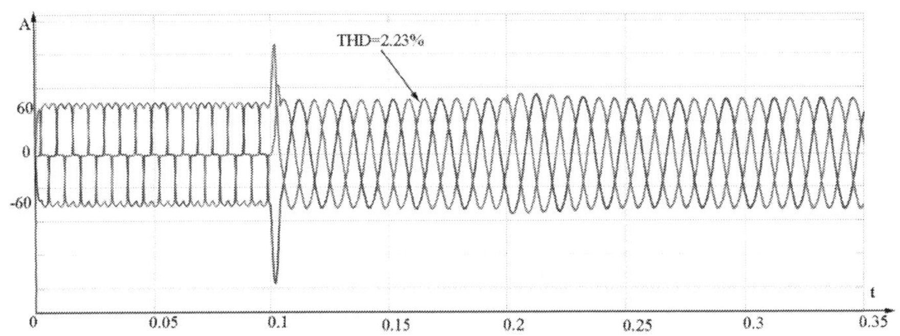

Figure 5. Compensation effect of APF under ANFIS-PI control and repetitive control.

As shown in figure 5, under the strategy of ANFIS-PI and repetitive control, THD is 2.2% after APF compensation, which is better than the first two methods.

Table 1. Compensation effect of APF under different control algorithms.

Control algorithm	THD
PI control	10.47%
PI control and repetitive control	3.41%
ANFIS-PI and repetitive control	2.23%

According to figure 5, it can be seen that at 0.2s, the load has a sudden change, and the current has a small fluctuation, which has little effect on the system. Therefore, the algorithm proposed in this paper has good superiority in the process of APF compensation for sudden changes in load.

5. Conclusion

In this paper, the compound control strategy of ANFIS-PI and repetitive control is used to compensate the harmonic current of active power filter (APF), aiming at the disadvantage that the traditional control strategy of APF can not track the harmonic current without error. PI control has fast response speed, but the tracking compensation degree of harmonic current is not high, and the total harmonic distortion (THD) after compensation often can not meet the requirements of the system. Repetitive control can track and compensate the zero steady-state error of each harmonic in the system, but it has time delay and poor dynamic performance. Therefore, it is combined with PI control to form a compound control strategy. However, considering the influence of sudden load change on PI control, ANFIS is used to adaptively adjust P parameter and I parameter. The simulation results show that the composite control strategy has high accuracy and fast dynamic response. It is an ideal APF harmonic current tracking compensation scheme.

References

[1] Lobos, T., Kozina, T., Leonowicz, Z., Koglin, H.J. (2000) Advanced signal processing methods of harmonic analysis in power systems. Ifac Proceedings Volumes, 33(5): 147-152.

[2] Chen, J., Zhang, X., Wen, C. (2016) Harmonics attenuation and power factor correction of a more electric aircraft power grid using active power filter. IEEE Transactions on Industrial Electronics, 63(12): 1-1.

[3] Djeghloud, H., Benalla, H. (2007) Harmonics cancellation using svpwm shunt active power filter. Compel, 26(1): 114-126.

[4] Xu, Q., Zhong, X., Hu, J., Chen, G. (2015) Compensation current control of APF based on error iteration pi and improved repetitive control. Automation of Electric Power Systems, 39(3): 124-131.

[5] Wei, X. (2010) Shunt APF based on current loop repetitive control. Journal of Convergence Information Technology, 5(4): 29-37.

[6] Jang., J.-S., R. (1993) ANFIS: adaptive-network-based fuzzy inference system. IEEE Trans on Smc, 23(3): 665-685.

[7] Wang, C., Zhou, Q., Zhou, J. (2012) Research on selection method of parameters in main circuit for four-leg APF. Electric Drive, 42(11): 33-37.

[8] Guo, X., Lin, H., & Chen, G. (2021) SIC-MOSFET shunt active power filter based on half-cycle SDFT and repetitive control. Energy Reports, 7(12): 246-252.

[9] Chen, J., Shao, H., Liu, C. (2021) An improved deadbeat control strategy based on repetitive prediction against grid frequency fluctuation for active power filter. IEEE Access, PP(99): 1-1.

[10] Kalarathi, M., Prabha, N.R. (2020) Multilevel parallel active power filter based on ANFIS controller for harmonic alleviation. Iranian Journal of Science and Technology - Transactions of Electrical Engineering, (15).

[11] Manivasagam, R., Prabakaran, R. (2020) Power quality improvement by UPQC using ANFIS-based hysteresis controller. International Journal of Operational Research, 37(2): 174-.

Adaptive damping control strategy of virtual synchronous generator based on fuzzy control

Wenkai Li, Jie Xie, Mengzhao Zhang* and Chunlin Guo

School of Electrical and Electronic Engineering, North China Electric Power University, Changping District, Beijing, 102200, China

*Corresponding author's e-mail: 120202201184@ncepu.edu.cn

Abstract. Based on the topology and small signal model of virtual synchronous generator, the influence of virtual inertia and damping coefficient on transient process is analysed. In order to solve the problem of overshoot with small damping coefficient and slow response with large damping coefficient, a fuzzy control method is introduced, which can completely eliminate overshoot without affecting the response speed. Another fuzzy controller is designed to optimize the output of the virtual synchronous generator in the primary frequency modulation process according to the SOC of the energy storage system.

1. Introduction

New energy generation generally requires frequent changes in power instructions. Every time the power instruction is changed, it is equivalent to experiencing a transient process. The control mode of the inverter determines the impact of changing the power instruction on the grid. Compared with PQ control and PQ droop control, the superiority of VSG grid-connected control mode has been widely recognized. However, if the damping coefficient is too small, it will lead to the phenomenon of oscillation; If the damping coefficient is too large, the response will be too slow and the power requirements cannot be met in time. Fuzzy control can eliminate transient process overtones without affecting the response power command pigment.

Constant damping has major drawbacks. Too large damping can ensure the transient process without oscillation, but will make the output power of primary FM error. Low damping can reduce such errors, but again there are oscillations. In view of this contradiction, many scholars have proposed the methods of transient damping and inertia in recent years. Literature [1] proposed a solution of introducing a transient damping link that reflects the characteristics of the damping windings of synchronous generators, realizing VSG simulation of transient and steady-state characteristics of synchronous generators at the same time, and establishing a small signal dynamic model of VSG control with transient damping characteristics. Li Zhijun et al. The author from Hebei University of Technology proposed a VSG control strategy to improve the damping characteristics of inertia. By adding a differential correction link to the inertia damping control, VSG's inertia damping characteristics in different frequency bands are changed to achieve the purpose of considering the dynamic and static characteristics of the system.

These results solve the contradiction of transient damping and inertia to some extent. However, it relies too much on VSG's small signal model. Due to the linear relationship between the generator power Angle and the output power Pe in a very small range, the accuracy can be guaranteed when the system frequency changes and the power instruction is changed in a small range. However, in the case of wind farm custom change or photovoltaic panel illumination mutation, the output power fluctuates greatly, and the small-signal model loses its accuracy.

2. Virtual synchronous generator

2.1. Virtual synchronous generator topology

VSG method is only an improvement on the interface inverter control algorithm, does not need to adjust the grid structure on a large scale, has a strong practical utility in engineering. The structure of the VSG can be roughly divided into two layers, the first layer calculates the active power and reactive power instructions, and the second layer controls the inverter output of the corresponding current or voltage, so that the power output of the inverter meets the requirements of the power command[2]. In this paper, VSG is a current-type virtual synchronous generator, that is, the inverter output current is selected as the control amount, and the control mode is hysteresis control[3].

VSG's simulation of a real synchronous generator is reflected in the calculation of power instructions. Active power instruction P_{ref} equivalent to the real synchronous generator's primer input power, the inverter output power is equivalent to the real synchronous generator's output power to the external circuit.

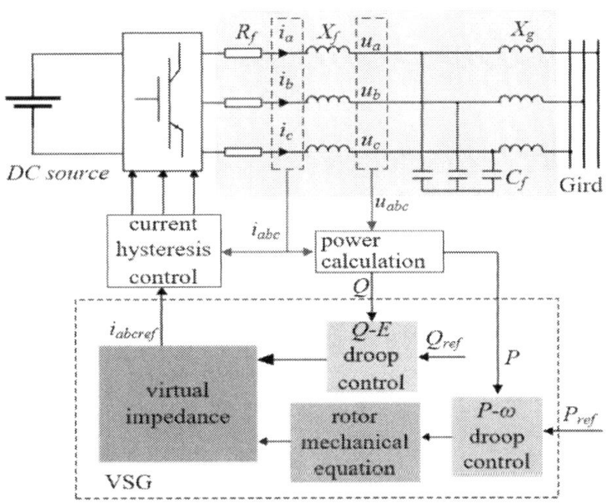

Figure 1. The topology of VSG control circuit

In the process of following the change of active power instruction with the active power output of the inverter, a transient module is artificially introduced. This transient module is essentially an equation of the mechanical motion of the rotor:

$$\begin{cases} J\dfrac{d\omega}{dt}=\dfrac{P_m}{\omega}-\dfrac{P_e}{\omega}-D_p(\omega-\omega_0) \\ \dfrac{d\delta}{dt}=\omega-\omega_0 \\ \delta=\int(\omega-\omega_0)dt=\varphi-\int\omega_0 dt \\ \varphi=\int\omega dt \end{cases} \tag{1}$$

In this equation, ω is system angle frequency, and ω_0 is rated angle frequency; J is the moment of inertia; D_p is the damping coefficient; P_m, P_e and D(ω-ω_0) respectively represent the corresponding power of mechanical power, electromagnetic power and damping torque. δ is the internal power angle of the synchronous generator, where φ is the phase of the port line voltage of the synchronous machine when the phase current is the reference phasor.

From the outside circuit, the inverter output is as if it had a rotational inertia and damping effect. Unlike a real synchronous generator, where the moment of inertia and damping are determined by the mechanical structure and are almost impossible to change after being manufactured, the moment of inertia and damping coefficients of a virtual synchronous machine can be adjusted as required.

In terms of reactive power and voltage amplitude, the synchronous and armature resistors are compared to the inverter filter and parasitic resistors[7]. The average voltage of the midpoint of the inverter bridge arm is compared to the excitation motor potential of the synchronizer. In a synchronous generator, the equations that describe the relationship between excitation electromotive force and port voltage are[5]:

$$\dot{E} = \dot{U} + \dot{I}R_f + j\dot{I}X_f \tag{2}$$

In (2), $\dot{E}$ is excitation electric potential, $\dot{U}$ is generator port voltage, R_f and X_f are respectively generator armature resistors and synchronous resistors, $\dot{I}$ is the stator current.

2.2. Paraments Setting

Convert J and D_p to a per-unit value, introduce variables H and D.

$$\begin{cases} H = \dfrac{J\omega_0^2}{S_n} \\ D = \dfrac{D_p\omega_0^2}{S_n} \end{cases} \tag{3}$$

In (3), H and D are respectively the inertial time constant and damping coefficients after converting to per-unit value, and Sn is rated power capacity.

The most important parameters of a virtual synchronizer are rotational inertia and damping. The rotor mechanical equation can be refined into a second-order small signal system, as shown in Figure 2.

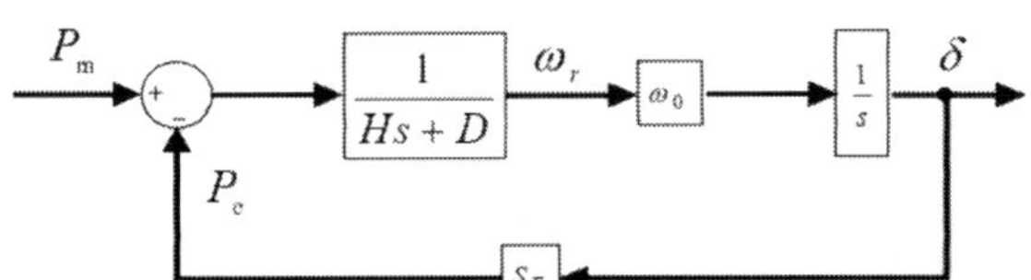

Figure 2. VSG Small-signal modeling

S_E is the per-unit value of synchronous power, and its expression is:

$$S_E = \frac{1}{S_n}\frac{\partial P}{\partial \delta}\bigg|_{\delta=\delta_0} = \frac{E_0 U_l}{S_n Z}\sin(\alpha - \delta_0) \tag{4}$$

The model is based on two assumptions. At first, we assume that the power instruction Pm is unchanged, and secondly, we assume that the inverter filter impedance is close to pure inductive circuit. In another word, $\alpha \approx 90$ degrees. Write down the meaning of the transfer function between P_e and P_m:

$$G_{VSG}(s) = \frac{P_e(s)}{P_m(s)} = \frac{\omega_0 \dfrac{S_E}{H}}{s^2 + \dfrac{D}{H}s + \omega_0 \dfrac{S_E}{H}} \tag{5}$$

The natural oscillation angular frequency and damping ratio of this second-order transfer function are:

$$\begin{cases} \omega_n = \sqrt{\omega_0 \dfrac{S_E}{H}} \\ \xi = 0.5D\sqrt{\dfrac{1}{\omega_0 S_E H}} \end{cases} \tag{6}$$

The final virtual synchronous generator parameters are shown in Table 1.

Table 1. VSG parameters

Parameter	Value
Rated active power P_n	5KW
Rated reactive power Q_n	0Kvar
Rated frequency f_n	50Hz
Rated line voltage U	380V
Line inductance L_f	0.001H
Parasitic resistance R_f	0.004Ω
The inertia time constant H	0.1s
Damping coefficient D	15
Primary frequency modulation coefficient K_f	10KW/Hz
Reactive power regulation factor k_q	0.00001
DC side rated voltage U_{dc}	1500V

3. Fuzzy Control

The Modification of the damping coefficient can be regard as applying two extra components, which can be described as equation:

$$D = D_0 + \Delta D_1 + \Delta D_2 \tag{7}$$

In (7), D is the final damping coefficient of VSG, D_0 is the initial damping coefficient of VSG, ΔD_1 and ΔD_2 are the amendment components from transient fuzzy controller and the first frequency modulation fuzzy controller. The value of the first component should make the transient process of power instruction change without overshoot, and the value of the second component is determined by the energy storage unit SOC value.

3.1. Transient Damping Fuzzy control rules

The output of the transient damping fuzzy controller is ΔD_1, The control principle of ΔD_1 is that the damping coefficient for regulation during the transient process is that the VSG power response is not overstated and zeroed after entering steady state. When the actual power value differs greatly from the power instruction, the VSG damping needs to be reduced appropriately to speed up the response, and when the actual power is closer to the instruction value, the damping becomes larger to suppress overshooting.

The transient damping fuzzy controller has two inputs. The first input is power error E, and the second input is the derivative of power error with respect to time Ec. The expression is as follows:

$$\begin{cases} E = P_{ref} - P \\ E_c = \dfrac{dE}{dt} \end{cases} \tag{8}$$

The fuzzy subset of power error E is {NB,NS,O,PS,PB}, the corresponding language set is {negative big, negative small, zero, positive small, positive big}, and the fuzzy domain is [-5,5]. The fuzzy subset of derivative of the power error input is {NB,NS,O,PS,PB}, the corresponding language set is the same as E. Its fuzzy domain is [-100,100]. The fuzzy subset of the VSG damping adjustment is {NB, NM, NS, O, PS, PM, PB}, and the fuzzy domain [-3.5,3.5]. The Membership degree function is shown in Figure 3.

The design of fuzzy control rules is briefly described here. In the transient process of power instruction change, the ideal situation is that when error E is large, the error change rate is also large, so as to eliminate the error quickly. When the error E is small, the error change rate is also small; As the error E goes to zero, the error rate Ec also goes to zero; There is no overshoot in the whole transient process.

The first step is to eliminate overshoot. The overshooting phenomenon has obvious characteristics, as shown in the case that the error E is already small, the error change rate Ec is still very large, which indicates that the damping is insufficient and the damping should be increased. If the error E is large and the error rate Ec is small, it indicates that the transient response process is slow. This is caused by oversized damping, so damping should be appropriately reduced. Specific fuzzy control rules are shown in Table 2.

Table 2. Formatting sections, subsections and subsubsections

E	***E_c***				
	NB	***NS***	***O***	***PS***	***PB***
NB	*PS*	*PS*	*NB*	*NM*	*PS*
NS	*PB*	*PB*	*O*	*O*	*PB*
O	*PB*	*PS*	*O*	*PM*	*PB*
PS	*PB*	*O*	*O*	*PB*	*PB*
PB	*O*	*NS*	*NM*	*PS*	*PS*

3.2. First Frequency modulation Fuzzy control rules

New energy generation is usually equipped with energy storage systems. Generally, the power of new energy itself cannot be adjusted, and energy storage should be combined to participate in a frequency modulation. Therefore, it is necessary to optimize the SOC of energy storage system. The energy storage device used in this paper is battery group, with a total capacity of 5000Ah.

Here are two input variables in the frequency modulation fuzzy controller. The first is the state of charge of the energy storage device , the second input is the frequency offset Δf, which is equal to $f - f_n$.

The variable f is the actual value of power system, and fn is system rated frequency.

The fuzzy subset of battery SOC is {VL,L,O,H,VH}, the corresponding language set is {very low, low, normal, high, very high}, and the fuzzy domain is [0,1]. The fuzzy subset of frequency input is {L,O,H}, the corresponding language set is {below power frequency, normal, above power frequency}, and the fuzzy domain is [-1,1]. The fuzzy subset of the VSG damping adjustment is {negative large, negative small, zero, positive small, positive big}, and the fuzzy domain [-1,1]. The Membership degree function is shown in Figure. 4.

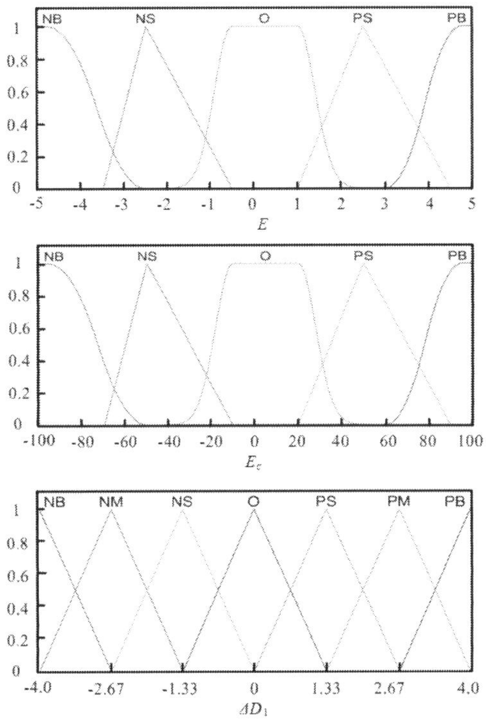

Figure 3. Transient damping fuzzy control rules

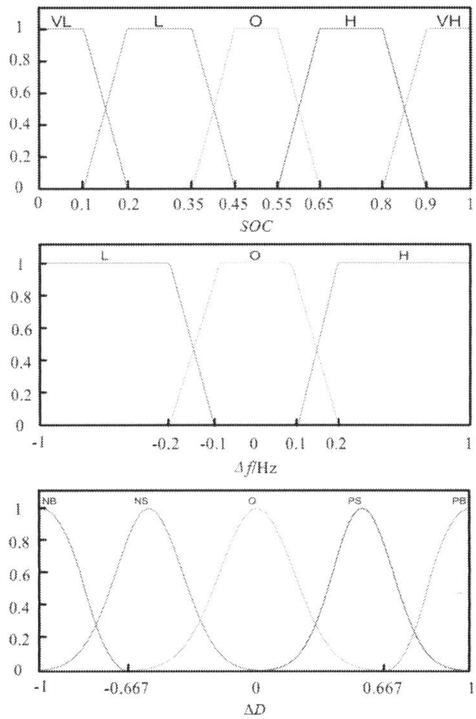

Figure 4. Secondary adjustment of battery membership function.

The principle of this fuzzy control is that, if the energy storage device has a high charge, and VSG senses that the frequency of the power grid decreases, then the energy storage system engages in a frequency modulation with multiple discharges, and the damping coefficient D should be increased accordingly. When VSG perceives that the power grid frequency is greater than the rated frequency, the energy storage device reduces or increases the charging amount according to the state of charge to participate in the primary frequency modulation, and correspondingly reduces or increases the damping coefficient D.

Based on the above principles, a specific table of fuzzy control rules is developed, as shown in Table 3.

Table 3. Parameter optimization results

P_{ref}	The SOC of Energy Storage System				
	VL	***L***	***O***	***H***	***VH***
D	NB	NB	O	O	O
O	NS	NS	O	PS	PB
C	O	O	O	PS	PB

4. Simulation And Verifications

4.1. Transient damping fuzzy control

The simulation experiment of transient damping fuzzy control is divided into two working conditions. In one case it's a step up, and in the other case it's a down step. It increased from 1KW to 6KW and decreased from 6KW to 1KW.

At t=0.3s, change the VSG active power instruction from 1KW to 6KW, and the change range is equivalent to 1.0PN.As can be seen from Figure 5, when the damping is too small, D=1.5, the transient process will have overharmonic oscillation.It enters a steady state at t=0.7s.The damping is slightly larger. When D=7.5, although there is no oscillation, there is still a slight overshoot, and it enters the steady state at about 0.9s.When D=15 and 20, there is no overharmonic oscillation, but the time to enter the steady state is more delayed at 1.0s and 1.2s respectively.

After the damping of fuzzy control is adopted, the overharmonic oscillation is eliminated and the steady state is entered at t=0.7s. It is proved that the damping fuzzy control strategy is correct and effective.

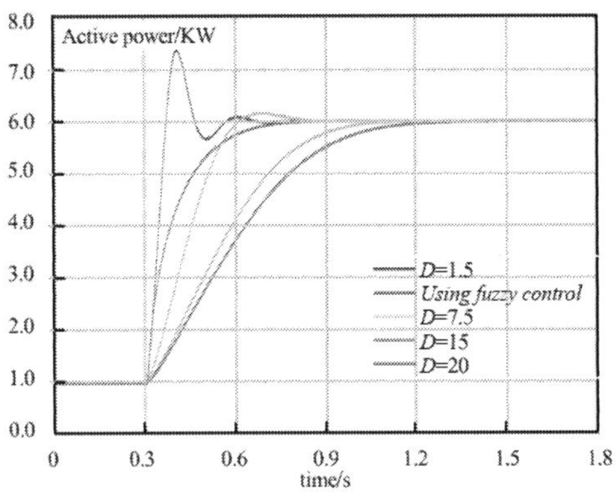

Figure 5. VSG damping using fuzzy control effect (rising step)

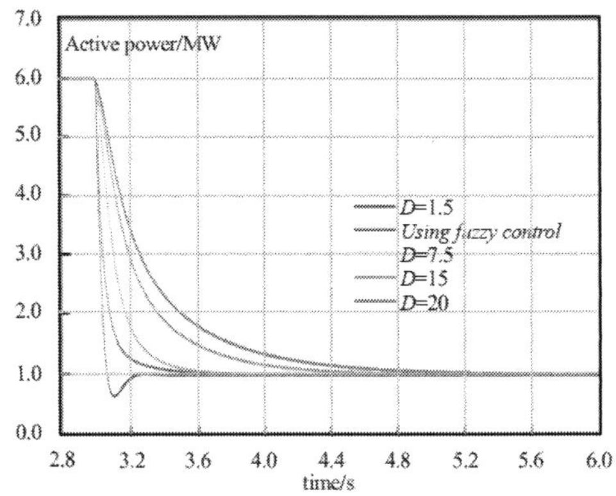

Figure 6. VSG damping using fuzzy control effect (drop step)

When the power drops by 5KW, the transient response process is asymmetric even if the damping is the same.This is also the reason for the asymmetry of fuzzy control rules.After fuzzy control, the overshoot can be eliminated and the response can be achieved quickly. In the case of power falling, the transient curves of fuzzy control damping and constant damping are compared in Figure 6.

4.2. First frequency modulation damping control

After introducing fuzzy control, the damping coefficient of VSG that will be a constant without fuzzy control changes with the SOC of the energy storage unit during the period when the frequency is not equal to the active frequency.

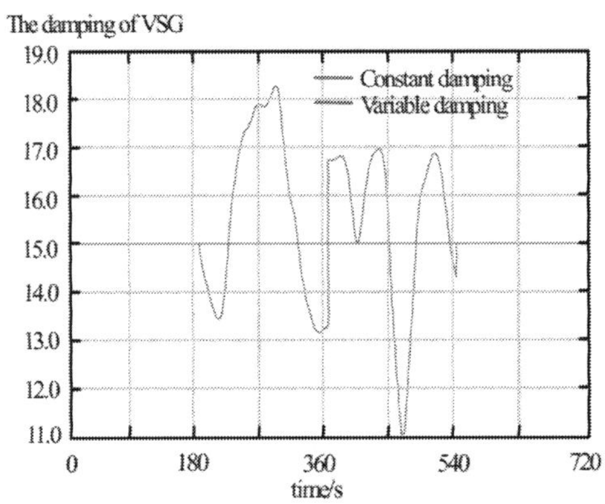

Figure 7. VSG damping coefficient optimization before and after comparison

Assume that the frequency of the grid decreases by 0.25Hz from the rated frequency during 180s to 360s, and increases by 0.25Hz from the rated frequency during 360s to 540s.The comparison of damping coefficient before and after fuzzy control is shown in fig. 7. During 180s to 540s, the damping coefficient of VSG always fluctuates with the change of SOC. The damping change process is shown in Figure 7.

The initial SOC of the energy storage device is 0.8, the energy is sufficient, according to the first frequency modulation damping fuzzy control rules, the energy storage system appropriately increases the discharge depth, and the first frequency modulation contribution to the system is increased. Therefore, the SOC minimum value has changed from 0 .528 to 0.507 after fuzzy control optimization. The comparison before and after SOC optimization is shown in Figure 8.

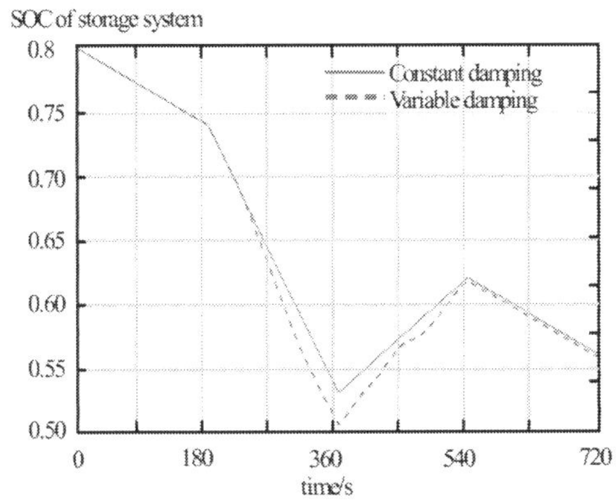

Figure 8. SOC comparison before and after optimization.

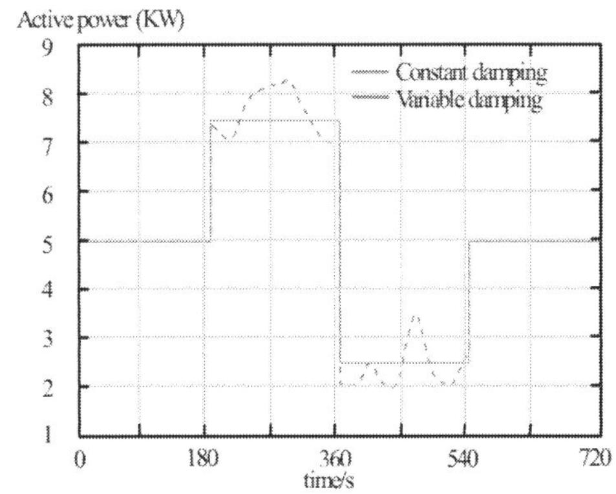

Figure 9. Frequency modulation power comparison.

The extra power in the first frequency modulation process before and after optimization is shown in Figure 9. Before the introduction of fuzzy control, if the frequency offset is unchanged, the power change of VSG is unchanged. after the introduction of fuzzy control, the power change is changed with the SOC.

5. Conclusion

In this paper, a transient damping fuzzy controller is designed, so that the damping coefficient is always in a state of adjustment during the transient process of power instruction change, and the following conclusions are obtained:

(1) If the fuzzy control rules are appropriate, overshooting can be completely eliminated in the transient process and the speed of entering the steady state is not affected at all, which proves the effectiveness of the control strategy.

(2) Based on SOC of energy storage system, another set of appropriate fuzzy control rules is designed to optimize the output of primary frequency modulation of energy storage system in the period when the system frequency is different from the rated frequency.

(3) In this paper, the design of fuzzy control rules using heuristic method, more time-consuming and laborious, the follow-up work will focus on the research of general, intelligent fuzzy control rules.

References

[1] Dongfeng X, Mingxing T. (2020) Relationship between frequency characteristics of virtual synchronous generator and parameters of energy storage equipment[J]. Power System Technology. pp:1-13.

[2] Lin W, Yuntao J, Wenchuan W. (2021) Optimal Design of Inertia and Damping Parameters of Virtual Synchronous Microgrid for Improving Frequency Stability. Proceedings of the CSEE, 41(13): 4479-4490.

[3] Zheng Z, Weihua S, Li R. (2015) Mathematical Model and Strategic Energy Storage Selection of Virtual Synchronous Generators. Automation of Electric Power System, 39(13): 22-31.

[4] Zhitao Z. (2018) Research on Grid--connected Control of Photovoltaic Power Generation System Based on Virtual Synchronous Generator. Xi'an: Chang'an University.

[5] Xin L I, Guoliang L, Ranchen Y, (2018) VSG Control Strategy With Transient Damping Term and Seamless Switching Control Method. Power System Technology, 42(7): 2081-2088.
[6] XU Chongfu, LI Wanru, XU Ningyi, WANG Yifei, LIU Yonghui, (2020) Analysis and Improvement of Damping Characteristics of Virtual Synchronous Generator Control Applied to Energy Storage Converter. Power System Technology, 44(5): 1656-1663.
[7] Xufeng L, Limin L, Lexiang C, (2019) Power Decoupling Control Strategy in Virtual Synchronous Generator With Improved Reactive Power Loop Based on Adaptive Virtual Impedance. Power System Technology, 43(10): 3752-3760.

Research on Insulation State of 10kV XLPPE Fault Cable Based on Qt Technology

Tengfei Zhao[1,2*], Yongjie Nie[1], and Min Cao[1]

[1.] Power Science Research Institute, Yunnan Power Grid Co., Ltd., Kunming 650217, China

[2.] School of Electrical and Electronic Engineering, North China Electric Power University, Beijing 102206, China

[*]Corresponding author's e-mail: zhaotengfei@ncepu.edu.cn

Abstract. The cross-linked polyethylene cable will undergo thermal aging during service, which will seriously affect the service life of the cable. The current integration method can analyze the charge injection and conduction parameters from the dynamic change rate of the charge over time, and then analyze the insulation state of the polymer. Aiming at the field 10KV fault breakdown cable, this paper uses the current integration method to compare the dynamic change characteristics of the charge of the faulty cable and the new cable, and uses the ratio of Q(t) at t=600s and t=4s to study Charge injection and accumulation situation. Through analysis, it is found that the fault breakdown of the cable, the maximum charge accumulation under high voltage is 9000 Q/nC, the charge change rate is about 2, and the dielectric constant is 3.1, which is much greater than that of the new cable, indicating that the breakdown of the three-phase cable is mainly caused by the cable. It is caused by aging and eventually leads to insulation deterioration and breakdown.

1. Introduction

With the acceleration of urbanization, XLPE cables are widely used in power systems because of their good dielectric properties and physical and chemical properties. However, when the power cable has been in operation for more than 25 years, under the combined action of electricity, heat, mechanical force and environmental factors, it enters the aging failure period [1]. When the cable is in the process of aging and failure, the cross-linked polyethylene molecular chain gradually breaks under the action of multiple stresses, and the microscopic defects continue to expand. Under the action of a DC electric field, space charges are easy to accumulate in microscopic defects, causing local electric field distortion, further accelerating insulation aging, and even causing final insulation breakdown [2-4]], which will cause greater losses to the power grid. Therefore, evaluating the insulation status of power cables is of great significance to the safe operation of power grids.

At present, the evaluation of the insulation state mainly uses performance tests in terms of dielectric, physical and chemical, and space charge. Among them, the research on physical and chemical properties mainly uses differential scanning calorimetry analysis, X-ray diffraction analysis, infrared spectroscopy and other methods to study the chemical groups, oxidation, thermal decomposition, crystallization, and crosstalk of XLPE cable insulation before and after electrical-thermal aging. The difference in the state of the link and the molecular chain, and then the insulation state is evaluated. Chen Zhiyong et al. studied the effects of electrical aging and accelerated water tree aging on the physical and chemical properties of cross-linked polyethylene insulation through physical and chemical experiments, and established a microscopic model of water tree aging (AWTT). The mechanism has been analyzed in depth [5]. Xu Jun et al. conducted physical and chemical experiments on heat-aged samples at different temperatures and different times, and studied the effects of thermal cracking activation energy, crystal morphology, and molecular structure on the physical and chemical structure of cross-linked polyethylene cable insulation [6]. Regarding the study of dielectric properties, it mainly uses polarization/depolarization current method (PDC), broadband dielectric spectrometer, and theoretical models to analyze the increase in dielectric constant and loss, increase in conductivity, and breakdown field of XLPE before and after aging. Strong and other characteristic parameters, and then obtain the analysis basis of XLPE aging degree. Zhang Qiaofeng et al. studied the influence of the degree of crosslinking on the time domain dielectric properties of crosslinked polyethylene by using the polarization/depolarization current method. The results show that the crosslinking reaction can effectively improve the insulation of crosslinked polyethylene. Performance, and the greater the degree of crosslinking [7]. Ye Gang et al. conducted polarization/depolarization current (PDC) experiments on XLPE accelerated water tree aging samples at different aging stages, and used the IT-LT curve and the extended Debye model to study the time-frequency domain The relationship between electrical characteristics and insulation aging [8]. Regarding the research of space charge, the pulsed electro-acoustic method (PEA) is mainly used to test the space charge. Through the experiment of pressurization and short-circuit process, the space charge distribution, average bulk charge density and Characteristic parameters such as volume charge density attenuation rate. Yin Yi and others of Shanghai Jiaotong University used the isothermal relaxation current method (IRC) to detect the aging conditions of unaged and water-aged, and applied the full-scale full-scale space charge measurement method to study the space charge distribution of the cable body. The research results show that the aging factor and deep trap energy level increase after aging in water, space charge accumulates during polarization, and shallow trap charge disappears during short circuit [9]. Jing Qiang et al. used gas chromatography-mass spectrometry (GC-MS) and space charge method to study the relationship between the type and content of crosslinked by-products remaining in the insulation after different degassing times and the space charge. Studies have shown that the content of cross-linking by-products such as α-methylstyrene, acetophenone, and benzyl alcohol decreases with the extension of the degassing time. With the increase of the degassing time, the average bulk charge density and the attenuation rate of the bulk charge density in the samples with the same short-circuit time decrease [10].

Most of the above detection methods are carried out for sheet-like patterns and cannot be used to test the whole cable as a whole. The space charge test can test the whole cable, but its test electric field strength and temperature range are limited; in addition, its resolution is low, and it is greatly affected by the interference and noise of the test site. [11,12]. In recent years, Takadad and others in Japan have proposed a new method for evaluating the insulation state of cables, which can evaluate the insulation state of cables through the rate of change of charge before and after aging [13-15], but it has not been applied in the field of overall cable inspection in China. Therefore, this paper prepares XLPE cable sheet samples with different aging degrees in the laboratory, and uses the current integration method to study the charge dynamic characteristics of the XLPE samples under different aging degrees under different electric fields. The evaluation of the cable insulation status has a certain meaning.

2. Introduction to the principle of the method

2.1. Overall cable Q(t) test principle

The overall cable test principle of the current integration method is shown in Figure 1. The overall cable test is composed of three parts: a high-voltage DC voltage source, an integral capacitor, an internal analog-to-digital conversion signal detection unit, and a data receiving and processing unit. The high-voltage DC voltage source is connected to the overall cable through the integrating capacitor, and forms a closed loop through the grounding wire. The amount of charge accumulated in the overall cable can be calculated by integrating the current flowing through the high-voltage circuit, as shown in equation (1). At the same time, the accumulated charge can be evaluated by the voltage induced across the integrating capacitor. An impedance amplifier is used to convert the obtained Q(t) into a stable voltage VQ(t). After ADC (analog-to-digital converter), the voltage analog signal is converted into a digital signal, and finally the signal is transferred to the computer software for processing through the zigbee signal transmitter and receiver, as shown in equation (2).

$$Q(\mathrm{t}) = \int_0^t I(\mathrm{t})dt \tag{1}$$

$$VQ(t) = \frac{Q(t)}{C\mathrm{int}} = \frac{1}{C\mathrm{int}}\int_0^t I(\mathrm{t})dt \tag{2}$$

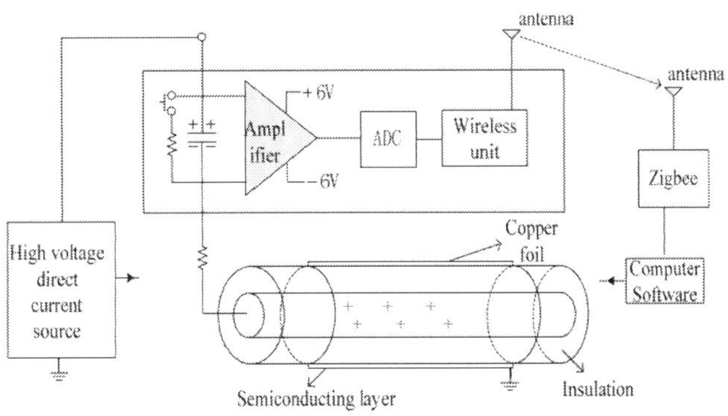

Figure 1. Schematic diagram of the overall cable test.

2.2. Q(t) aging assessment method

The aging state of the cable is related to the accumulation state and level of carriers in the material. Therefore, the aging state of the material can be determined by the amount of charge accumulated in the material. The current integration method uses the initial charge and the change in the charge after the end of the pressure as the insulation state evaluation method, as shown in Figure 2.

$$\frac{Q(t)}{Q_0} = 1 + \frac{Q_{spac}(t)}{Q_0} + \frac{Q_{leak}(t)}{Q_0} \tag{3}$$

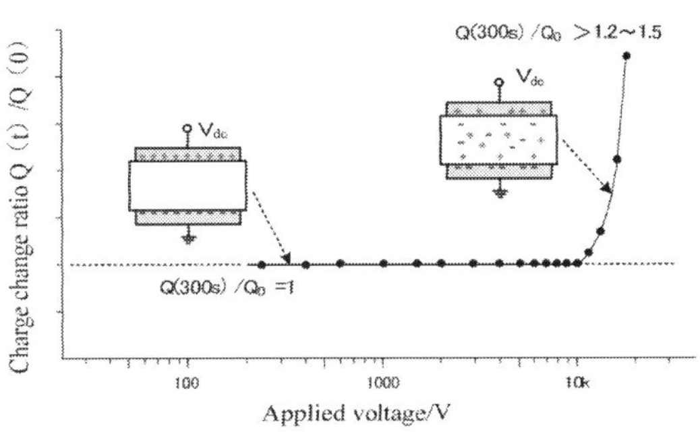

Figure 2. The rate of change of the amount of charge under different voltages.

When Qt/Q0=1, $\frac{Q_{spac}(t)}{Q_0}+\frac{Q_{leak}(t)}{Q_0}=0$, it indicates that there is only induced charge generated by pressurization at this time, no charge accumulation inside, only instantaneous charging current.

When Qt/Q0>1, $0<\frac{Q_{spac}(t)}{Q_0}+\frac{Q_{leak}(t)}{Q_0}<1$, it indicates that charge accumulation (charge conduction) occurs, space charge is dominant, and absorption (polarization) or electrical conduction current exists inside the sample.

When Qt/Q0>1, $1<\frac{Q_{spac}(t)}{Q_0}+\frac{Q_{leak}(t)}{Q_0}$, it indicates that the leakage current charge is dominant at this time.

3. Acquisition and processing of samples

Obtain the cable style from the 10kv three-phase cable faulty cable running on site. The cable has been in service for 5 years and was damaged due to breakdown. At a distance of 20 cm from the breakdown point of the whole cable, the three-phase single-core cable is obtained by cutting off the cross section and stripping off the surface layer and armor layer of the cable. Each single-core cable head and tail of the semi-conductive layer of 10cm each stripped and wiped with alcohol and wrapped in copper foil.At the same time, select a new cable of the same model that has not been put into operation, perform the same steps, and set up a control experiment. The information of the overall cable style is shown in Table 1

Table 1. Basic parameters of the cable.

Cable number	length	the inside diameter of	Outer diameter	Copper foil length
New cable	76	24	34	45
A phase cable	76	24	34	45
B phase cable	76	23	32	47
C phase cable	80	23	35	49

4. Experimental results

4.1. Qt test results

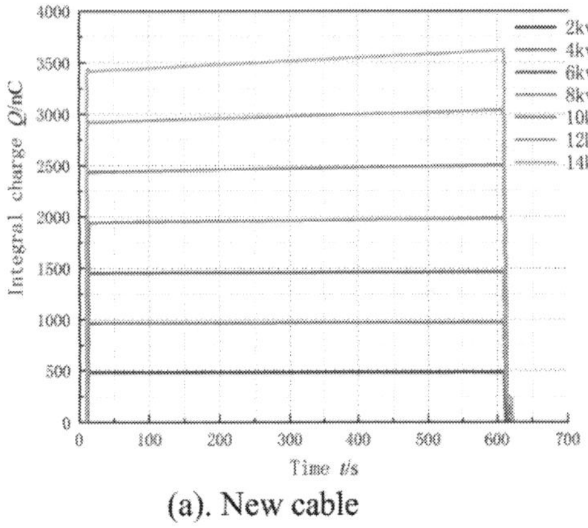

(a). New cable

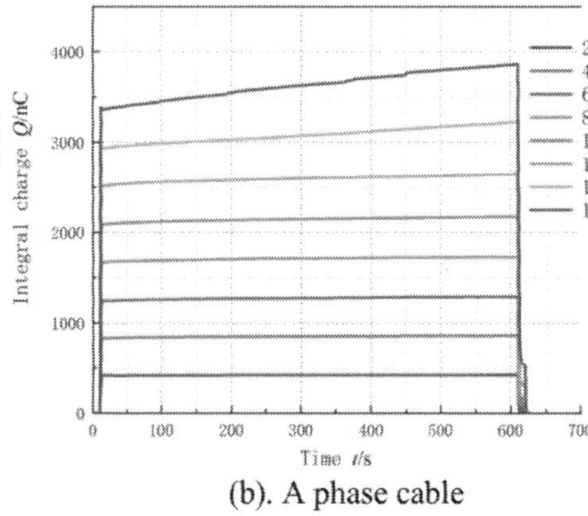

(b). A phase cable

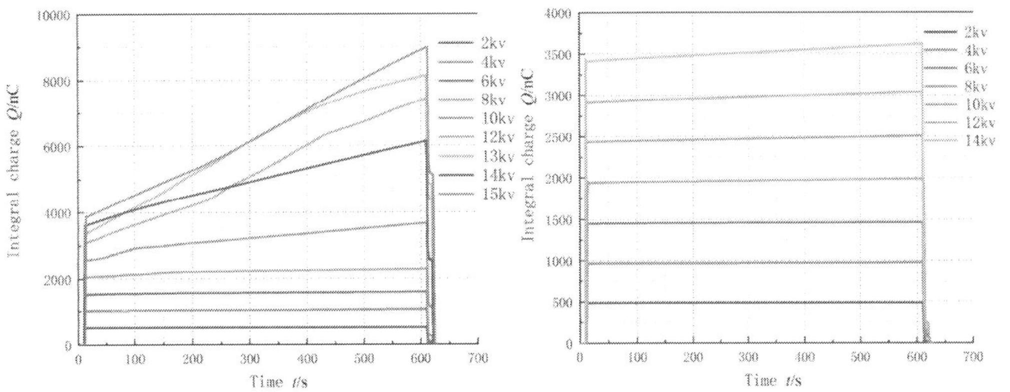

(c). B phase cable (d). C phase cable

Figure 3. The charge change curve of the new cable and the fault breakdown cable.

Figure 3 shows the results of the charge test results of the new cable and the fault-breakdown cable under different voltages. It can be seen from the figure that the amount of charge increases as the voltage increases. Among them, the charge of the new cable and the C-phase cable remains almost constant with the increase of time. For the A and B-phase cables, when the voltage is less than 8kV, the increase of the charge with time remains constant. When the voltage is greater than 10kV, the amount of charge increases with time. Increasing, of which the B-phase cable has the largest change trend with time.

4.2. Current test result

Figure 4 shows the current test results of the new cable and the fault breakdown cable under different voltages. It can be observed from the figure that the XLPE cable will generate an instantaneous charging current at the moment of pressurization, and then reach a steady state, forming an internal steady state or leakage current. The volatility of the current curve is relatively large at low voltage, and the volatility of the current curve decreases with the increase of voltage. Among them, the current change curves of the new cable and the C-phase cable are both below the order of 10-9, the maximum current of the A-phase cable is about 10-9, and the maximum current of the C-phase cable reaches the order of 10-8.

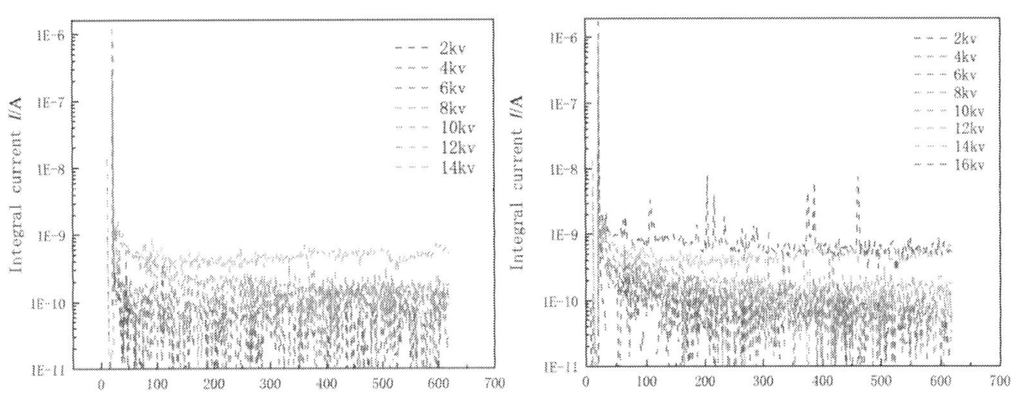

(a). New cable (b). A phase cable

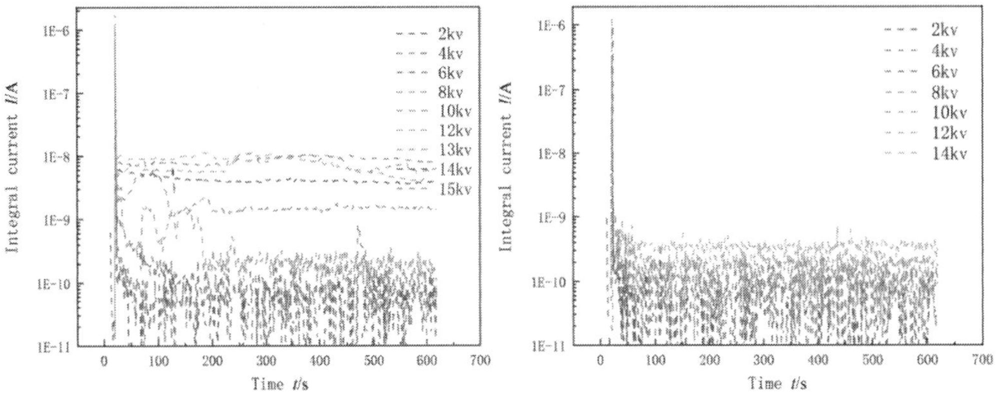

(c). B phase cable (d). C phase cable

Figure 4. The current change curve of the new cable and the fault breakdown cable.

4.3. Dielectric constant test result

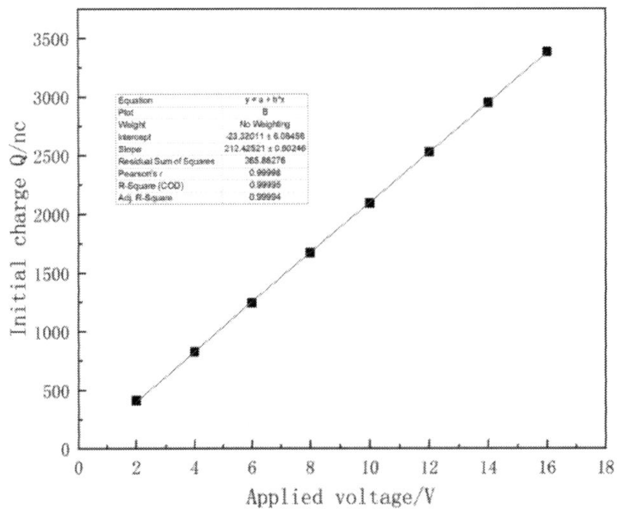

Figure 5. The relationship between the applied voltage of the A-phase cable and the instantaneous interface charge Q0.

Figure 5 is the relationship curve between the applied voltage and the instantaneous charging charge on the phase A cable. According to the relationship between capacitance, charge and voltage, the capacitance can be calculated. In this paper, the Origin2019b software is used to linearly fit the voltage of the phase A cable to the initial charge, and the capacitance of the sample is calculated to be about 0.8μF. The relative dielectric constant of the cable is calculated according to formula (4). The new cables are 2.3, A-phase cable 2.4, B-phase cable 3.1, and C-phase cable 2.5. Among them, the dielectric constant of the B-phase cable is higher than that of the new cable and the A and C-phase cables.

$$C_S = \frac{2\pi\,\varepsilon_0\varepsilon_r l}{\ln\dfrac{b}{a}} \tag{4}$$

Among them: Cs is the capacitance of the cable, L is the length of the cable, a is the radius of the cable, and b is the diameter of the cable.

5. Discuss

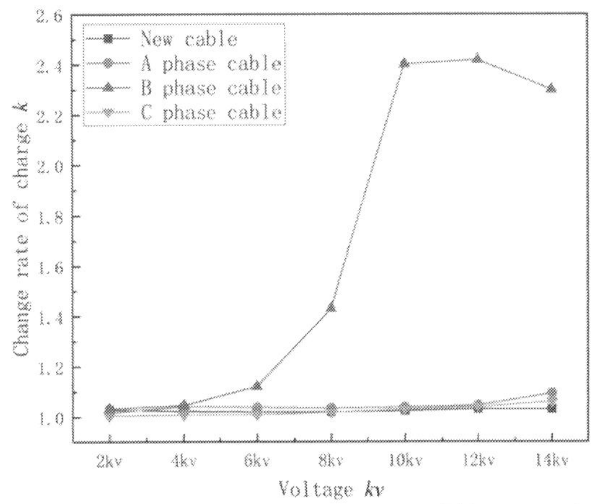

Figure 6. Change rate of charge under different voltage.

Figure 6 shows the charge change rate of the new cable and the fault breakdown under different voltages. Take K=1.2 as the threshold of space charge injection. From the figure, it can be found that the charge change ratio of the new cable and the phase B cable is close to 1, and there is no obvious change with the increase in voltage, indicating that there is no charge injection at this time. The cable insulation performance is good. With the increase of this voltage, the change rate of the charge amount of the B-phase cable increases significantly, and it is greater than 1.2. At this time, charge accumulation occurs, indicating that the insulation performance of this phase cable is seriously degraded. This is because in actual operation of the cable, it will be affected by factors such as electricity, heat, mechanical stress, moisture, etc., which will cause the cable to age during service, resulting in changes in the molecular structure of the XLPE material, and changes in the energy level density and depth of the trap. The long-term electric charge accumulates under high-voltage direct current. Deterioration of insulation performance most leads to deterioration of XLPE's dielectric properties, such as changes in parameters such as dielectric constant and current. Among them, the dielectric constant and current of the B-phase cable are greater than that of the new cable, and the A and C-phase cables. The final result causes the B-phase cable insulation breakdown.

6. Conclusion

In this paper, the Q(t) characteristics of the 10kV overall fault breakdown cable are studied, and the following conclusions are finally obtained.

1. When the overall cable ages, the molecular structure and energy level trap density of the XLPE cable change, and then charge accumulation occurs under high DC field strength. According to the results of the Q(t) test, it can be found that the K value of phase A is the largest, and $K_B>K_A>K_C>$new cable.

2. When the overall cable is in an aging state, the insulation performance of XLPE deteriorates, which is manifested in the increase of dielectric constant, current and other parameters, which will eventually lead to the breakdown of the B-phase cable.

3. The overall cable parameters obtained by Q(t) technology can well reflect the insulation performance of 10kV XLPE cables, and Q(t) technology can be used to provide a basis for evaluating the insulation status of cables with higher voltage levels.

Acknowledgments
Authors wishing to acknowledge assistance or encouragement from colleagues, special work by technical staff or financial support from organizations should do so in an unnumbered Acknowledgments section immediately following the last numbered section of the paper.

References

[1] Liao R.J., Zhou T.C., Liu., et al. Electrical tree branching test and partial discharge characteristics of cross-linked polyethylene power cables [J] China Electrical Engineering Journal, 2011, 31(28).
[2] YING Q.L., WEI D., GAO X.Q., et al. Development of high voltage XLPE power cable system in China[C] // Proceedings of the 6th International Conference on Properties and Applications of Dielectric Materials. Xi'an, China: IEEE, 2000: 247-253.
[3] Zhao S.L., Zhou K., He M., et al. Dielectric response characteristics and insulation state evaluation of cables under impulse voltage[J]. High Voltage Technology, 2019, 45(4).
[4] ZHANG Y.W., LEWINE J., ALQUIE C., et al. Evidence of strong correlation between space-charge buildup and breakdown in cable insulation[J]. IEEE Transaction on Dielectrics and Electrical Insulation, 1996, 3(6): 778-783.
[5] Chen Z.Y., Luo Z.X, Zhang J., et al. The influence of electrical aging and accelerated water tree aging on the physical and chemical properties of cross-linked polyethylene insulation[J]. Journal of Xi'an Jiaotong University, 2015, 49(04)): 32-39.
[6] Xu J., Wang X.D., , et al. The effect of thermal aging on the physical and chemical structure of XLPE cable insulation[J]. Insulating Materials, 2013, 46(02): 33 -37.
[7] Zhang Q.F., Zhou K., Li K.L., et al. The effect of crosslinking degree on the time domain dielectric properties of crosslinked polyethylene[J]. Insulating Materials, 2020, 53(12): 27-31.
[8] Ye G., Yao X.C., Li T., Zou B.C., et al. Time-frequency domain dielectric characteristics of XLPE cable insulation aging[J]. High Voltage Technology, 2018, 44(11): 3713-3719.
[9] Qian K.Y., Zhang Y., Liu D.Y., et al. Application of full-scale space charge measurement in the evaluation of the condition of aging cables in water[J]. High Voltage Technology, 2018, 44(05): 1706-1714.
[10] Tu Y.J., Zhang H.L., Gao Z. Jing Q., et al. The effect of degassing treatment on cross-linking by-products and space charge characteristics in cross-linked polyethylene[J]. Insulating Materials, 2018, 51(04): 58- 63.
[11] Liu D.Y., Zhang Y., Deng Y.K., et al. Development of a full-size cable space charge detection system under power frequency electric field[J], High Voltage Technology, 2019, 45(10): 3327-3334.
[12] Mazzanti G., Chen G., Fothergill J.C., et al. A Protocol for Space Charge Measurements in Full-size HVDC Extruded Cables [J], Ieee Transactions on Dielectrics and Electrical Insulation, 2015, 22(1): 21-34.
[13] Fujii M, Fukuma M, S.M., et al. Deterioration diagnosis of epoxy resin evaluated by current integration meter[C]. in 2017 Interna-tional Symposium on Electrical Insulating Materials (ISEIM). 2017: 319-322.
[14] Takada T., Fujitomi T., Mori T., et al. New diagnostic method of electrical insulation properties based on current integration[J], Ieee Transactions on Dielectrics and Electrical Insulation, 2017, 24(4): 2549-2558.
[15] Iwata S., Kitani R., and Takada T. Diagnostic technique for electrical tree by current integration method[C]. in 2019 IEEE Electrical In-sulation Conference (EIC). 2019: 317-320.1

The Research of GIS Discharge Influence Based on Particle Properties in Installation Environment

Leicheng Lin [1], Dan Chen[1]*

(1. College of Electrical and Automation, Fuzhou University, Fuzhou, Fujian, 350108, China;)

*Corresponding author's e-mail: T03019@fzu.edu.cn

Abstract: In order to explore the influence of atmospheric free particles that exit during GIS installation on the stable operation of GIS, the size, shape and inner properties of the particles contained in the air are analyzed. And based on 500kV GIS equipment, the GIS gas chamber model and the model of free particles were established. By using the finite element simulation software, the surface electric field distribution when the free particles remain on the insulator surface is solved. The results show that the electric field distribution of particles which exist along the surface of insulator is directly related to the relative permittivity of the material of the particles themselves, and the intensity of the electric field along the surface increases with the increase of the relative permittivity. with the fact that the relative permittivity of air particles is almost below 20, as a result, the existence of 10μm grade non-metallic free particles will not cause flashover discharge. But, when a small number of Fe_3O_4 particles which size over 1mm or when them accumulate regularly, it will lead to flashover discharge. Therefore, the suspended atmospheric particles in the installation environment have little impact on the stable operation of GIS. When considering the installation environmental conditions, the installation can be carried out on a clear, windless day which can save costs.

1. Introduction

The Gas Insulated Switchgear has been widely used in high voltage transmission system because of their good insulation performance, high reliability and small footprint[1]. However, it is inevitable that many equipment faults has turn up in the long run of GIS equipment. Of all the faults of GIS equipment, the rate of insulation faults is the highest which reaches 58%. Among them, the faults of basin-type insulator accounted for a large proportion, accounting for 26.6%[2]. A lot of researches on the simulated discharge caused by particles along the surface of basin-type insulators have been done and achieved fruitful results by scholars. Hongxin Ji studied the discharge behavior and discharge characteristics of linear and sheet metal particles, and proposed the conditions that different types of metal particles meet the runout under the operating voltage[3]. Youhua Gao established a three-dimensional finite electric field analysis and calculation model for basin-type insulators with metal particles above in 800kV GIS, and conducted surface flashover experiments accordingly. It is concluded that the maximum field strength on the surface of the basin-type insulator increases with the increase of the field strength along the surface where the metal particles are located before[4].

Studies have shown that the presence of particles along the surface of the basin-type insulator causes electric field distortion along the surface of the basin-type insulator, as a result, forming a through channel in the surface of basin-type insulator. Although insulation faults may not necessarily be induced in the experiment, the amount of partial discharge changes greatly, and it is likely to cause surface flashover discharge after long-term accumulation[5]. The particles existing in the GIS mainly come from various particles which enter during the installation process and the metal particles generated due to abrasion during the operation of the equipment. On the premise of ensuring that the GIS products are qualified, the source of particles is mainly the particles that enter the cavity during the installation process. Considering that during the actual installation of the GIS, the installation steps such as vacuuming, standing, and injecting SF_6 gas are still required, so the number of particles remaining inside the GIS is still relatively small.

To explore the impact of the properties of free particles, which enter the GIS cavity in the installation, on the stable operation of the GIS, the finite element simulation software is used to simulate the flashover discharge of free particles existing along the basin-type insulator. The relationship between the particle properties of the installation environment and the discharge of the GIS is obtained.

2. Air free particles analysis

Researches shows that the sedimentation characteristics of free particles and dusts in the air are different due to the different particle size under the action of gravity. The free particles which has a small size less than 10μm can float in the air. While the particles with a great size more than 10μm that will settle on the ground quicklly. Therefore, it can be considered that under no windy weather conditions, most of the atmospheric particles with a particle size greater than 10μm in the air have settled on the ground and have no affect in the installation process, while the size of most atmospheric particles floating in the air that will affect the installation process remain below 10μm. During the installation of the GIS, only particles smaller than 10μm may remain in the cavity of the GIS after installation.

Through the investigation and research, we can know that except for particularly inactive metals such as gold and silver, other metals generally exist as salts, oxides. According to the study of the morphology of atmospheric particles, we can know that the particles with the size less than 0.5μm accounted for 72.3%, the particles with the size between 0.5μm and 1μm accounted for 12.9%, the particles with the size between 1μm and 2.5μm accounted for 10.4% and the particles with the size between 2.5μm and 10μm accounted for 4.4%. Fossil fuel combustion, soil dust and motor vehicle exhaust were the main causes of atmospheric particulate pollution in Fuzhou city, and the effect of sea salt aerosol was less than the other three. The number of particles decreases with the increase of particle size and the shape of the main particles of PM10 include spherical, chain, cotton-shaped, strip, flake, irregular shape and special shape. And in Fuzhou, the main types of atmospheric particles are soot aggregate, fly ash particles, mineral particles, biological particles[6]. So on this basis, the simulation model of air particles and the relationship between air cleanliness and GIS internal discharge are established, and the mutual influence between them is discussed.

3. GIS gas chamber simulation model

The state Grid 500kV voltage level GIS equipment is selected as the research reference object in this study. Through Ansys software, a simple simulation model of GIS gas chamber and insulator is established. Through simulation, the relationship between airborne particulate matter and GIS stable operation during GIS installation is explored, the structure of GIS equipment is shown in Figure 1.

From Figure 1, it can bee seen that the GIS is composed of six parts include center guide rod, basin-type insulator, gas chamber, the case, and the covers which are basin-type insulator actually. Considering the size of the actual equipment, the overall height and width of the GIS are set to 6m and 720mm.

4. Simulation about influence of discharge free particles exist on surface of insulator

4.1. Simulation parameter setting
Imported the established 3D Model of GIS gas chamber into Ansys software and set the materials of each component of the model. Among them, the materials of the central guide rod, the case and the basin-type insulator were metal aluminum, metal aluminum and epoxy resin. And 0.4MPa SF_6 gas was filled in the gas chamber. The relative permittivity distributions of aluminum, epoxy resin and SF_6 gas are infinity, 6.000 and 1.002.

In the simulation experiment, the rated short-time power frequency withstand voltage is used to carry out the simulation experiment. The central guide rod is loaded with 740kV voltage, the metal case is grounded.

4.2. The distribution of electric field along the insulator surface
Firstly, the distribution of surface electric field of basin-type insulators in GIS should be explored when GIS is in stable operation and the result of simulation is as the Figure 2 shown. When there are no particles on the surface of the basin-type insulator, the maximum creeping electric field appears near the central guide rod as the maximum is 4.7456kV/mm. And the minimum value appears at the edge of the basin-type insulator with the numeric about 0.2182kV/mm. The distribution of the electric field of the basin-type insulator is radial, and the intensity of the electric field decreases with the increase of the length of the surface.

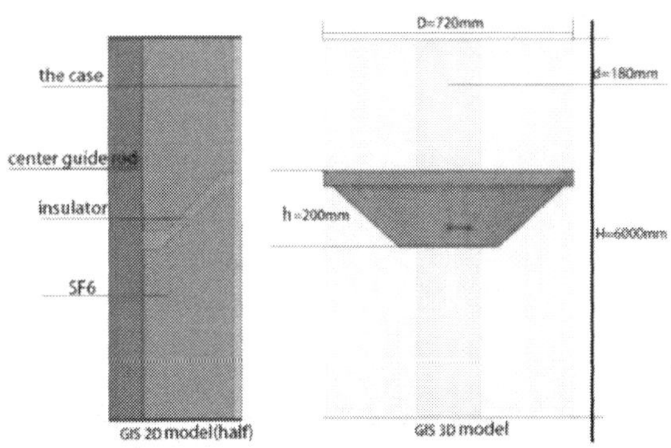

Figure 1. Simple model of GIS gas chamber

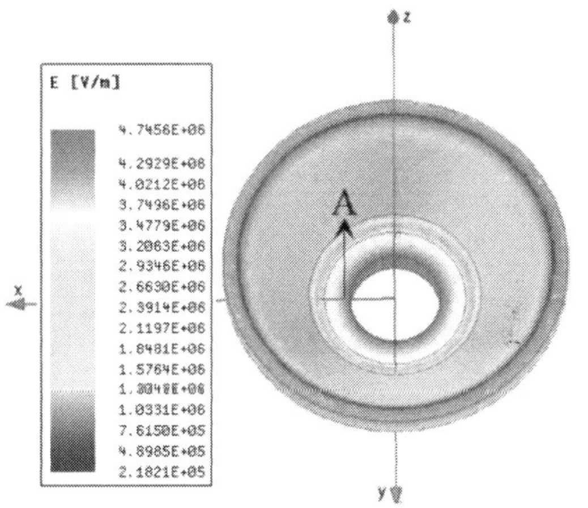

Figure 2. Electric field distribution on surface of basin-type insulator without metal particles

According to literature[4], when particles exist in high electric field, they have the greatest influence on insulators. Therefore, according to Figure 2, location A of the high electric field area is selected.

4.3. Influence of particle type
It is considered that some atmospheric particles remain in the GIS and settle to the insulator surface. Explore the influence of particles existing in location A. Taking Fuzhou as a reference city, it is considered that the coarse particle size is 10μm and the fine particle size is 2.5μm. The types, shapes and particle sizes of each particle are shown in Table 1:

Table1. Properties of different particles

Property / Types	Shape	Size	relative permittivity
Fe_3O_4	strip	a=10μm, b=h=1μm	20
	flake	a=b=10μm, h=1μm	
SiO_2	strip	a=10μm, b=h=1μm	4.42
$CaSO_4$	strip	a=10μm, b=h=1μm	5.1
	flake	a=b=10μm, h=1μm	
C	sphere	r=2.5μm	5.87
$CaCO_3$	sphere	r=2.5μm	8.67

Establish the 3D model of particles by using Ansys software, and divide the mesh was properly. The simulation results were obtained as shown in Figure 3:

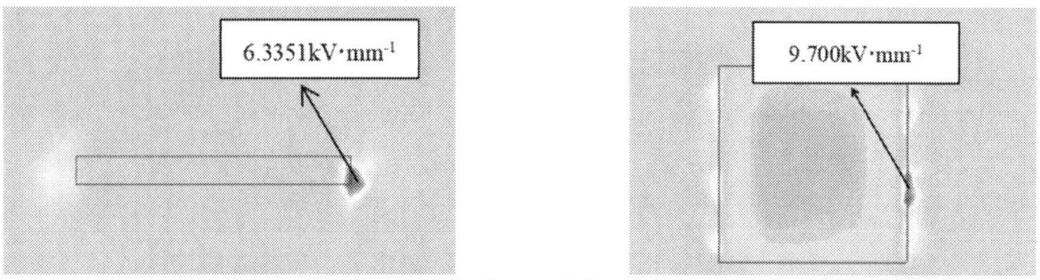

Figure 3. Effects of Fe_3O_4 particles on the electric field

When different kinds of particles in Table 1 exist along the surface of the basin-type insulator, the variation of the electric field intensity along the surface of the insulator is shown in Table 2:

Table 2 . The electric field strength of insulator in the presence of different particles

Property / Types	Shape	E_{max}/kV·mm^{-1}
Fe_3O_4	strip	6.3551
	flake	9.7000
SiO_2	strip	4.8006
$CaSO_4$	strip	4.8655
	flake	5.6183
C	sphere	4.7442
$CaCO_3$	sphere	4.7852

As is shown the table 2, in the process of GIS installation, the particles will impact on the electric field of insulator in different degrees. The presence of particles makes the charge along the surface accumulate in a small amplitude, and the electric field around the particles increases in a small range compared with that without the presence of particles. Metal oxide has the greatest influence on the surface electric field intensity of the insulator, but it does not reach the degree of flashover and will not cause the insulator to flashover.

4.4. Influence of particle relative permittivity

From Table 2, when particles exist along the surface of insulators, charge accumulates around the particles. Which is no only related to the shape and size of the particles to a certain extent, but also due to relative permittivity of the particles.

Select strip particles with length, width and height of 10μm, 1μm and 1μm placed it in location A. Change the relative permittivity of particles to explore the influence of the internal properties of particles on the electric field distribution along insulator surface. The simulation result is shown as Figure 4.

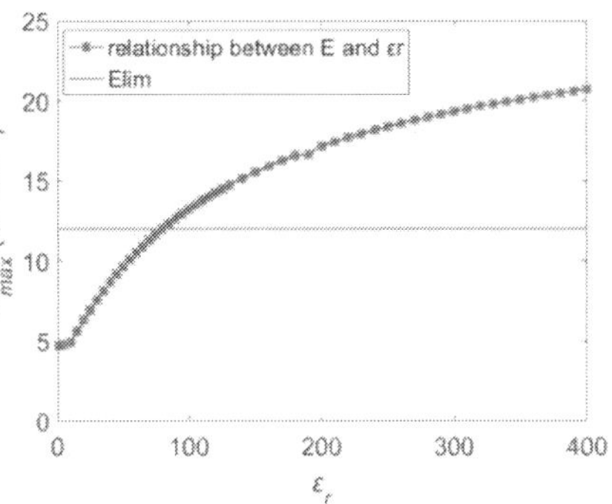

Figure 4. Relationship between ε_r and E_{max}

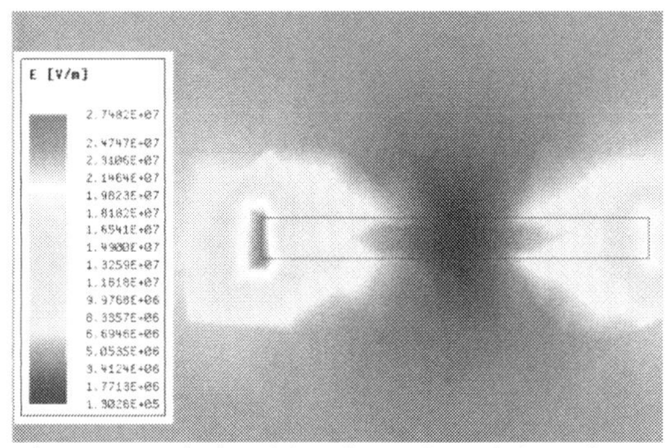

Figure 5. Relationship of metal particle and E_{max}

As can be seen, with the increase of ε_r, the surface electric field strength of the insulator E_{max} also increases, but the increasing amplitude is decreasing constantly, and finally there is a trend of convergence. On the premise that the strip particles are placed at position A, when the value of ε_r reaches 80, the surface electric field strength of the insulator reaches the limit value which may lead to the damage of the insulator.

The relative permittivity of the conductor is infinity. Conduct a group of simulation experiments to verify the result above, the results are shown in Figure 5. When the granular material is metal, the maximum of the electric field along the surface of the insulator is 27.482kV·mm^{-1}, which is consistent with the law in Figure 4.

4.5. Influence of particle size

In installation, sometimes in order to meet the deadline, the GIS equipment is installed in windy weather. As a result, the settled particles float into the air again and may makes the number of particles with large size increase in the air and thus enter the GIS during installation finally. Therefore, on the basis of Table 2, select Fe_3O_4 which has the greatest influence on the surface electric field strength of insulators as the research object to explore the distribution of the surface electric field strength of insulators when the particle size increases.

Establish 3D models of Fe_3O_4 with different sizes and set the length of the longest side of the strip and flake as *a*, maintain the aspect ratio of 10:1:1 for the strip and 10:10:1 for the flake particle. Through finite element simulation, the results are shown in Figure 6:

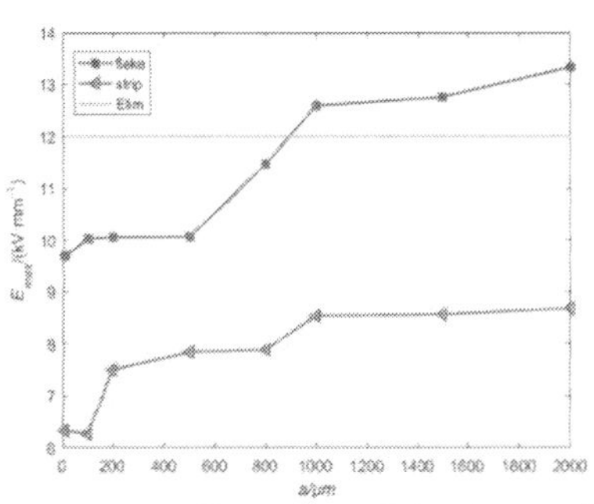

Figure 6. Effect of particle size on field strength

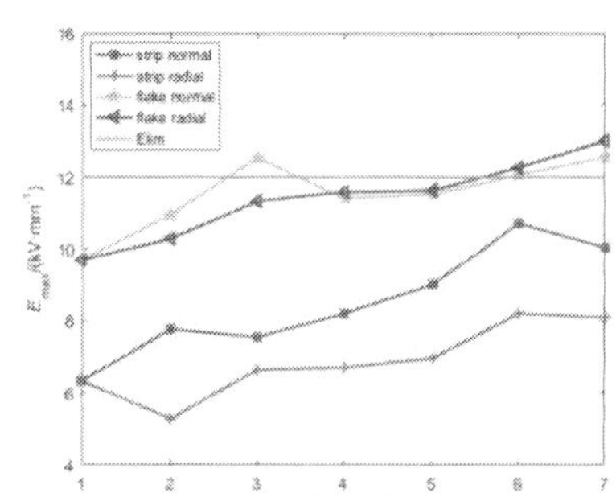

Figure 7. Effect of particle aggregation on field strength

As can be seen from Figure 6, when the shape of Fe_3O_4 is flake or approximately flake, it may exceed the surface electric field strength limit of insulators with the increase of particle size, leading to the occurrence of surface flashover discharge of insulators. Moreover, according to the simulation results, the particle size which may cause the flashover discharge is more than 1mm, while the number of this size is very little among the particles floating in the atmosphere.

4.6. The influence of particle aggregation

When the particles exist in GIS, the particles sedimentation may occur which may lead to particle aggregation because there is no gas flow. Therefore chose the Fe_3O_4 particle to analyse the influence of particle aggregation on the surface electric field strength.

The discharge condition of Fe3O4 particles clustered along radial and normal directions was simulated, and the results were shown in Figure 7:

As can be seen from Figure 7, the influence of Fe3O4 particle aggregation on the surrounding electric field is related to the shape of the particle itself. The flashover discharge of insulators along the surface will not occur if the strip particles gather, but when the flake particles are distributed along the normal and radial directions, the electric field strength around the particles will increase with the increase of the particles. When the number of particles is greater than 5, the electric field strength around the particle group will exceed the maximum allowable electric field strength limit along the surface of the insulator, leading to flashover discharge.

5. Conclusion

Taking GIS basin-type insulator as the analysis object, the 3D simulation model of GIS and free particles are established by using finite element analysis software. On this basis, the electric field distribution along the insulator surface when dust remains on the insulator surface is simulated, and the relationship between the presence of particles and the flashover discharge along the GIS insulator is explored. The specific conclusions are as follows:

(i) The distribution of the electric field of the basin-type insulator is radial, and the intensity of the electric field along the surface is within the limit value, and it can operate stably when there is no foreign matter.

(ii) When different kinds of non-metallic particles appear on the surface of the insulator, flashover discharge does not occur. It can be found that the influence of particles on the surface electric field distribution of insulators is related to the property, size and shape of particles. The maximum of electric field increases with the increase of relative permittivity and finally tends to be stable.

(iii) When the Fe_3O_4 particles which size is greater than 1mm exist on the surface of basin-type insulator or the particles gathered with the number over 5, the flashover discharge around the particles will occur and the stable operation of insulators will be damaged. When considering the installation environmental conditions, the installation can be carried out on a clear, windless day which can save costs.

Reference

[1] Li X, Liu W D, Xu Y, et al. (2019) Surface charge accumulation and pre-flashover characteristics induced by metal particles on the insulator surfaces of 1100 kV GILs under AC voltage. High Voltage, 5: 134-142.
[2] Zhang X X, Xiao S, Su N, et al. (2014) GIS partial discharge pattern recognition based on the chaos theory. IEEE Transactions on Dielectrics and Electrical Insulation, 21:783-790.
[3] Ji H X, Li C R, Pang Z K, et al. (2017) Influence Factors of the Lifting Voltage of Linear Particle in GIS[J]. Proceedings of the CSEE, 37:301-312.
[4] Gao Y H, Wang C Y, Liu X M, et al. (2015) Analysis of electric field of basin-type insulator existing metal particles and its influence on surface flashover. Advanced Technology of Electrical Engineering and Energy, 34:56-61.
[5] Zhuang C, Zeng J B, Yuan C Z, et al. (2019) Effect of Surface Foreign Matters on the Insulation

Property of 252kV Gas Insulated Switchgear Basin Insulators. Transactions of China Electrotechnical Society, 34:4208-4216.
[6] Wang P. (2021) Morphology analysis of PM10 single particle in Fuzhou urban road. Fujian Analysis & Testing, 30:37-41.

Research on fine tuning parameter insulator identification based on YOLOV4 algorithm

Meng Wang[1a], Yuan Sun[1b*], Feize Xia[1c]

[1] School of Mechanical Engineering, Shanghai DianJi University, Shanghai , China

*Corresponding author's e-mail: suny@sdju.edu.cn

Abstract—Aiming at the problem that the training network time of YOLOV4 algorithm is too long due to the large data set of aerial insulator images, a method based on YOLOV4 algorithm is proposed to shorten the training time by fine-tuning parameters without affecting the positioning detection accuracy. Based on the development of UBANTU virtual machine, through CUDA and CUDNN environment configuration, and through the detection and verification of insulator aerial photo data set, the feasibility of accurate positioning of insulators under the condition of fine tuning parameters of YOLOV4 algorithm is successfully proved.

1. Introduction

With the increasing mileage of high-voltage transmission lines in China, regular inspection of transmission lines has become the most important to maintain the stable operation of the power system. High-voltage transmission lines are often laid across provinces and cities, and often cannot avoid passing through areas with complicated terrain conditions. Insulators, as an important part of transmission lines, are not only abundant in quantity but also very easy to be damaged under complex climatic conditions. Insulator detection has become an important target of transmission line detection. The traditional manual inspection of insulators requires inspection workers to carry heavy inspection tools to travel in the rugged areas where transmission lines are set up, which is not only dangerous but also inefficient. In the context of intelligent inspection of modern power grid, it has become a trend to use aerial images of UAV for insulator fault detection. Aerial images of insulators often have many irrelevant background interference, and how to optimize the algorithm for aerial insulator positioning is the main research direction of researchers in China [1].

Traditional target detection and recognition methods are represented by feature extraction method based on SIFT algorithm and HOG algorithm, as well as classifier method based on SVM algorithm. However, these traditional target detection methods can only be applied to specific situations under manual setting, and are not capable of self-learning to adapt to different backgrounds [2]. With the development of deep learning, a target detection method based on CNN convolution network emerged, but the detection speed is slow due to the increasing number of network layers and large parameters of CNN convolution network [3].

Therefore, after considering the detection speed and accuracy, YOLOV4 algorithm is selected as the algorithm of insulator positioning detection. In this study, an improved method of fine-tuning YOLOV4 algorithm was proposed in UBANTU virtual machine to detect and locate insulators.

2. Ubuntu and YoloV4 algorithms

Compared with the Windows system we often use, more and more image processing scholars prefer to process images in Ubuntu VIRTUAL machine. When configuring the virtual environment for Windows system, the environment may fail because a certain software is not installed or different software conflicts may occur. Creating virtual environments in Ubuntu is much easier and more successful than in Windows. Ubuntu also has many advantages such as no need to defragment files, a very efficient file management system, a powerful command line, and a great user experience.

Yolov4 algorithm is improved by improving model training skills and stacking methods on the basis of YOLOv3 algorithm. The combination of Darknet-53 residual network and CSP can not only reduce the amount of computation, but also make the image extract deeper feature data. Mosaic sample enhancement method and the addition of SPP network modules enable the algorithm to adapt to recognition under different environmental backgrounds [4]. Yolov4 model is divided into four parts: input, trunk, neck and head. Its network structure is shown in Figure 1.

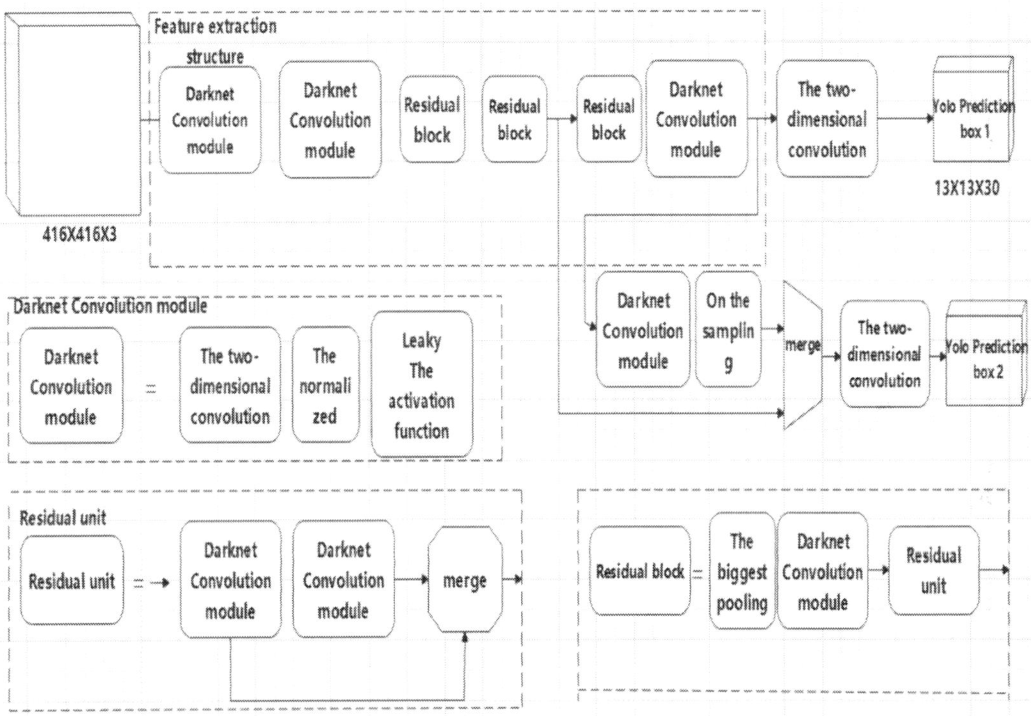

Fig.1 Yolov4 network structure

3. Experiment and result analysis

3.1. Model parameters and fine tuning

Set batch parameter yolov4-tiny-Insulato. CFG in the CFG folder under the root directory. The default value is 64. The meaning of subdivision is how many images can be divided. For example, during training, if multiple images cannot be input simultaneously, split the input training in batches and set it to 2. Width and height parameters are set to 416 to prevent errors caused by excessive image size, and the maximum number of iterations of MAX_batches is set to 2000.

3.2. Model evaluation

Yolov4 algorithm was used to train the model, among which the most important is the loss function value in the case of multiple iterations, whose calculation formula is as follows [5] :

$$f_{LOSS} = \lambda_{coord} \sum_{i=1}^{s^2} \sum_{j=1}^{b} I_{ij}^{obj} \left[\left(x_i - \hat{x}_i \right)^2 + \left(y_i - \hat{y}_i \right)^2 \right]$$
$$+ \sum_{i=1}^{s^2} \sum_{j=1}^{b} I_{ij}^{obj} \left(c_i - \hat{c}_i \right)^2 + \lambda_{noobj} \sum_{i=1}^{s^2} \sum_{j=1}^{b} I_{ij}^{noobj} \left(c_i - \hat{c}_i \right)^2$$
$$+ \sum_{i=1}^{s^2} \sum_{j=1}^{b} I_{ij}^{obj} \left(c_i - \hat{c}_i \right)^2 + \lambda_{noobj} \sum_{i=1}^{s^2} \sum_{j=1}^{b} I_{ij}^{noobj} \left(c_i - \hat{c}_i \right)^2$$
$$+ \sum_{i=1}^{s^2} I_{i}^{obj} \sum_{c \in classes} \left(p_i(c) - \hat{p}_i(c) \right)^2 \quad (1)$$

The meaning of each formula in the formula:f_{LOSS} is the loss function,λ_{coord} is the loss weight of the target position coordinates, λ_{noodj} is the loss weight of target frame confidence,Whether the JTH anchor frame of grid I contains a target is expressed as I_{ij}^{obj} or I_{ij}^{noodj} ,The prediction and actual target frame are represented by (x_i, y_i, w_i, h_i) and $(\hat{x}_i, \hat{y}_i, \hat{w}_i, \hat{h}_i)$ respectively,I the prediction category and actual category of grid are represented by c_i and $\hat{c}_i$ respectively,The predicted target probability and actual target probability of grid I are represented by $p_i(c)$ and $\hat{p}_i(c)$ respectively,B is the number of borders, S^2 is the number of grids.

3.3. Experimental environment and data set

In this paper, 600 aerial images published by State Grid on Github were used as data sets. Before the training data set, test sets and training sets were separated in a ratio of 1:9, and then the image format was converted into XML format [6]. The experiment was conducted on the Ubuntu VIRTUAL machine platform with two thousand iterations.

3.4. Comparative analysis of test results

The results of aerial aerial insulator detection by YOLOV4 network with original parameters and the network with improved parameter fine-tuning are shown in Figure 2 and figure 3. The YOLOV4 network with fine-tuning parameters can still achieve good detection results under different lighting conditions and complex backgrounds.

(a)

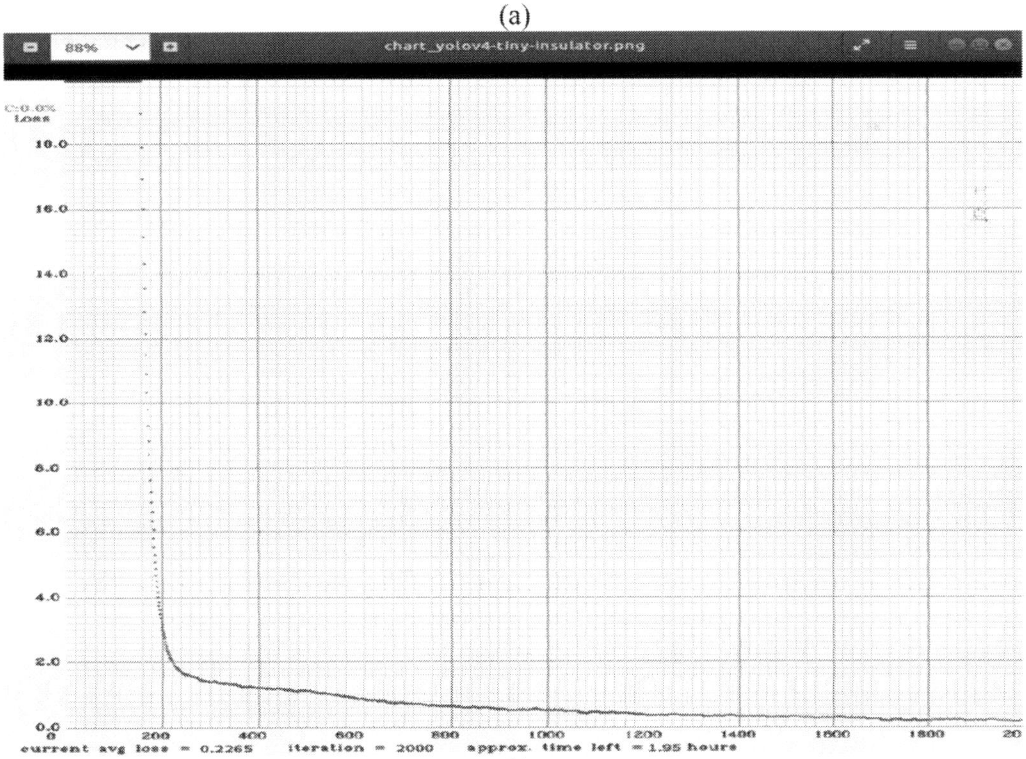

(b)

Fig.2 Original network detection result

(a)

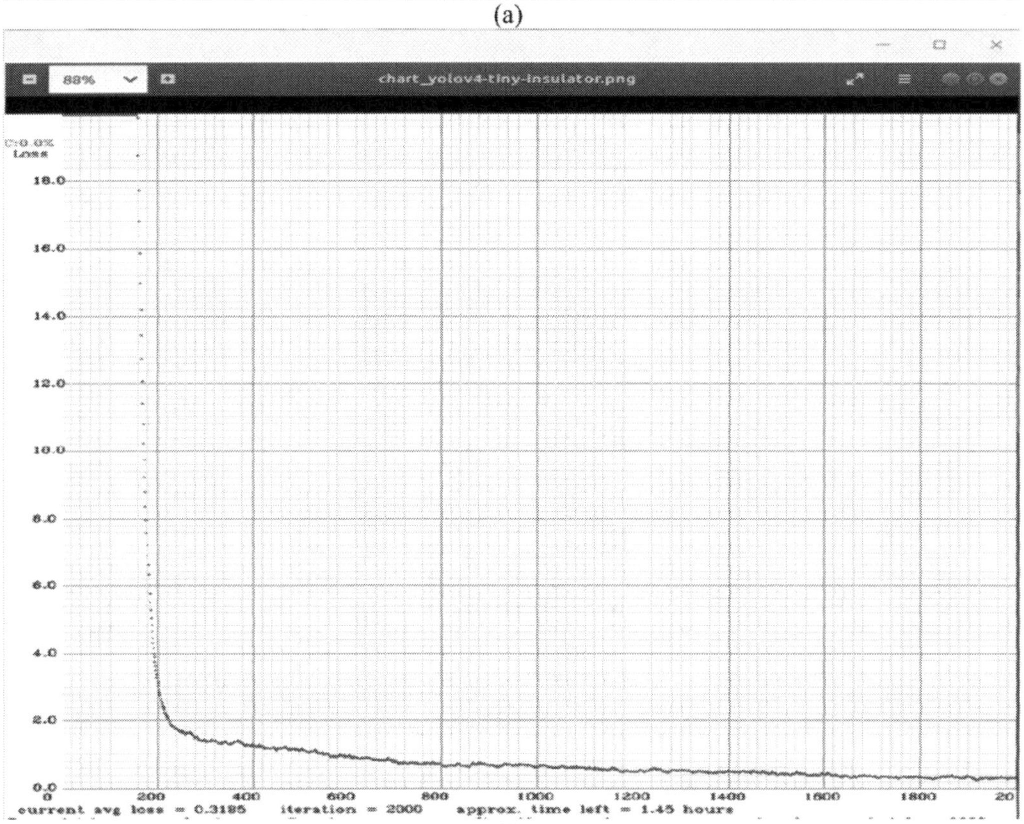

(b)

Fig.3 Network detection results after parameter tuning

Compared with the original algorithm, the training time of the fine-tuning improved algorithm is significantly improved. Before the improvement, the training time reaches 93 hours. After the fine-tuning improved algorithm, it only needs 31 hours to complete the training. Compared to two network loss function of the image at the same time found the training after the completion of the loss function were similar, the set of tests on the test can also be reflected, two networks of insulator recognition accuracy reached 86.6% and 85.8%, respectively, in the subsequent experiments under the condition of increasing data sets, fine-tuning of the network will have obvious advantages.

4. Conclusion

The experiment in this paper was conducted in Ubuntu based on yOLOV4 algorithm, and good ideal results were obtained through fine-tuning experiments and analysis. However, this paper did not test in the case of more complex background and other interference conditions, and its stability in the case of larger data sets remains to be studied.

References

[1] Liu CHANG-xin, LI Mu. Aerial insulator detection based on Deep detachable convolutional Network [J]. Mechanical and Electrical Information,2021(2):43-45,48. (in Chinese) DOI:10.3969/j.issn.1671-0797.2021.02.019.

[2] Li Xiaoyong, WEI Zexian, Yang Yulin. Research on application of HOG and SVM in Dynamic Weighing Wheel Axle Recognition [J]. Highway Traffic Science and Technology, 201,38(7):145-151. (in Chinese) DOI:10.3969/j.issn.1002-0268.2021.07.019.

[3] He Fengfeng, He You, Liu Zhuanggadao, et al. Research progress of convolutional neural network in radar automatic target recognition [J]. Journal of electronics & information technology,2020,42(1):119-131. DOI:10.11999/JEIT180899.

[4] ABDURAHMAN FETULHAK, FANTE KINDE ANLAY, ALIY MOHAMMED. Malaria parasite detection in thick blood smear microscopic images using modified YOLOV3 and YOLOV4 models[J]. BMC Bioinformatics,2021,22(1). DOI:10.1186/s12859-021-04036-4.

[5] Digital cinema (D-cinema) packaging. Stereoscopic picture track file:BS ISO 26429-10-2009[S].

[6] Tian Ersheng, Li Chunlei, Zhu Guodong, et al. Transmission lines broken hazard identification algorithm based on YOLOv4 [J]. Computer system application, 2021, 30 (7) : 190-196. The DOI: 10.15888 / j.carol carroll nki. Csa. 008082.

Robust path-following control of underactuated AUVs with multiple uncertainties and state constraints

Kankan Deng[1,*], **Jianming Miao**[1,2], **Xingyu Sun**[1]

[1]School of Marine Engineering and Technology, Sun Yat-sen University, Zhuhai 519082, China

[2]Southern Marine Science and Engineering Guangdong Laboratory (Zhuhai), Zhuhai 519082, China

[*]dengkk@mail2.sysu.edu.cn

Abstract—This paper proposes a novel robust controller for horizontal path-following problem of an underactuated AUV subject to multiple uncertainties and state constraints. Firstly, four reduced-order extended state observes (ESOs) are designed to estimate the multiple uncertainties, and the estimated values are adopted in the design of kinematic and dynamic controller. Secondly, to address the state constraints, the barrier Lyapunov function is incorporated with the kinematic controller. To resolve the problem of input saturation, the auxiliary design system is utilized in the dynamic controller. To address the problem of "explosion of complexity" inherent in the conventional back-stepping method, a nonlinear tracking differentiator is utilized to obtain the derivative of the desired yaw speed. Finally, the results of numerical simulation are performed to demonstrate the effectiveness of the proposed controller.

1. Introduction

In recent years, autonomous underwater vehicle (AUV) has been widely utilized in all kinds of underwater applications, such as deep sea exploration, underwater target tacking and ocean sampling [1, 2]. In order to better fulfil the above-mentioned tasks, it is essential to achieve accurate control for an AUV. However, due to the existence of model nonlinearity, model parameters perturbations and unknown environment disturbances, it is a difficult task to precise control the AUV [3].

Path-following is main technical basis of an AUV to accomplish various complex underwater missions, and it has been attracted widespread attention from researchers. In [4], a compound controller, which is based on reduced-order ESOs, was proposed for the path-following of the underactuated AUV subject to multiple uncertainties, where the reduced-order ESOs were designed to estimate the multiple uncertainties. However, the problem of input saturation is not taken into consideration. In [5], a novel controller was presented for the path-following of an underactuated unmanned surface vehicle (USV) with disturbances. A disturbance observer was utilized to compensate the unknown dynamics, and an auxiliary system was adopted to eliminate the effect of input saturation. However, the state constraints are not considered. In [6], a trajectory tracking controller, which was incorporated with barrier Lyapunov function, was proposed for an USV with input saturation and full-state constraints. The barrier Lyapunov function was employed to address the full-state constraints, and an anti-windup compensator was utilized to compensate the input saturation.

Inspired by the afore-mentioned considerations, this paper proposes a robust controller for horizontal path-following of an underactuated AUV subject to multiple uncertainties and state constraints. The

Journal of Physics: Conference Series **2121** (2021) 012042 doi:10.1088/1742-6596/2121/1/012042

multiple uncertainties, including the model parameters perturbances, the environment disturbances, and unmodeled dynamics, are estimated by the constructed reduced-order ESOs, the state constraints are addressed by using the barrier Lyapunov function, and the input saturation is compensated by the anti-windup compensator.

2. Preliminaries

2.1 Underactuated AUV model

Based on the previous study [7], the kinematic model is described as:

$$\begin{cases} \dot{x} = u\cos\psi - v\sin\psi \\ \dot{y} = u\sin\psi + v\cos\psi \\ \dot{\psi} = r \end{cases} \tag{1}$$

Referring the previous study [8], the dynamics of an underactuated AUV in the horizontal plane is modified as:

$$\begin{cases} \dot{u} = \dfrac{m_{22}}{m_{11}}vr - \dfrac{X_u}{m_{11}}u - \dfrac{X_{u|u|}}{m_{11}}u|u| + \dfrac{\tau_u}{m_{11}} + d_u \\ \dot{v} = -\dfrac{m_{11}}{m_{22}}ur - \dfrac{Y_v}{m_{22}}v - \dfrac{Y_{v|v|}}{m_{22}}v|v| + d_v \\ \dot{r} = \dfrac{m_{11} - m_{22}}{m_{33}}uv - \dfrac{N_r}{m_{33}}r - \dfrac{N_{r|r|}}{m_{33}}r|r| + \dfrac{\tau_r}{m_{33}} + d_r \end{cases} \tag{2}$$

Taking into account the effect of input saturation, the saturated control inputs are defined as:

$$\tau_i = \begin{cases} \tau_{i\max}, & \tau_{i0} > \tau_{i\max} \\ \tau_{i0}, & \tau_{i\min} \le \tau_{i0} \le \tau_{i\max}, \quad i = u, r \\ \tau_{i\min}, & \tau_{i0} < \tau_{i\min} \end{cases} \tag{3}$$

2.2 Path-following error dynamics

Referring the previous study [8], the path-following error dynamics is represented as:

$$\begin{cases} \dot{x}_e = -\dot{s}\left(1 - c_c y_e\right) + v_t\cos\psi_e \\ \dot{y}_e = -c_c\dot{s}x_e + v_t\sin\psi_e \\ \dot{\psi}_e = r + \dot{\beta} - c_c\dot{s} \end{cases} \tag{4}$$

In order to overcome the dependency on an accurate model, the unknown time-varying side-slip angular velocity in the path-following error dynamics is regarded as the kinematic uncertainty, and the reduced-order ESO is utilized to estimate it. Therefore, the third equation of the error dynamics is modified as:

$$\dot{\psi}_e = r + d_\psi - c_c\dot{s} \tag{5}$$

The line-of sight (LOS) law is given by:

$$\psi_{los} = -\tan^{-1}\left(\frac{y_e}{\Delta}\right) \tag{6}$$

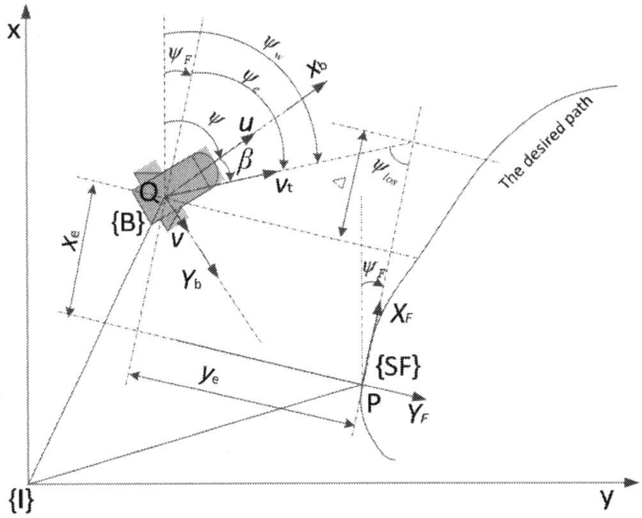

Fig. 1. The description of frames in the AUV path-following issue.

2.3 Problem formulation

The horizontal path-following problem can be described as follows:

Considering an underactuated AUV governed by Eqs. (1) and (2), design a robust controller for the control inputs, so that the underactuated AUV can follow a predefined path under the given desired speed.

3. Path-following controller design

3.1 The reduced-order ESO design

In this section, the reduced-order ESOs are designed to estimate the kinematic and dynamic uncertainties. Four reduced-order ESOs are given by:

$$\begin{cases} \dot{p}_1 = -\omega_1 p_1 - \omega_1^2 u - \omega_1 \left[\dfrac{m_{22}}{m_{11}} vr - \dfrac{X_u}{m_{11}} u - \dfrac{X_{u|u|}}{m_{11}} u|u| + \dfrac{\tau_u}{m_{11}} \right] \\ \hat{d}_u = \omega_1 u + p_1 \text{ , } \omega_1 > 0 \end{cases} \tag{7a}$$

$$\begin{cases} \dot{p}_2 = -\omega_2 p_2 - \omega_2^2 v - \omega_2 (-\dfrac{m_{11}}{m_{22}} ur - \dfrac{Y_v}{m_{22}} v - \dfrac{Y_{v|v|}}{m_{22}} v|v|) \\ \hat{d}_v = \omega_2 v + p_2 \text{ , } \omega_2 > 0 \end{cases} \tag{7b}$$

$$\begin{cases} \dot{p}_3 = -\omega_3 p_3 - \omega_3^2 r - \omega_3 \left[\dfrac{m_{11} - m_{22}}{m_{33}} uv - \dfrac{N_r}{m_{33}} r - \dfrac{N_{r|r|}}{m_{33}} r|r| + \dfrac{\tau_r}{m_{33}} \right] \\ \hat{d}_q = \omega_3 r + p_3 \text{ , } \omega_3 > 0 \end{cases} \tag{7c}$$

$$\begin{cases} \dot{p}_4 = -\omega_4 p_4 - \omega_4^2 \psi_e - \omega_4 \left(r + c_c \dot{s} \right) \\ \hat{d}_\psi = \omega_4 \psi_e + p_4 \text{ , } \omega_4 > 0 \end{cases} \tag{7d}$$

3.2 Kinematic controller design

3.2.1 Attitude control law design

Consider the following Lyapunov function candidate:

$$V_1 = \frac{1}{2}(\psi_e - \psi_{los})^2 \tag{8}$$

The derivative of Eq. (8) is given by:

$$\dot{V}_1 = (\psi_e - \psi_{los})(r + \dot{\beta} - c_c(s)\dot{s} - \dot{\psi}_{los}) \tag{9}$$

With the estimated value of kinematic uncertainty $\hat{d}_\psi$, the desired yaw velocity is designed as:

$$r_d = c_c(s)\dot{s} - \hat{d}_\psi + \dot{\psi}_{los} - k_1(\psi_e - \psi_{los}) \tag{10}$$

3.2.2 The virtual target movement control law design

To address the state constraints, choose the following barrier Lyapunov function:

$$V_2 = \frac{1}{2}x_e^2 + \frac{1}{2}y_e^2 + \frac{1}{2}\ln(\frac{k_{c1}^2}{k_{c1}^2 - x_e^2}) + \frac{1}{2}\ln(\frac{k_{c2}^2}{k_{c2}^2 - y_e^2}) \tag{11}$$

Differentiating the Eq. (11), we have:

$$\dot{V}_2 = -x_e(c_1\dot{s} - v_t\cos\psi_e + c_1 x_e c_c y_e - c_2 x_e c_c y_e) + c_2 y_e v_t \sin\psi_e \tag{12}$$

The virtual target movement control law is designed as:

$$\dot{s} = \frac{k_2}{c_1}x_e + v_t\cos\psi_e - (c_1 - c_2)x_e c_c y_e \tag{13}$$

where $c_1 = (\frac{1}{k_{c1}^2 - x_e^2} + 1)$, $c_2 = (\frac{1}{k_{c2}^2 - y_e^2} + 1)$,

3.3 Dynamic controller design

3.3.1 Yaw angular velocity control law design

Choose the following Lyapunov function candidate:

$$V_3 = k_4 V_1 + \frac{1}{2}(r - r_d)^2 + \frac{1}{2}\chi_1^2 \tag{14}$$

In order to address the effect of the input saturation, an anti-windup compensator is employed as:

$$\dot{\chi}_1 = \begin{cases} -k_{\chi_1}\chi_1 - \dfrac{|(r - r_d)\Delta\tau_r| + 0.5\gamma_1^2\Delta\tau_r^2}{\chi_1^2}\chi_1 + \gamma_1\Delta\tau_r & |\chi_1| > \overline{\chi} \\ 0 & |\chi_1| \le \overline{\chi} \end{cases} \tag{15}$$

Differentiating Eq. (15), we have:

$$\begin{aligned} \dot{V}_3 = {} & k_4\left[(\psi_e - \psi_{los})(\tilde{r} + r_d + \dot{\beta} - c_c(s)\dot{s} - \dot{\psi}_{los})\right] + \tilde{r}\left(\frac{m_{11} - m_{22}}{m_{33}}uv - \frac{N_r}{m_{33}}r - \frac{N_{r|r|}}{m_{33}}r|r|_r + \frac{\tau_{r0} + \Delta\tau_r}{m_r} + d_r - \dot{r}_d\right) \\ & + \chi_1(-k_{\chi_1}\chi_1 - \frac{|\tilde{r}\Delta\tau_r| + 0.5\gamma_1^2\Delta\tau_r^2}{\chi_1^2}\chi_1 + \gamma_1\Delta\tau_r) \end{aligned} \tag{16}$$

With the estimated value of dynamic uncertainty $\hat{d}_r$, design the control moment τ_{r0} as:

$$\tau_{r0} = m_r\left[\dot{r}_d - k_3(r - r_d) - k_4(\psi_e - \psi_{los}) + k_3\chi_1 - \frac{m_{11} - m_{22}}{m_{33}}uv + \frac{N_r}{m_{33}}r + \frac{N_{r|r|}}{m_{33}}r|r| - \hat{d}_r\right] \tag{17}$$

In order to overcome the problem of "explosion of complexity" inherent in the traditional back-stepping method, the following NTD is utilized to obtain the derivative of the desired yaw speed [9]:

$$\begin{cases} \text{fh} = \text{fhan}\left(r_c(\text{k}) - r_d(\text{k}), \dot{r}_c(\text{k}), R, h\right) \\ r_c(\text{k}+1) = r_c(\text{k}) + T \cdot \dot{r}_c(\text{k}) \\ \dot{r}_c(\text{k}+1) = \dot{r}_c(\text{k}) + T \cdot \text{fh} \end{cases} \tag{18}$$

Therefore, the control moment τ_{r0} is modified as:

$$\tau_{r0} = m_r \left[\dot{r}_c - k_3(r - r_c) - k_4(\psi_e - \psi_{los}) + k_3\chi_1 - \frac{m_{11} - m_{22}}{m_{33}} uv + \frac{N_r}{m_{33}} r + \frac{N_{r|r|}}{m_{33}} r|r|_r - \hat{d}_r \right] \tag{19}$$

3.3.2 Surge velocity control law design

Consider the following Lyapunov function candidate:

$$V_4 = \frac{1}{2}(u - u_d)^2 + \frac{1}{2}\chi_2^2 \tag{20}$$

Similar to the design of yaw angular velocity control law, an anti-windup compensator is employed as:

$$\dot{\chi}_2 = \begin{cases} -k_{\chi_2}\chi_2 - \dfrac{|(u - u_d)\Delta\tau_u| + 0.5\gamma_2^2 \Delta\tau_u^2}{\chi_2^2} \chi_2 + \gamma_2 \Delta\tau_u & |\chi_2| > \bar{\chi} \\ 0 & |\chi_2| \le \bar{\chi} \end{cases} \tag{21}$$

Differentiating Eq. (21), we have:

$$\dot{V}_4 = \tilde{u}\left(\frac{m_{22}}{m_{11}} vr - \frac{X_u}{m_{11}} u - \frac{X_{u|u|}}{m_{11}} u|u| + \frac{\tau_{u0} + \Delta\tau_u}{m_u} + d_u - \dot{u}_d \right) + \chi_2(-k_{\chi_2}\chi_2 - \frac{|\tilde{u}\Delta\tau_u| + 0.5\gamma_2^2\Delta\tau_u^2}{\chi_2^2}\chi_2 + \gamma_2\Delta\tau_u) \tag{22}$$

With the estimated value of dynamic uncertainty $\hat{d}_u$, design the control force τ_{u0} as:

$$\tau_{u0} = m_u \left[\dot{u}_d - k_5(u - u_d) + k_5\chi_2 - \frac{m_{22}}{m_{11}} vr + \frac{X_u}{m_{11}} u + \frac{X_{u|u|}}{m_{11}} u|u| - \hat{d}_u \right] \tag{23}$$

4. Simulation results

In this section, numerical simulations are performed to validate the effectiveness of the proposed controller, and the parameters of the underactuated AUV are the same as in [8]. The parameters of the proposed controller are chosen as $k_1 = 10$, $k_2 = 5$, $k_3 = 5$, $k_4 = 2$, $k_5 = 3$. The desired path is parameterized by $x_F(\mu) = \sum_{i=0}^{4} a_i \mu^i$ and $y_F(\mu) = \sum_{i=0}^{4} b_i \mu^i$, and the corresponding path parameters are shown in [10]. The multiple uncertainties are generated as: $d_u = 0.1\sin(0.1\text{t})$, $d_v = 0.1\sin(0.05\text{t})$, $d_r = 0.1\cos(0.1\text{t})$.

The path-following results are shown in Fig. 2(a)-(d). In Fig. 2(a), the controller can force the underactuated AUV to accurately converge to the desired path under the multiple uncertainties. The velocity evolutions and control inputs are drawn in Fig. 2(b) and Fig. 2(c), respectively. It is obvious that the surge speed can converge to the given desired speed, and the control inputs are limited in a valid rage. The path-following errors are shown in Fig. 2(d). It is evident that the tracking errors quickly tend to narrow field near zero.

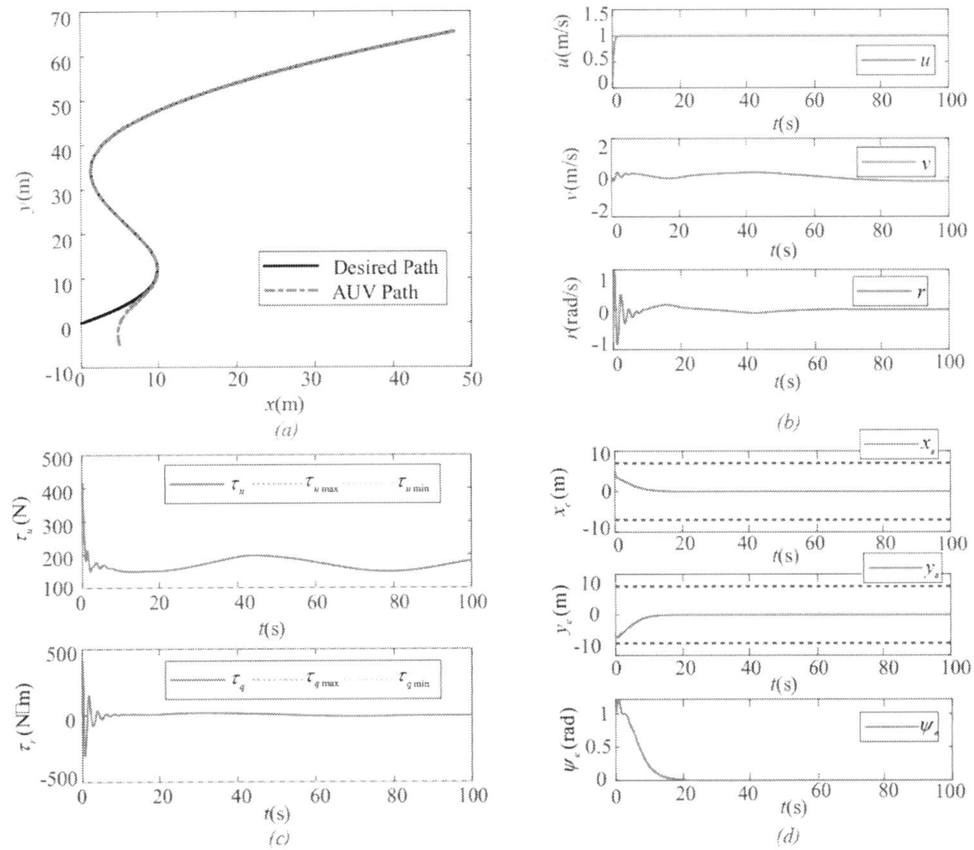

Fig. 2. (a) path-following performance. (b) velocity evolutions. (c) control inputs. (d) path-following errors.

5. Conclusion

In this paper, a novel controller is proposed to address the path-following problem of an underactuated AUV subject to multiple uncertainties and state constraints. Four reduced-order ESOs are utilized to estimate the kinematic and dynamic uncertainties, the barrier Lyapunov function is developed to address state constraints, and the auxiliary design system is adopted to eliminate the effect of input saturation.

Acknowledgments

This work is supported by the Natural Science Foundation of Guangdong Province of China (2019A1515011405), the Science and Technology Planning Project of Guangdong Province of China (311020011), and the Key-Area Research and Development Program of Guangdong Province (2020B1111010004).

Reference

[1] Y. Zhang, X. Liu, M. Luo, and C. Yang, "MPC-based 3-D trajectory tracking for an autonomous underwater vehicle with constraints in complex ocean environments," *Ocean Engineering,* vol. 189, p. 106309, 2019.

[2] Y. Xia, K. Xu, W. Wang, G. Xu, X. Xiang, and Y. Li, "Optimal robust trajectory tracking control of a X-rudder AUV with velocity sensor failures and uncertainties," *Ocean Engineering,* vol. 198, p. 106949, 2020.

[3] S. He, S.-L. Dai, and F. Luo, "Asymptotic trajectory tracking control with guaranteed transient behavior for MSV with uncertain dynamics and external disturbances," *IEEE Transactions on*

Industrial Electronics, vol. 66, no. 5, pp. 3712-3720, 2018.
[4] J. Miao, S. Wang, Z. Zhao, Y. Li, and M. M. Tomovic, "Spatial curvilinear path following control of underactuated AUV with multiple uncertainties," *ISA transactions,* vol. 67, pp. 107-130, 2017.
[5] Y. Zhao, B. Qiu, G. Wang, and Y. Fan, "Robust Path Following Control of Underactuated Unmanned Surface Vehicle With Disturbances and Input Saturation," *IEEE Access,* vol. 9, pp. 46106-46116, 2021.
[6] H. Qin, C. Li, Y. Sun, and N. Wang, "Adaptive trajectory tracking algorithm of unmanned surface vessel based on anti-windup compensator with full-state constraints," *Ocean Engineering,* vol. 200, p. 106906, 2020.
[7] C. Shen, Y. Shi, and B. Buckham, "Path-following control of an AUV: A multiobjective model predictive control approach," *IEEE Transactions on Control Systems Technology,* vol. 27, no. 3, pp. 1334-1342, 2018.
[8] X. Xiang, C. Yu, Q. Zhang, and G. Xu, "Path-following control of an auv: Fully actuated versus under-actuated configuration," *Marine Technology Society Journal,* vol. 50, no. 1, pp. 34-47, 2016.
[9] J. Han, "From PID to active disturbance rejection control," *IEEE transactions on Industrial Electronics,* vol. 56, no. 3, pp. 900-906, 2009.
[10] L. Lapierre and D. Soetanto, "Nonlinear path-following control of an AUV," *Ocean engineering,* vol. 34, no. 11-12, pp. 1734-1744, 2007.

Calibration system design for multi-beam sonar

Yicheng Liu[1,2a], Song Ye[3b*],Anmin Zhang[1c], Mingyuan Li[4d]

[1] School of Marine Science and Technology, Tianjin University, Tianjin 300072, China

[2] Tianjin Research Institute of Water Transport Engineering,Tianjin 300000,China

[3] National University of Defense Technology, Changsha 410000, China

[4] College of Underwater Acoustic Engineering, Harbin Engineering University, Harbin 150001, China

[a]liuyicheng1990s@163.com, [b*]nauyml@foxmail.com,[c]zhanganmin@sina.com, [d]nauyml@163.com

Abstract—A set of calibration system for multi-beam sonar is designed by the principle and method of underwater acoustic metrology. Using the multi-dimensional operation control device, the calibration for acoustical index is completed in the anechoic pool, including the sound source level and beam width. The calibration for geometrical index is carried out in the large scale and deep-water prototype pool, including the bathymetric accuracy. Components and method of the calibration system are introduced, as well as the expanded uncertainty of the calibration system is presented. By comparing the calibrating value and the indicating or nominal value of the detected multi-beam sonar, test results show that the sound source level error is less than 0.7dB, the beam width indication error is less than 10% and the bathymetry value error is less than 0.2%.

1. Introduction

Multi-beam sonar has become the main sonar equipment for high precision bathymetry detection by its high efficiency and full coverage. The performance improvement, the frequency range extension and the type increase of sonar equipment put forward many new requirements for underwater acoustic measurement. In order to ensure the accuracy of ocean measurement data, it has practical research and application value to measure its detection performance regularly. Sound source level and beam width are the basic acoustic parameters of multi-beam sonar. They are closely related to the backscattering intensity and bathymetry accuracy. The above two parameters should be the acoustical indexes of multi-beam sonar calibration. Bathymetric accuracy affect the quality and efficiency of underwater topographic survey. The above parameters should be the geometrical indexes of the multi-beam sonar calibration. K.G. Foote et al. [1] developed the protocols for calibrating multi-beam sonar by means of the standard-target method. A calibration procedure to determine high resolution, three dimensional transmit and receive beam patterns of a Reson SeaBat 7125 MBES was performed using the fresh water calibration tank at the University of New Hampshire[2].

2. Methods

This part describes the calibration system design, including the experimental tank, rotary and lifting device, standard measurement Instrument. Calibration method for acoustical index and geometrical index of multi-beam sonar is also validated in this part.

2.1. Calibration System Design

The multi-dimensional operation control device for acoustical index is mounted on the anechoic water tank, which can realize precise rotation, lifting, horizontal displacement and other functions. It is the main actuator of water acoustic measurement experiment. Multi-beam sonar transducer and standard hydrophone can be installed in the multi-dimensional operation control device (Figure 1), the angle and distance between them can be precisely controlled for the calibration of acoustic indicators. The maximum permissible error of angle control is ±0.05°, and the maximum permissible error of displacement control is ±0.1cm.

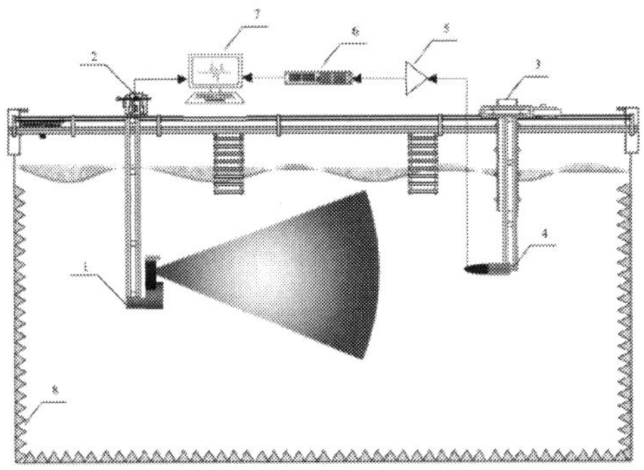

Fig.1 Diagram of multi-dimensional operation control device and auxiliary equipments on anechoic tank. 1-Transducer;2-Rotary bracket; 3-Lifting bracket; 4-Standard hydrophone; 5- Pre-amplifier;6-Signal collector; 7-Display control computer; 8-Anechoic material.

The multi-dimensional operation control device for geometrical index is mounted on the prototype deep-water port pool. Multi-dimensional operation control device on the prototype deep-water pool (Figure 2) can install the multi-beam sonar transducer, and it can accurately control the distance between the transducer and the standard reflecting wall, which is used for the calibration of geometric index. The maximum permissible error of angle control is ±0.1°, and the maximum permissible error of displacement control is ±0.5 cm.

Fig.2 Prototype deep-water pool and calibration equipment

2.2. Calibration Method

The experimental procedure for acoustic index is briefly described as follows:

(1) The tested transducer is installed at the bottom of the bracket of the multi-dimensional operation control device. Make its launching fan parallel to the water surface, and adjust the working parameters of sonar equipment to make sure it transmit pulse signal normally and stable.

(2) The standard hydrophone is connected to the signal collector. The signal collector records the open circuit voltage amplitude of the hydrophone in real time. The transmitting transducer is placed at the same depth underwater as the receiving hydrophone and meet the far field distance [3].

(3) Vertical direction test. Adjust the hydrophone to the lowest end, the transducer is measured roughly alignment the direction of the hydrophone and emit an acoustic signal with certain pulse width. The transducer remained stationary, and the hydrophone bracket is lifted at 1cm. Record the open

circuit voltage of the hydrophone at the same time. In the process of moving hydrophone, a voltage peak appears, which is the horizontal plane of acoustic axis of multi-beam sonar.

(4) Horizontal test. After finding the horizontal plane of the acoustic axis, fix the hydrophone. At this time, rotate the tested transducer in the horizontal plane through the precision rotary device, and control the servo motor to rotate the transducer at 0.1° angular spacing. After one revolution, an open circuit voltage of hydrophone can be collected at each angular position, and the emission directivity pattern is drawn. In the process of rotating the transducer, the position where the maximum voltage is found is the direction of the transducer's acoustic axis. At this point, the direction of the acoustic axis is found and pulse signals received.

(5) After the signal is filtered, the effective value of open circuit voltage for steady state part of pulse is calculated. According to the relation curve of sensitivity and frequency of standard hydrophone, the corresponding voltage sensitivity of each detection frequency range can be found. Combined with the voltage sensitivity M of the standard hydrophone, the free field sound pressure at each position can be calculated. The performance parameters such as sound source level and beam width are analyzed and calculated.

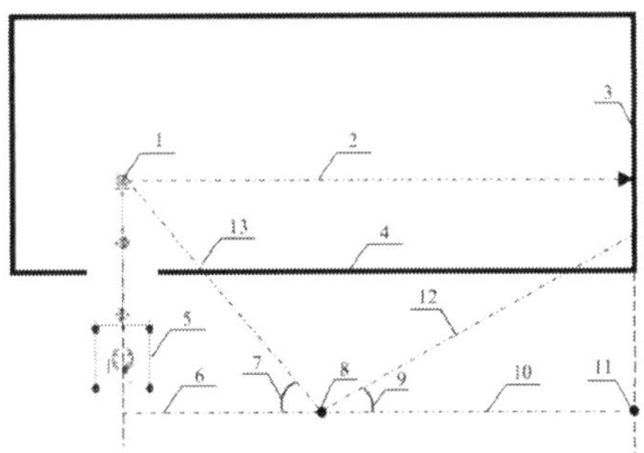

Fig.3 Calibration Principle of Bathymetry. 1-Transducer. 2-Central beam acoustic axis. 3-Short side wall of the tank. 4-Long side wall of the tank. 5-Multi-dimensional operation control mechanism. 6-Horizontal distance of transducer. 7-Horizontal angle of transducer. 8-Total station control point. 9-Horizontal angle of reflection point. 10-Horizontal distance of reflective surface. 11-Backsight point. 12-Slant distance in horizontal plane of reflective point. 13-Slant distance in horizontal plane of transducer.

The calibration process of bathymetry is as follows: the transducer of multi-beam sonar is installed on the connecting flange at the lifting rod of the multi-dimensional operation control mechanism, making the angle of multi-beam sonar transducer between 90°±0.1°. Adjust the turning mechanism to place multi-beam sonar transducer at 4 m underwater and 4 m offshore through the multi-dimensional operation control mechanism. Underwater transverse ranging is replaced by vertical bathymetry. A sound velocity profiler is used to measure the standard sound velocity at the water interface of the multi-beam sonar transducer. Input this sound velocity value into the multi-beam sonar display and control software. A straight line is arranged with the calibrated total station, and it is perpendicular to the surface of the short side wall of the pool. A point on the line is selected as the backsight point of total station measurement. Total station measures the horizontal distance between the control point and the standard reflector L_0, the horizontal distance between the control point and the multi-beam transducer array L_p, and difference of two distance $L_s = L_0 - L_p$, is regarded as the standard value of multi-beam sonar bathymetry. By rotating the transmitting angle of the transducer, all beams in the transmitting sector of the multi-beam sonar are irradiated to the reflection plate. Collect and analyze the data of multi-beam sonar, and make the difference between the standard value and the sounding indication value. Calculate the bathymetric indication error of all beam points. The principle schematic of bathymetry calibration is shown in Figure 3.

3. Results

Table 1 shows the test results of iBeam8120 multi-beam sonar made in China underwater acoustic experiment. The multi-beam operating frequency of this model is 200 kHz. The beam Angle is 1.5°×1.5°, and the maximum angle of the beam sector is 160°. Here's what needs to be said: the first 1.5° refers to the emission directivity of the multi-beam sonar along the track line, the second 1.5° refers to the receiving directivity along the vertical track line, and 160° refers to the emission directivity in the direction of the vertical track line.

Table 1. Acoustic indexes test data of multi-beam sonar

Equipment Model	**Sound Source Level**		**Beam Width** ($2\theta_{-3dB}$)	
	Nominal Value	Test Value	Nominal Value	Test Value
iBeam Multi-beam sonar	220dB	219.350dB	Along track direction is 1.5°.	1.62°(60°) 1.51°(0°) 1.64°(-60°)
			Vertical track line direction (reception) is 1.5° (center beam).	1.55°
			Vertical track line direction (emission) is 160°.	161.8°

The abnormal water depth was eliminated according to Dixon's Criterion. According to the requirements of IHO Standards for Hydrographic Surveys [4] and JJG 139-2017 Multi-beam Echo Sounder – Shallow Water, the indication error of multi-beam sonar is evaluated. Seabat8125 multi-beam sonar is used as the equipment been tested. At the position 116.41m, central beam No.120 collects 50 ping bathymetric data. The minimum value is 116.22m, the maximum value is 116.41m, the range is 0.19m, the average is 116.32m, and the standard deviation is 0.05m. The calibration results of bathymetry are shown in Table 2, and the maximum relative error is better than 0.2%.

Table 2. Calibration Results of Bathymetry

Indicating value of sonar /m	18.04	20.85	25.32	33.54	41.39	51.05	58.79	61.49	116.35
Standard value of total station /m	18.13	20.84	25.32	33.49	41.15	50.85	59.08	61.26	116.53
Indication error /m	-0.09	0.02	0.00	0.05	0.24	0.20	-0.29	0.23	-0.18

3.1. *Uncertainty evaluation of beam width measurement.*

The main sources and data of measurement uncertainty of beam width of multi-beam sonar are shown in Table 3 and Table 4.

Table 3. Comprehensive analysis for the components of along-track beam width uncertainty

Source	**Standard Uncertainty Component /Hz**		**Sensitivity Ratio**
	Type A	**Type B**	
Measuring repeatability	0.003	/	1
The uncertainty component introduced by lifting devices	/	0.02	-1
The uncertainty component introduced by installation error	/	0.01	-1

Table 4. Comprehensive analysis for the components of cross-track beam width uncertainty

Source	**Standard Uncertainty Component /Hz**		**Sensitivity Ratio**
	Type A	**Type B**	
Measuring repeatability	0.01	/	1
The uncertainty component introduced by rotation devices	/	0.06	-1
The uncertainty component introduced by installation error	/	0.01	-1

Combined standard uncertainty is,

$$\mu_c = \sqrt{0.01^2 + 0.06^2 + 0.01^2} = 0.062° \tag{1}$$

Expanded uncertainty of beam width cross -track direction. Coverage factor k=2.

$$U = 2\mu_C = 0.13^\circ \tag{2}$$

3.2. Uncertainty evaluation of bathymetric measurement.

Comprehensive analysis of uncertainty component is in Table 5.

Table 5 Comprehensive analysis table for standard uncertainty component of bathymetry calibration

Standard uncertainty component	Source of uncertainty	$\mu(x_i)$	Sensitivity ratio
$\mu(L_{mb})$	Repeatability and resolution of multi-beam sonar	10mm	1
$\mu(L_0)$	Total station error	25mm	-1
$\mu(\Delta L_1)$	Sound velocity meter error	5mm	1
$\mu(\Delta L_2)$	Instrument installation error	12mm	1
$\mu(\Delta L_3)$	The uneven reflection wall deviates from aiming point	17mm	1

Because each standard uncertainty component is not related, so combined standard uncertainty is:

$$\mu_c = \sqrt{\mu^2(L_{mb}) + \mu^2(L_0) + \mu^2(\Delta L_1) + \mu^2(\Delta L_2) + \mu^2(\Delta L_3)} = 34\text{mm} \tag{3}$$

Coverage factor $k = 2$, the extension uncertainty is,

$$U = 2 \times \mu_c = 68\text{mm} \tag{4}$$

The total station measurement result is 65.141m, the arithmetic mean of 10 water depth data measured by multi-beam sonar is 65.123m. The calibration value of multi-beam sonar is,

$$L = L_s + \Delta L = 65.123\text{m} + 0.018\text{m} = 65.141\text{m},\ U = 68\text{mm},\ k = 2. \tag{5}$$

4. Conclusions

Calibration system design of multi-beam sonar refers to JJG 139-2017 Multi-beam Echo Sounder—Shallow Water. It can realize the calibration of multi-beam sonar source level, beam width and bathymetry, the measurement uncertainty of each calibration parameter meets the requirements of regulation. By comparing the calibrating value and the indicating or nominal value of the detected multi-beam sonar, test results show that the sound source level error is less than 0.7dB, the beam width indication error is less than 10% and the bathymetry value error is less than 0.2%.

Acknowledgments

This work was supported in part by the Special Fund for Basic Scientific Research Business Expenses of Central Public Welfare Scientific Research Institutes under Grant TKS20200202, National Key R&D Program of China (grant Nos. 2018YFF0212202) and Tianjin Research Institute for Water Transport Engineering, MOT Key R&D Program (grant No. TKS160102, TKS170201).

References

[1] Foote, K.G.; Chu, D.; Hammar, T.R.; Baldwin, K.C., and Jech, J.M. Protocols for calibrating multi-beam sonar. The Journal of the Acoustical Society of America,2005, 117(4 Pt 1), 2013-2027.

[2] Lanzoni, J.C., and Weber, T.C.High-resolution calibration of a multi-beam echo sounder. MTS/IEEE Oceans 2010, 1-7.

[3] Foote, K.G., and Martini, M.A.Standard-target calibration of an acoustic backscatter system. IEEE Oceans, 2010,58(1), 1-5.

[4] Standards for Hydrographic Surveys. Special Publication No.44(5th Edition). International Hydrographic Bureau （IHO）, MONACO.

[5] Foote, K.G., and Moustier, C.D. Standard-target calibration of sonars for imaging and scattering measurement: an ISO-standard development. The Journal of the Acoustical Society of America, 2015, 138(3), 1822-1822.

The path-following control of the underactuated AUV with input saturation and multiple disturbances

Xingyu Sun[1], Jianming Miao[1,2*], Kankan Deng[1], Yanyun Wang[1]

[1]School of Marine Engineering and Technology, Sun Yat-sen University, Zhuhai 519082, China

[2]Southern Marine Science and Engineering Guangdong Laboratory (Zhuhai), Zhuhai 519082, China

[*]miaojm@mail.sysu.edu.cn

Abstract—Autonomous underwater vehicle (AUV) in marine resource surveys plays an important role. This paper proposes a new path-following control frame for the underactuated AUV with input saturation and multiple disturbances. The disturbances include external disturbances, model parameter uncertainties, unmodeled dynamics and other random disturbances. Compared to most of previously published literatures, which treat disturbances as lumped disturbances, a composite hierarchical anti-disturbance control (CHADC) strategy is adopted to achieve higher precision path following. A disturbance observer (DOB) is constructed to estimate and eliminate the disturbances with partial known information, while the H_∞ theory is used to optimize the path-following controller to attenuate the other disturbances satisfying the L_2-norm bound condition and improve the robustness of system. Besides, Lyapunov direct method and back-stepping method are used to design the path-following controller, where the input saturation is considered, the extended state observer (ESO) is used to estimate the uncertainty of kinematic controller and the nonlinear tracking differentiator (NTD) is used to simplify the controller. Finally, simulations are given to demonstrate the effectiveness of the proposed control law.

1. Introduction

In recent years, the study of underactuated AUVs in theory and practice taking had preliminary in progress. Taking into account nonlinearity and external disturbances, [1] shown an adaptive nonlinear second-order sliding mode controller to achieve precise path following control for AUV. In [2], a Lyapunov-based model predictive control framework for AUV is designed, which can solve actual constraints and use computing resources to improve the trajectory tracking performance of AUV. [3] studied the multiple unmanned underwater vehicle cooperative control, in order to achieve synergy target tracking. [4] proposed a backstepping-based nonlinear controller for AUV, and [5] adopted a backstepping-based state feedback control method in the path following of underactuated AUV. However, the backstepping method often involves the problem of "explosion of complexity", therefore, [6] adopted dynamic surface control (DSC) to solve the problem, [7] used NTD to avoid the problems. Uncertainties and external disturbances play an important role in AUV path-following control. In most studies, the disturbances to the aircraft are not considered, or the disturbances are assumed as lumped disturbances, which is not in line with reality. Therefore, [8] proposed that the external disturbances, unmodeled dynamics and the system uncertainties are regarded as equivalent. Uncertainty and disturbance estimator (UDE) is used to estimate and provide feedforward compensation. Although this method can solve the interference problem, it is too conservative to achieve high-precision path tracking. A composite anti-disturbances method was proposed in [9] to deal with external disturbances, then, the uncertainty of the exogenous system was further considered in[10], making this processing method more likely to be applied to the actual system.

In this paper, with the studies of previous advanced methods, we propose a novel path-following control frame for underactuated AUVs to address the problems of input saturation and multiple disturbances.

2. Problem statement

2.1. AUV Modeling

The schematic diagram of underactuated AUV path following system can be shown in Fig.1, the kinematics equation can be described as[4, 11]

$$\begin{cases} \dot{x} = u\cos(\psi_B) - v\sin(\psi_B) \\ \dot{y} = u\sin(\psi_B) + v\cos(\psi_B) \\ \dot{\psi}_B = r \end{cases} \tag{1}$$

The dynamic equations can be described as[12]

$$\begin{cases} \dot{u} = \dfrac{m_{22}}{m_{11}}vr - \dfrac{X_u}{m_{11}}u - \dfrac{X_{u|u|}}{m_{11}}u|u| + \dfrac{\tau_u}{m_{11}} + d_u \\ \dot{v} = -\dfrac{m_{11}}{m_{22}}ur - \dfrac{Y_v}{m_{22}}v - \dfrac{Y_{v|v|}}{m_{22}}v|v| + d_v \\ \dot{r} = \dfrac{m_{11} - m_{22}}{m_{33}}uv - \dfrac{N_r}{m_{33}}r - \dfrac{N_{r|r|}}{m_{33}}r|r| + \dfrac{\tau_r}{m_{33}} + d_r \end{cases} \tag{2}$$

where u , v and r are the surge speed, sway speed and yaw speed of the underactuated AUV, τ_u and τ_r are the control input, m is the mass of AUV, I_Z is the moment of inertia, X_u, $X_{\dot{u}}$, $X_{u|u|}$, Y_v, $Y_{\dot{v}}$, $Y_{v|v|}$, N_r, $N_{\dot{r}}$ and $N_{r|r|}$ are the nominal hydrodynamic parameters, which $m_{11} = m - X_{\dot{u}}$, $m_{22} = m - Y_{\dot{v}}$, $m_{33} = I_Z - N_{\dot{r}}$ represent the combined inertia and added mass terms. d_u, d_v and d_r are the total disturbances, including environmental disturbance, unmodeled errors and system uncertainties.

The error dynamics equations are

$$\begin{cases} \dot{x}_e = -\dot{s}(1 - c_c y_e) + v_t\cos\psi_e \\ \dot{y}_e = -c_c \dot{s} x_e + v_t\sin\psi_e \\ \dot{\psi}_e = r + \dot{\beta} - c_c\dot{s} \end{cases} \tag{3}$$

where s is the abscissa of the virtual target point P on the desired path, Cc is the curvature of the desired path, $\beta = \tan^{-1}(v/u)$ is the sideslip angle, $v_t = \sqrt{u^2 + v^2}$ is the total speed. (x_e, y_e, ψ_e) are the follow errors in the Serret-Frenet frame.

Taking into account the effect of the input saturation of the control inputs in the practical application, the actual control force τ_u and τ_r moment can be formulated as:

$$\tau_i = \begin{cases} \tau_{i\max}, & \tau_i^* > \tau_{i\max} \\ \tau_i^*, & \tau_{i\min} \le \tau_i^* \le \tau_{i\max}, \quad i = u, r \\ \tau_{i\min}, & \tau_i^* < \tau_{i\min} \end{cases} \tag{4}$$

where τ_i^* $(i=u,r)$ represents the control input without considering the problem of input saturation. $\tau_{i\max}$ and $\tau_{i\min}$ $(i=u,r)$ are the upper and lower bounds of the control input, respectively.

The dynamic model of the underactuated AUV can be expressed in matrix form:

$$\dot{v} = Mv + F_{01}f_{01}(v) + M_1\tau + d \tag{5}$$

the disturbances can be written as $d = M_1 d_0 + H_1 d_1$ by layering. Hence, the model (5) can be rewritten as

$$\dot{v} = Mv + F_{01}f_{01}(v) + M_1[\tau + d_0] + H_1 d_1 \tag{6}$$

2.2. Problem Statement

The control objectives of this paper are that design the controller τ_u , τ_r and $\dot{s}$ to make

$$\sup_{t\in[t_0,\infty]}\|x_e\|\le\delta_1,\quad \sup_{t\in[t_0,\infty]}\|y_e\|\le\delta_2,\quad \sup_{t\in[t_0,\infty]}\|\psi_e-\psi_{los}\|\le\delta_3,\quad \sup_{t\in[t_0,\infty]}\|u-u_d\|\le\delta_4,$$

under the joint action of multiple disturbances (d_u, d_v with d_r) and input saturation.

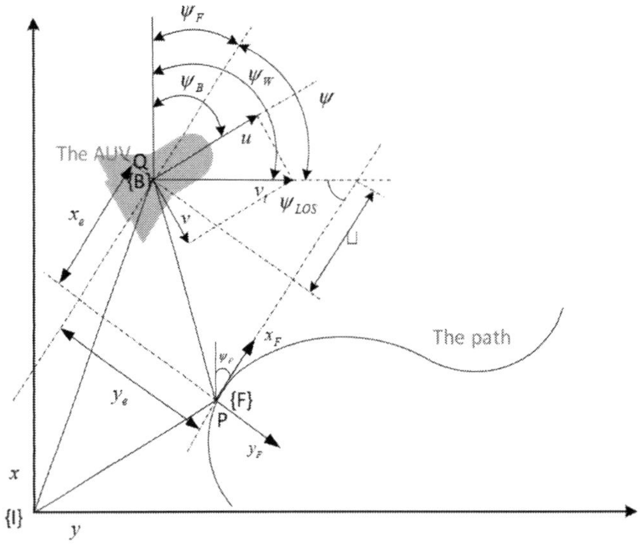

Fig.1 Frame and variable definitions of AUV path following

3. Controller design

3.1. Path-following controller design

Using the line-of-sight (LOS) guidance algorithm described in [13], the LOS guidance law is selected as:

$$\psi_{LOS} = \tan^{-1}\left(\frac{-y_e}{\Delta}\right) \tag{7}$$

where ψ_{LOS} is the LOS angle, Δ is the look-ahead distance.

To avoid long periods of full rudder, we chose to add constraints at the virtual of heading angle velocity, therefore, the control law can be described as follows:

$$r = c_c(s)\dot{s} - \dot{\beta} + \dot{\psi}_{los} - k_1(\psi_e - \psi_{los}) + k_1\chi_1 \tag{8}$$

where χ_1 is auxiliary system state which has been design as[14]:

$$\dot{\chi}_1 = \begin{cases} -k_{\chi_1}\chi_1 - f_1\chi_1 + \gamma_1\Delta r & \|\chi_1\| > \bar{\chi}_1 \\ 0 & \|\chi_1\| \le \bar{\chi}_1 \end{cases}$$

among the function, $f_1 = \frac{|\tilde{\psi}\Delta r| + 0.5\gamma_1^2\Delta r^2}{\chi_1^2}$, $k_{\chi_1} > \frac{1}{2} + \frac{k_1}{2}$, $\gamma_1 > 0$, $\tilde{\psi} = \psi_e - \psi_{LOS}$, $\Delta r = r - r_d$, $\bar{\chi}_1 > 0$.

Due to the complexity of the calculation of $\dot{\beta}$, choose ESO to estimate the uncertainty d_ψ which represent $\dot{\beta}$ [12].

$$\begin{cases} \dot{\xi}_1 = -\omega_1\xi_1 - \omega_1^2\psi_e - \omega_1(r - c_c\dot{s}) \\ d_\psi = \omega_1\psi_e + \xi_1, \omega_1 > 0 \end{cases} \tag{9}$$

As a result, the control law of virtual yaw speed becomes:

$$r_d = c_c(s)\dot{s} - \hat{d}_\psi + \dot{\psi}_{los} - k_1(\psi_e - \psi_{los}) + k_1\chi_1 \tag{10}$$

The virtual target movement control law is designed as

$$\dot{s} = k_2 x_e + v_t \cos\psi_e \tag{11}$$

The control input τ_u can be designed as follow, notice the saturation of τ_u.

$$\tau_u = m_{11}\left[\dot{u}_d - k_5(u - u_d) - \frac{m_{22}}{m_{11}}vr + \frac{X_u}{m_{11}}u + \frac{X_{u|u|}}{m_{11}}u|u| - d_u + k_5\chi_2\right] \tag{12}$$

where χ_2 is the is auxiliary system state which has been designed as χ_1, $k_5 > 0$ is the control gain.

Design control input τ_r as follow:

$$\tau_r = m_{33}\left[\dot{r}_d - k_3(r - r_d) - k_4(\psi_e - \psi_{los}) - \frac{m_{11} - m_{22}}{m_{33}}uv + \frac{N_r}{m_{33}}r + \frac{N_{r|r|}}{m_{33}}r|r| - d_r\right] \tag{13}$$

where $k_3 > 0$ and $k_4 > 0$ are the control gains. Due to the form $\dot{r}_d$ is too complicated, using NTD can simplify the above design controller (16) as follows[12]:

$$\tau_r = m_{33}\left[\dot{r}_c - k_3(r - r_c) - k_4(\psi_e - \psi_{los}) - \frac{m_{11} - m_{22}}{m_{33}}uv + \frac{N_r}{m_{33}}r + \frac{N_{r|r|}}{m_{33}}r|r| - d_r\right] \tag{14}$$

3.2. Composite controller design:

First, design the disturbance observer:

$$\begin{cases} d_0 = V\omega \\ \omega = \xi_2 - Lv \\ \dot{\xi}_2 = (W + LM_1V)\omega + L(Mv + F_{01}f_{01} + M_1\tau) \end{cases} \tag{15}$$

where $\hat{d}_0$ is the estimation of disturbances d_0, $\hat{\omega}$ is the estimation of ω, ξ_2 is auxiliary state of the observer. The estimation error is $e_\omega = \omega - \omega$, then

$$\dot{e}_\omega = (W + LMV)e_\omega + LH_1d_1 + H_2\delta \tag{16}$$

Then, make $\tau = -d_0 + \tau_0$, where τ_0 is the control input when the disturbances are absent, where $\tilde{\psi}_e = \psi_e - \psi_{LOS}$. Substituting τ into (5) yields

$$\dot{v} = Mv + F_{01}f_{01} + M_1(d_0 - d_0 + \tau_0) + H_1d_1 \tag{17}$$

make $\tilde{\boldsymbol{v}} = \boldsymbol{v} - \boldsymbol{v}_d$

$$\dot{\tilde{v}} = (G_0 + K)\tilde{v} + F_{01}\bar{f}_{01} + M_1Ve_\omega + H_1d_1 + H_3d_2 \tag{18}$$

where $G_0 = diag([0 \;\; -\frac{Y_{\dot{v}}}{m_{22}} \;\; 0])$, $\overline{f_{01}} = (0 \;\; f_v \;\; 0)^T$, $d_2 = \left[k_5\chi_2 \;\; -k_4\tilde{\psi}_e \;\; (-Y_v/m_{22})v_d - \dot{v}_d\right]^T$,

$K = \begin{bmatrix} k_5 & 0 & 0 \\ 0 & 0 & k_3 \end{bmatrix}^T$. Combine (16) and (18), we can gain that

$$\begin{cases} \dot{\bar{v}} = G\bar{v} + F\bar{f}_{01} + Hd \\ z = C\bar{v} \end{cases} \tag{19}$$

where $\bar{v} = \begin{bmatrix} \tilde{v} \\ e_\omega \end{bmatrix}$, $G = \begin{bmatrix} G_0 + K & M_1V \\ 0 & W + LM_1V \end{bmatrix}$, $F = \begin{bmatrix} F_{01} \\ 0 \end{bmatrix}$, $H = \begin{bmatrix} H_1 & 0 & H_3 \\ LH_1 & H_2 & 0 \end{bmatrix}$, $d = \begin{bmatrix} d_1 & \delta & d_2 \end{bmatrix}^T$, $C = \begin{bmatrix} C_1 & 0 \end{bmatrix}$.It is noted that we denote by a matrix $M < 0$ the M is a negative definite matrix[10]. Therefor, for given parameters $\lambda > 0$, $\gamma > 0$, if there are real symmetric matrices $Q_1 > 0$, $P_2 > 0$, R_1, R_2, such that

$$\begin{bmatrix} \Pi_1 & F & H_1 & 0 & H_3 & Q_1U_1^T & M_1V \\ * & -\frac{1}{\lambda^2}I & 0 & 0 & 0 & 0 & 0 \\ * & * & -\gamma^2 I & 0 & 0 & 0 & H_1^T R_2^T \\ * & * & * & -\gamma^2 I & 0 & 0 & H_2^T P_2^T \\ * & * & * & * & -\gamma^2 I & 0 & 0 \\ * & * & * & * & * & -\gamma^2 I & 0 \\ * & * & * & * & * & * & \Pi_2 \end{bmatrix} < 0 \tag{20}$$

where $*$ represents the corresponding elements in the symmetric matrix, $R_1 = KQ_1$, $R_2 = P_2L$, $\Pi_1 = G_0Q_1 + Q_1^T G_0^T + KQ_1 + Q_1^T K^T$, $\Pi_2 = P_2W + W^T P_2^T + P_2LM_1V + V^T M_1^T L^T P_2$, then the nonlinear system represented by equation (19) is robust and asymptotically stable when the disturbance d is absent, and satisfies $\|z\|_2 < \gamma\|d\|_2$.If formula (20) is true, it needs to be satisfied $\Pi_1 < 0$, $\Pi_2 < 0$,and K and L can be obtained by linear matrix inequality (LMI).

4. Numerical Simulation

Table 1 shows the nominal parameters of the AUV model. The initial conditions of the AUV, the designed parameters for the proposed control law and the bounds of control forces and virtual yaw speed are set as **Table 2**.

Table 1 The nominal parameters of the underactuated AUV [11]

$m = 185kg$, $I_z = 50kgm^2$, $X_{\dot{u}} = -30kg$, $Y_{\dot{v}} = -80kg$, $N_{\dot{r}} = -30kgm^2$, $X_u = 70kg/s$,
$X_{u\|u\|} = 100kg/m$, $Y_v = 100kg/s$, $Y_{v\|v\|} = 200kg/m$, $N_r = 50kgm^2/s$, $N_{r\|r\|} = 100kgm^2$

Table 2 Kinds of parameters of the underactuated AUV

Initial position and posture of the underactuated AUV	$x(0) = 15\text{m}$, $y(0) = -15\text{m}$, $u(0) = 0.1\text{m/s}$, $v(0) = 0\text{m/s}$, $r(0) = 0\text{rad/s}$, $\psi(0) = \frac{\pi}{2}\text{rad}$, $s(0) = 0\text{m}$, $u_d = 1\text{m/s}$
The bounds of control force and virtual yaw speed	$\tau_{u\max} = 200\text{N}$, $\tau_{u\min} = 0\text{N}$, $r_{d\max} = 0.6(rad/s)$, $r_{d\min} = -0.6(rad/s)$

The proposed controller	$k_1 = 10$, $k_2 = 5$, $k_4 = 2$, $k_{\chi_1} = k_{\chi_2} = 1.6$, $h = 0.1\text{s}^{-1}$, $R = 100\text{m/s}^2$ $T = 0.01\text{s}^{-1}$

The underactuated AUV is required to follow a desired path described by (21) with the parameters shown in **Table 3.**

$$x_F(\mu) = \sum_{i=0}^{4} a_i \mu^i ,\ y_F(\mu) = \sum_{i=0}^{4} b_i \mu^i \tag{21}$$

Table 3 Desired path parameters

Parameter	***i*=0**	***i*=1**	***i*=2**	***i*=3**	***i*=4**
a_i	0	0.87	-0.02	10^{-5}	1.5×10^{-6}
b_i	0	0.5	-5×10^{-4}	10^{-5}	10^{-7}

Select $W = \begin{bmatrix} 0 & 0.5 \\ -0.5 & 0 \end{bmatrix}$, $V = \begin{bmatrix} 0.8 & 0 \\ 0 & 0.5 \end{bmatrix}$ to make d_0 can be used to describe the harmonic disturbances, and the L and K can be calculated by LMI.

$$L = \begin{bmatrix} -134.375 & 0 & -0.2261 \\ 0.3799 & 0 & -80 \end{bmatrix},\ \mathrm{K} = \begin{bmatrix} -k_5 & 0 & 0 \\ 0 & 0 & -k_3 \end{bmatrix} = \begin{bmatrix} -0.5 & 0 & 0 \\ 0 & 0 & -0.5 \end{bmatrix}$$

The rest disturbances d_1 can be set as

$$\begin{cases} d_{1u} = 0.01\sin(0.1t) + 0.01 \\ d_{1v} = 0.01\cos(0.1t) - 0.01 \\ d_{1r} = 0.01\sin(0.2t) + 0.01 \end{cases} \tag{22}$$

The simulation results are shown in Fig. 2(a)-(d). It can be seen that the novel controller can make the AUV follow the desired path with high precision in Fig. 2(a). The Fig. 2(b) shows that the path following error asymptotically converge to neighborhood of zero. In Fig. 2(c), the surge speed u converges to u_d with great speed, and the sway speed and yaw speed are bounded. The Fig. 2(d) illustrate that the control input τ_r can be limited to an appropriate range by placing restrictions on the virtual of virtual yaw speed r_d, and the control input τ_u also in an appropriate range.

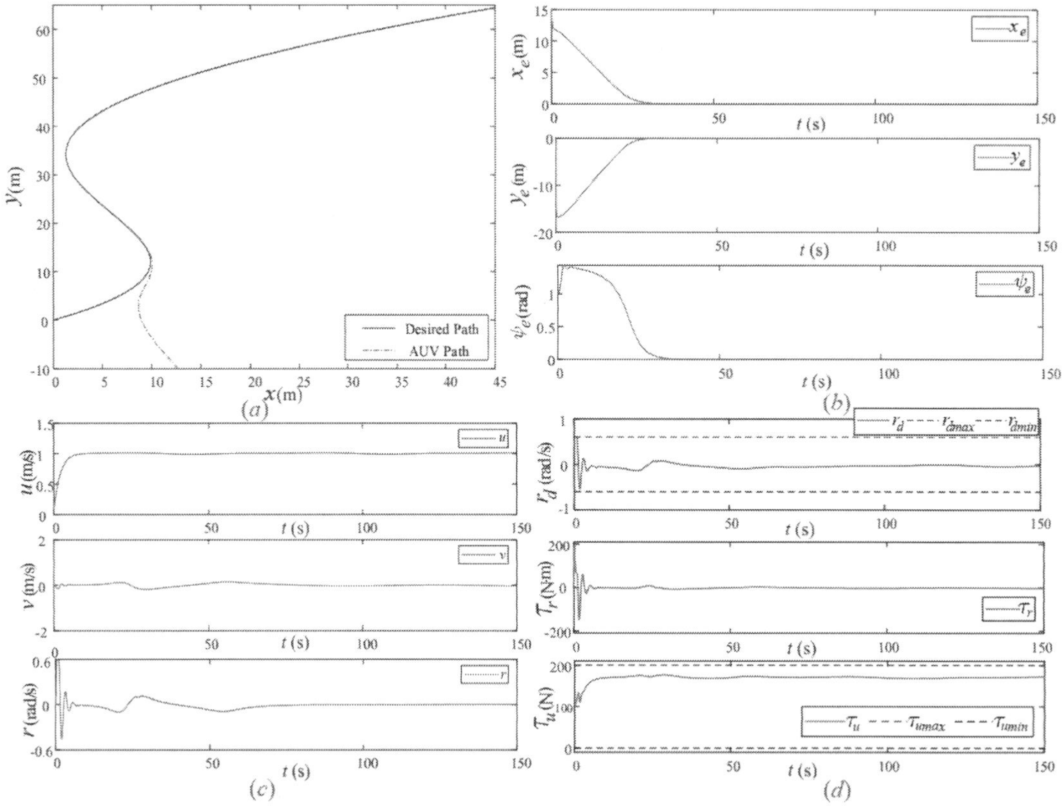

Fig. 2 (a) Desired path and AUV actual path. (b) Follow errors of the AUV. (c) Actual speeds of AUV. (d) Virtual yaw speed and control inputs of AUV.

5. Conclusions

This paper proposes a novel path following control scheme for the underactuated AUV with input saturation and multiple disturbances. By considering the characteristic of disturbances, which include external disturbances, model parameter uncertainties, unmodeled dynamics and other random disturbances, we assumed that the disturbances were divided into two types. The first type with partial known information was estimated by the disturbance observer and the other satisfied the L_2-norm bounded conditions was minimized by an H_∞ based path-following controller. Besides, when the path-following controller was designed, the input saturation was considered, and the $\dot{\beta}$ was regarded as uncertainty which was on-line estimated by ESO, which reduced the reliance on models and improved the practicability of the AUV. Numerical simulations validate the effectiveness of the proposed controller.

Acknowledgments

This work is supported by the Natural Science Foundation of Guangdong Province of China (2019A1515011405), the Science and Technology Planning Project of Guangdong Province of China (311020011), and the Key-Area Research and Development Program of Guangdong Province (2020B1111010004).

References

[1] Zhang *et al.*, "A novel adaptive second order sliding mode path following control for a portable AUV," 2018.

[2] C. Shen, Y. Shi, and B. J. I. T. o. I. E. Buckham, "Trajectory Tracking Control of an Autonomous Underwater Vehicle Using Lyapunov-Based Model Predictive Control," 2017.
[3] Z. Yan, X. Liu, J. Zhou, and D. J. I. A. Wu, "Coordinated Target Tracking Strategy for Multiple Unmanned Underwater Vehicles with Time Delays," pp. 1-1, 2018.
[4] L. Lapierre and D. Soetanto, "Nonlinear path-following control of an AUV," *Ocean Engineering,* vol. 34, no. 11-12, pp. 1734-1744, 2007.
[5] J. L. Zhang, X. B. Xiang, L. Lapierre, Q. Zhang, and W. J. Li, "Approach-angle-based three-dimensional indirect adaptive fuzzy path following of under-actuated AUV with input saturation," (in English), *Applied Ocean Research,* Article vol. 107, p. 13, Feb 2021, Art. no. 102486.
[6] K. D. von Ellenrieder, "Dynamic surface control of trajectory tracking marine vehicles with actuator magnitude and rate limits," *Automatica,* vol. 105, pp. 433-442, 2019.
[7] J. Miao, S. Wang, Z. Zhao, Y. Li, and M. M. Tomovic, "Spatial curvilinear path following control of underactuated AUV with multiple uncertainties," *ISA Trans,* vol. 67, pp. 107-130, Mar 2017.
[8] B. Ren, Q.-C. Zhong, and J. Dai, "Asymptotic Reference Tracking and Disturbance Rejection of UDE-Based Robust Control," *IEEE Transactions on Industrial Electronics,* vol. 64, no. 4, pp. 3166-3176, 2017.
[9] X. Yao and L. Guo, "Composite anti-disturbance control for Markovian jump nonlinear systems via disturbance observer," *Automatica,* vol. 49, no. 8, pp. 2538-2545, 2013.
[10] X. Wei and L. Guo, "Composite disturbance-observer-based control and H∞ control for complex continuous models," *International Journal of Robust and Nonlinear Control,* vol. 20, no. 1, pp. 106-118, 2010.
[11] X. Xiang, L. Lapierre, and B. Jouvencel, "Smooth transition of AUV motion control: From fully-actuated to under-actuated configuration," *Robotics and Autonomous Systems,* vol. 67, pp. 14-22, 2015.
[12] J. Miao, S. Wang, M. M. Tomovic, and Z. Zhao, "Compound line-of-sight nonlinear path following control of underactuated marine vehicles exposed to wind, waves, and ocean currents," *Nonlinear Dynamics,* vol. 89, no. 4, pp. 2441-2459, 2017.
[13] E. Borhaug, A. Pavlov, and K. Y. Pettersen, "Integral LOS control for path following of underactuated marine surface vessels in the presence of constant ocean currents," in *IEEE Conference on Decision & Control*, 2008.
[14] Z. Zheng, L. Ruan, M. Zhu, and X. Guo, "Reinforcement learning control for underactuated surface vessel with output error constraints and uncertainties," *Neurocomputing,* vol. 399, pp. 479-490, 2020.

AUTHOR INDEX

Institute of Physics
Dirac House, Temple Back
Bristol BS1 6BE UK

ISSN: 1742-6588
ISBN 978-1-7138-5092-2